Antarctica
A Keystone in a Changing World

Proceedings of the 10th International Symposium on
Antarctic Earth Sciences
Santa Barbara, California
August 26 to September 1, 2007

Editors:
Alan K. Cooper, Peter Barrett, Howard Stagg, Bryan Storey, Edmund Stump, Woody Wise,
and the 10th ISAES editorial team

*John Anderson
John Barron
Philip Bart
Donald Blankenship
Frederick Davey
Michael Diggles
Carol Finn
Paul Fitzgerald
Fabio Florindo
Jane Francis

Dieter Futterer
John Gamble
John Goodge
William Hammer
Patricia Helton
Erik Ivins
Philip Kyle
Wesley LeMasurier
Paul Mayewski

Timothy Naish
Sandra Passchier
Stephen Pekar
Carol Raymond
Carlo Alberto Ricci
Michael Studinger
David Sugden
Vanessa Thorn
Terry Wilson

This special International Polar Year volume is a joint effort of the Polar Research Board of the National Academies and the U.S. Geological Survey. The Polar Research Board is the U.S. National Committee to the Scientific Committee on Antarctic Research, which is the official sponsor of the 10th ISAES. Publication of printed papers is by the National Academies Press and publication of electronic media is by the U.S. Geological Survey.

THE NATIONAL ACADEMIES PRESS
Washington, D.C.
www.nap.edu

THE NATIONAL ACADEMIES PRESS   500 FIFTH STREET, N.W.   WASHINGTON, DC 20001

Note: This publication was overseen by two members of the Polar Research Board (PRB) who served as members of the 10th ISAES local organizing committee: Robin Bell, chairperson of the 10th ISAES Local Organizing Committee and Mahon C. Kennicutt II, U.S. delegate to the Scientific Committee on Antarctic Research (SCAR) and a vice president of SCAR.

Funding support for this collaborative project was provided by the U.S. National Science Foundation's Office of Polar Programs, the U.S. Geological Survey and SCAR. This book is a companion to the Online Proceedings of the 10th ISAES hosted by the USGS at http://pubs.usgs.gov/of/2007/1047/.

International Standard Book Number-13: 978-0-309-11854-5
International Standard Book Number-10: 0-309-11854-9

Additional copies of this book are available from the National Academies Press, 500 Fifth Street, N.W., Washington, DC 20055; (800) 624-6242 or (202) 334-3313 (in the Washington metropolitan area); Internet, http://www.nap.edu.

Printed in the United States of America

Front cover: Depicting change from greenhouse to icehouse times in Antarctica. *Upper:* Snowstorm on Seymour Island, Antarctic Peninsula. Image provided by Jane Francis. *Lower:* Middle Cretaceous forest on Antarctica; painting by Robert Nicholls, with permission for use by the artist and by the British Antarctic Survey who commissioned the painting and have it on display there.
Back cover: *Upper:* Estimates skin-depth temperatures derived from the thermal IR channel of historical AVHRR data. Image is from NASA website: http://svs.gsfc.nasa.gov/vis/a000000/a003100/a003188/index.html. *Lower:* Predictions of ice-sheet volume changes and the effect of these changes on global sea levels. Models are from the paper by Miller et al. in this book, and are in turn based on work of Deconto and Pollard (*Nature* 421:245-249, 2003).

Suggested citation: Cooper, A. K., P. J. Barrett, H. Stagg, B. Storey, E. Stump, W. Wise, and the 10th ISAES editorial team, eds. (2008). *Antarctica: A Keystone in a Changing World.* Proceedings of the 10th International Symposium on Antarctic Earth Sciences. Washington, DC: The National Academies Press.

THE NATIONAL ACADEMIES
Advisers to the Nation on Science, Engineering, and Medicine

The **National Academy of Sciences** is a private, nonprofit, self-perpetuating society of distinguished scholars engaged in scientific and engineering research, dedicated to the furtherance of science and technology and to their use for the general welfare. Upon the authority of the charter granted to it by the Congress in 1863, the Academy has a mandate that requires it to advise the federal government on scientific and technical matters. Dr. Ralph J. Cicerone is president of the National Academy of Sciences.

The **National Academy of Engineering** was established in 1964, under the charter of the National Academy of Sciences, as a parallel organization of outstanding engineers. It is autonomous in its administration and in the selection of its members, sharing with the National Academy of Sciences the responsibility for advising the federal government. The National Academy of Engineering also sponsors engineering programs aimed at meeting national needs, encourages education and research, and recognizes the superior achievements of engineers. Dr. Charles M. Vest is president of the National Academy of Engineering.

The **Institute of Medicine** was established in 1970 by the National Academy of Sciences to secure the services of eminent members of appropriate professions in the examination of policy matters pertaining to the health of the public. The Institute acts under the responsibility given to the National Academy of Sciences by its congressional charter to be an adviser to the federal government and, upon its own initiative, to identify issues of medical care, research, and education. Dr. Harvey V. Fineberg is president of the Institute of Medicine.

The **National Research Council** was organized by the National Academy of Sciences in 1916 to associate the broad community of science and technology with the Academy's purposes of furthering knowledge and advising the federal government. Functioning in accordance with general policies determined by the Academy, the Council has become the principal operating agency of both the National Academy of Sciences and the National Academy of Engineering in providing services to the government, the public, and the scientific and engineering communities. The Council is administered jointly by both Academies and the Institute of Medicine. Dr. Ralph J. Cicerone and Dr. Charles M. Vest are chair and vice chair, respectively, of the National Research Council.

www.national-academies.org

10TH ISAES EDITORIAL TEAM

Preface

Gondwana geologists have a timely metaphor: "Antarctica—the heart of it all" or from the geologic perspective: Antarctica—the center from which all surrounding continental bodies separated millions of years ago. The title of our book *"Antarctica: A Keystone in a Changing World"* reinforces the importance of continual changes in Antarctica's multifaceted history and the impact of these changes on global systems. In 2007, the Scientific Committee on Antarctic Research (SCAR) sponsored the 10th International Symposium on Antarctic Earth Sciences (10th ISAES) in Santa Barbara, California, to give researchers from 34 countries an opportunity to share and discuss recent discoveries in the Antarctic region. Such discoveries help decipher the prior and current roles of Antarctica in manifesting the global climatic changes, now seemingly accelerating.

The 10th ISAES coincides with the International Polar Year (IPY) that falls on the 50th anniversary of the International Geophysical Year (IGY). In recognition of these events, the symposium format and topics of keynote papers in the book envelop a broad spectrum with six themes covering key topics on evolution and interactions of the geosphere, cryosphere, and biosphere and their cross-linkages with past and historic paleoclimates. Emphasis is on deciphering the climate records in ice cores, geologic cores, rock outcrops, and those inferred from climate models. New technologies for the coming decades of geoscience data collection are also highlighted.

The 10th ISAES also marks the 44th year of such symposia, and denotes the first significant change in presentation and publication formats. Prior ISAES have a valued history of impressive printed symposia volumes, with a total of nearly one thousand printed papers (Table 1) that were solicited and printed after the symposia. In recognition of IPY and the desire to quickly document and disseminate Antarctic research results to the science community, the 10th ISAES changed to a new format of online and book publication. Presenters, other than keynote speakers, were asked to submit either a short research paper for peer review or an extended abstract without peer review before the symposium. Over 950 co-authors from 34 countries submitted 326 manuscripts. Prior to the symposium, 34 co-editors, over 200 peer reviewers and authors processed manuscripts into final publication format so that 92 percent were published in the Online Proceedings (http://pubs.usgs.gov/of/2007/1047/) before the symposium commenced and the remaining 8 percent were made available online to meeting participants and authors; these were then published in the Online Proceedings within the month following the symposium.

Ten keynote speakers were invited to contribute overview talks at the symposium and contribute a full-length paper. The keynote papers are printed in this book along with a paper that summarizes highlights of the 10th ISAES. Several reports from meetings and workshops held in conjunction with the symposium were also submitted. The DVD in the back of the book contains the keynote and summary papers and a complete copy of the 10th ISAES Online Proceedings (see also http://pubs.usgs.gov/of/2007/1047/), all in PDF format for access, search, and printing. The DVD can be accessed and used on either a MAC or PC.

This special IPY volume for the 10th ISAES is a 100-year milestone for Antarctic publications, and a first for any symposium publication. One hundred years ago, Ernest Shackleton's expedition members created and printed the first book ever published in Antarctica "on the ice in cryospace," with their scientific discoveries and personal vignettes—they titled the book *Aurora Australis*. One hundred years later, the 10th ISAES authors and editors created the first Antarctic pre-symposium proceedings "online in cyberspace," with

TABLE 1 The History of ISAES Symposia and Their Publication Volumes

No.	Location	Year	Symposium volume
I	Cape Town, South Africa	1963	Adie, R. J., ed. (1964), *Antarctic Geology*—Proceedings of the First (SCAR) International Symposium on Antarctic Geology, North Holland Publishing Co, Amsterdam, 758 pp.
II	Oslo, Norway	1970	Adie, R. J., ed. (1972), *Antarctic Geology and Geophysics*—Proceedings of the Second (SCAR) Symposium on Antarctic Geology and Solid Earth Geophysics, International Union of Geological Sciences, B1, Universitetsforlaget, Oslo, 876 pp.
III	Madison, Wisconsin, USA	1977	Craddock, C., ed. (1982), *Antarctic Geoscience*—Proceedings of the Third (SCAR) Symposium on Antarctic Geology and Geophysics, International Union of Geological Sciences, B4, University of Wisconsin Press, Madison, 1172 pp.
IV	Adelaide, Australia	1982	Oliver, R. L., P. R. James and J. B. Jago, eds. (1983), *Antarctic Earth Science*—Proceedings of the Fourth (SCAR) International Symposium on Antarctic Earth Sciences, Australian Academy of Science, Canberra, 697 pp.
V	Cambridge, UK	1987	Thompson, M. R. A., J. A. Crame and J. W. Thompson, eds. (1991), *Geological Evolution of Antarctica*—Proceedings of the Fifth (SCAR) International Symposium on Antarctic Earth Sciences, Cambridge University Press, Cambridge, 722 pp.
VI	Tokyo, Japan	1991	Yoshida, Y., K. Kaminuma and K. Shiraishi, eds. (1992), *Recent Progress in Antarctic Earth Science*—Proceedings of the Sixth (SCAR) International Symposium on Antarctic Earth Sciences, Terra Scientific Publishing, Tokyo, 796 pp.
VII	Siena, Italy	1995	Ricci, C. A., ed. (1997), The Antarctic Region: Geological Evolution and Processes—Proceedings of the Seventh (SCAR) International Symposium on Antarctic Earth Sciences, Terra Antarctica Publication, Siena, 1206 pp.
VIII	Wellington, New Zealand	1999	Gamble, J. A., D. N. B. Skinner, and S. Henrys, eds. (2002), *Antarctica at the Close of the Millennium*—Proceedings of the Eighth (SCAR) International Symposium on Antarctic Earth Sciences, The Royal Society of New Zealand Bulletin No. 35, Wellington, Terra Scientific Publishing, Tokyo, 652 pp.
IX	Potsdam, Germany	2003	Futterer, D. K., D. Damaske, G. Kleinschmidt, H. Miller, and F. Tessensohn, eds. (2006), *Antarctica: Contributions to Global Earth Sciences*—Proceedings of the Ninth (SCAR) International Symposium on Antarctic Earth Sciences, Springer-Verlag, Berlin-Heidelberg, 477 pp.
X	Santa Barbara, California, USA	2007	Cooper, A. K., P. J. Barrett, H. Stagg, B. Storey, E. Stump, W. Wise, and the Tenth ISAES editorial team, eds. (2008), *Antarctica: A Keystone in a Changing World*—Proceedings of the 10th (SCAR) International Symposium on Antarctic Earth Sciences, The National Academies Press, Washington, D.C. 162 pp. with DVD containing website http://pubs.usgs.gov/of/2007/1047/. The 10th ISAES Proceedings Volume is the first in the ISAES series to include both book and electronic publication formats.

new research findings and interpretations. The book and online proceedings are yet another way in which Antarctica and its scientists are effecting changes in the dissemination of geoscience and other information globally.

The 10th ISAES Proceedings volumes showcase the great breadth of Antarctic geoscience research at the time of IPY, and the importance of Antarctica in deciphering changes in global systems. The volumes illustrate the positive impact of this research in successfully preserving the spirit of collaboration, data-sharing, and use of Antarctica as a "continent for science" as intended by the Antarctic Treaty that was implemented in 1959 at the close of IGY.

Alan Cooper
Lead editor

Acknowledgments

The 10th ISAES was held under the auspices of the Scientific Committee on Antarctic Research (SCAR) and we thank SCAR for their ongoing support of these Antarctic geoscience symposia.

Creating the special International Polar Year volumes for the 10th ISAES, with book and DVD (and Online Proceedings) over the last three years has involved over 1200 people, most on a volunteer basis, in more than 34 countries, including over 950 authors, 34 editors and editorial staff, more than 200 peer reviewers, 23 members of the International and Local Organizing Committees and their staff, and many managers and staff from the Scientific Committee on Antarctic Research, The National Academies Polar Research Board, U.S. Geological Survey, National Science Foundation, The National Academies Press, Conference Exchange, Stanford University, and University of California at Santa Barbara. These people and institutions are warmly thanked for their varied contributions to this successful undertaking. There is not adequate space here to list all names, but see the DVD (Online Proceedings) for expanded lists of those who have assisted. We especially thank Stephen Mautner, Executive Editor, and Rachel Marcus, Managing Editor, of the National Academies Press, for their dedicated efforts in helping to make this book possible and in producing it on schedule.

U.S. Organizing Committee
Robin Bell, Chair
Alan Cooper, Publications manager
Ian Dalziel, International coordination
Carol Finn, Publications co-editor
Paul Fitzgerald, Program committee
James Kennett, Symposium co-manager
Bruce Luyendyk, Symposium manager
Samuel Mukasa, Development
Ross Powell, Workshops
Carol Raymond, Publications co-manager
Christine Siddoway, Field excursions
Terry Wilson, Program Committee chair
Chuck Kennicutt, SCAR ex-officio

Funding agencies for 10th ISAES
U.S. National Science Foundation
   Division of Polar Programs
Scientific Committee on Antarctic Research
   - Geosciences Standing Group
   - Antarctic Climate Evolution Program
U.S. Geological Survey

Collaborating agencies
The National Academies
      - Polar Research Board
U.S. Committee on International Polar Year
University of California at Santa Barbara
Stanford University

International Steering Committee
Peter Barrett, New Zealand
Alessandro Capra, Italy
Gino Casassa, Chile
Fred Davey, New Zealand
Ian Fitzsimons, Australia
Jane Francis, United Kingdom
Marta Ghidella, Argentina
Joachim Jacobs, Germany
Hubert Miller, Germany
Carlo Alberto Ricci, Italy

Contents

Cooper, A. K., P. J. Barrett, H. Stagg, B. Storey, E. Stump, W. Wise, and the 10th ISAES editorial team, eds. (2008). *Antarctica: A Keystone in a Changing World.* Proceedings of the 10th International Symposium on Antarctic Earth Sciences. Washington, DC: The National Academies Press.

Summary and Highlights of the 10th International Symposium on Antarctic Earth Sciences

T. J. Wilson,[1] R. E. Bell,[2] P. Fitzgerald,[3] S. B. Mukasa,[4] R. D. Powell,[5] C. Finn[6]

INTRODUCTION

The 10th International Symposium on Antarctic Earth Sciences (10th ISAES) was convened at the University of California, Santa Barbara, in August 2007. At the symposium about 350 researchers presented talks and posters with new results on major topics, including climate change, biotic evolution, magmatic processes, surface processes, tectonics, geodynamics, and the cryosphere. The symposium resulted in 335 papers and extended abstracts (Cooper et al., 2007, and this volume). Many science discoveries were presented spanning the last 2 billion years, from times when Antarctica was part of former supercontinents Rodinia and Gondwana to the present when Antarctica is an isolated, ice-covered land mass surrounded by seafloor spreading centers. In this summary we highlight some of the new results presented at the symposium.

TECTONICS IN THE SOUTH: A VIEW FROM THE SOLID EARTH

Antarctica occupies a key position for a greater understanding of the evolution of the supercontinents Rodinia and

[1]School of Earth Sciences, Ohio State University, Columbus, OH 43210-1522, USA.

[2]Lamont-Doherty Earth Observatory of Columbia University, Palisades, NY 10964-8000, USA.

[3]Department of Earth Sciences, Syracuse University, Syracuse, NY 13244-1070, USA.

[4]Department of Geological Sciences, University of Michigan, Ann Arbor, MI 48109-1005, USA.

[5]Department of Geology and Environmental Geosciences, Northern Illinois University, DeKalb, IL 60115-2854, USA.

[6]U.S. Geological Survey, Denver, CO 80225, USA.

Gondwana as well as the present-day global plate motion circuit. Challenges for Antarctic science include the presence of ice sheets, ice shelves, and annual sea ice, hence remote sensing over the continent and surrounding oceans is routinely employed. Extensive offshore seismic reflection studies provide information about the evolution of geologic structures and formation of the Antarctic ice sheets. For example, the "Plates and Gates" project (e.g., Maldonado et al., 2007) is examining links between the opening of the Drake and Tasman Passages, and the transition to an ice-covered continent. Onshore, techniques such as airborne geophysics provide insight into the geology and form of the continent under the ice. For example, aeromagnetic data along the western flank of the Transantarctic Mountains reveal a faulted margin with the Wilkes subglacial basin, hence casting doubt on a flexural uplift model for the mountains (Armadillo et al., 2007). Another approach to acquire information from interior Antarctica relies on proxies from sediments and glacial deposits. A 1440 Ma, A-type rapakivi granite boulder was discovered in glacial till in the Nimrod Glacier region (Goodge et al., 2007). This granite has a Nd-isotope age and detrital zircons that closely resemble granites from the Laurentian province of North America, demonstrating the presence of Laurentia-like crust in East Antarctica. This supports the postulated fit of East Antarctica and Laurentia over 1 billion years ago, initially suggested by the SWEAT hypothesis (i.e., **S**outh **W**est U.S. and **E**ast **A**nt-arctica connection). But the controversy continues, as new aeromagnetic data do not support the SWEAT reconstruction (Finn and Pisarevsky, 2007).

Correlation of the Cambro-Ordovician Ross and Delamerian orogenies of Antarctica and Australia are well established, but puzzling pieces of the Ross orogen occur in New Zealand and Marie Byrd Land. The Robertson Bay terrane

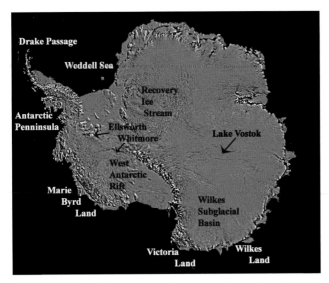

FIGURE 1 Mosaic image map of Antarctica derived from MODIS (Moderate-resolution Imaging Spectroradiometer) satellite data. SOURCE: See http://nsidc.org/data/nsidc-0280.html.

of northern Victoria Land is correlative with the western Lachlan fold belt of eastern Australia, and the Lachlan orogen rocks form in an extensional basin. The Cambrian rocks (Ross orogen correlatives) in New Zealand and Marie Byrd Land simply represent parts of a continental rift margin on the outboard side of the Lachlan fold belt (Bradshaw, 2007).

New models have been proposed for the origin of the West Antarctic rift system and the associated Transantarctic Mountains. Considerable debate at the symposium centered on the plateau hypothesis, in which the West Antarctic rift system and Transantarctic Mountains are thought to have been previously a high-topography plateau with thicker than normal crust. The proposed West Antarctic plateau is inferred to have collapsed in the Cretaceous during extension between East and West Antarctica. The Transantarctic Mountains are, in this hypothesis, the remnant western edge of the plateau modified by rift-flank uplift and glacial erosion. The first numerical model of the concept shows that plateau collapse could generate a remnant edge, depending on initial conditions (Bialas et al., 2007). The geological and thermochronologic evidence for the West Antarctic plateau formation and collapse, along with the tectonic implications, are discussed by Fitzgerald et al. (2007). Evidence for a drainage reversal in the Byrd Glacier region supports the presence of the West Antarctic plateau in the Mesozoic (Huerta, 2007). A synthesis of research from the exposed portion of the rift in Marie Byrd Land demonstrates that elevated crustal temperatures were attained by 140 Ma, causing voluminous melting, with lateral migration into wrench structures (Siddoway, 2008, this volume). Presence of melt aided the rapid evolution of the Cretaceous rift.

Other new geodynamic models of Antarctica were

advanced at the symposium, emphasizing the significance of Antarctica to studies of global geodynamics (see Figure 1). Sutherland (2008, this volume) presents a model for extension in the West Antarctic rift system, a model that fits well with global plate model circuits and the geology of New Zealand. An elegant "double-saloon-door seafloor spreading" model explains the breakup of Gondwana, magnetic anomalies in the Weddell Sea region, along with the rotation and translation of the Falkland Islands block and the Ellsworth-Whitmore Mountains crustal block (Martin, 2007).

LIFE IN ANTARCTICA: THE TERRESTRIAL VIEW

A definitive incremental change in our understanding of the evolution of life on Antarctic land emerged at the symposium, from the rich and diverse terrestrial presentations. Evidence of a vibrant world is preserved in nonglacial and glacial sedimentary deposits that rest on top of the tectonic basement structures.

Insights into Gondwana ecosystem dynamics are being gleaned from tracks of animals in Devonian deserts (Bradshaw and Harmsen, 2007); the climate records in Permian, Triassic, and Jurassic floras (Bomfleur et al., 2007; Miller and Isbell, 2007; Ryberg and Taylor, 2007); and the Triassic and Jurassic reptiles and dinosaurs of the Transantarctic Mountains (Collinson and Hammer, 2007; Smith et al., 2007).

For the last 100 million years, from the Late Cretaceous onward, the Antarctic continent has been situated over the South Pole in approximately its present location. In sharp contrast to current frigid polar conditions, abundant subtropical fossil plants are commonly found in Antarctic rocks. The subtropical nature of these fossil plants indicates warm, humid climates at high latitudes during the mid-Cretaceous (Francis et al., 2008, this volume). A variety of dinosaurs lived in these polar forests, as shown by the wealth of bones collected from the Antarctic Peninsula region. The Late Cretaceous dinosaur fauna is a relict of a cosmopolitan dinosaur assemblage that survived until the end of the Cretaceous in Antarctica after becoming extinct elsewhere (Case, 2007). Discoveries of juvenile marine reptile fossils indicate that Antarctica may have been a nursery for young marine reptiles (Martin et al., 2007).

The plant record reveals an interesting conundrum about the terrestrial response to the major climate transition from greenhouse to icehouse during the latest Eocene-earliest Oligocene. Analyses of fossil leaf collections from the Antarctic Peninsula show a temperature decline from warm temperate to cold climates through the Eocene. These cold Eocene climates may have had winter frosts coinciding with a decline in plant diversity (Francis et al., 2008, this volume). A similar cooling trend dominates the marine isotope record. In contrast, the offshore pollen record, recovered from Integrated Ocean Drilling Program (IODP) cores, indicates only short-term responses to individual cooling events and fairly stable land temperatures through the Oligocene (Grube and Mohr, 2007).

The newly identified fossil plant record shows that even during the Antarctic icehouse, the continent was not barren and supported a diverse ecosystem. Recent discoveries of fossil plants and insects in the Dry Valleys shows that small bushes of southern beech, *Nothofagus*, along with mosses and beetles, persisted in Antarctica during the mid-Miocene (Ashworth et al., 2007). A rich fossil assemblage of dwarf beeches, beetles, a fly, snails, fish, mosses, and cushion plants is also preserved in the Sirius Group high in the Beardmore region, sandwiched between glacial tillites (Ashworth et al., 2007; Francis et al., 2008). Although the marine record points to an increasingly ice-covered continent, life continued on land.

The convergence of paleontology and modern biology is producing advances in our understanding of biodiversity in Antarctica. Molecular studies of the scant living Antarctic biotas—mosses, mites, springtails—indicate an origin for these biotas in Antarctica over 40 million years ago (Convey et al., 2007). The Miocene mosses recently discovered in the Dry Valleys are identical to species living today in Antarctica. High-altitude nunataks and ice-free coastal niches must have existed during glacial times to act as refugia for these animals and plants.

Examining the biochemical role of early ice sheets and the development of Earth's atmosphere, Raub and Kirschvink (2008, this volume) discuss the link between intense global glaciations and atmospheric oxygen generation. Geochemical evidence associated with all three pre-Phanerozoic glacial events shows that "whiffs to gigantic bursts" of oxygen accompany deglaciation. The increasing oxygen in the atmosphere is attributed to UV photochemistry producing H_2O_2 that remains locked in the ice until deglaciation. Upon deglaciation the peroxide decomposes to molecular O_2. In their model the ice sheets are a crucial part of an inorganic mechanism that drives the evolution of the oxygen-mediating enzymes that predate oxygenic photosynthesis.

UNDERSTANDING THE CHANGING PLANET: THE PALEOCLIMATE VIEW

Significant progress has been made regarding Antarctica's Neogene-Pleistocene climate and its role in the global climate system, based on recently collected data. Studies of geologic proxies at various timescales from decadal-centurial to millennial and millions of years are under way to resolve the paleoclimatic events, including the important Paleogene "Greenhouse to Icehouse" transition for the Antarctic region. The many varied studies will continue throughout the International Polar Year and beyond.

Since the 9th ISAES in 2003, scientific rock drilling has made significant advances under the umbrella of the Scientific Committee on Antarctic Research's (SCAR's) ACE (Antarctic Climate Evolution) program. Several different initiatives are ongoing in the form of ANDRILL (Antarctic Geological Drilling) and SHALDRIL (Shallow Drilling). Drilling will continue with a long anticipated IODP cruise to the Wilkes Land margin that is now scheduled during the International Polar Year.

The first drilling season of ANDRILL in the McMurdo Ice Shelf Project yielded an unprecedented record of at least 60 ice-sheet fluctuations in the past 13 Ma, with indications of both warmer-than-present climate and ice sheets in the pre-Pleistocene period (Naish et al., 2008, this volume). ANDRILL and its predecessor Cape Roberts Project have successfully demonstrated that rock drilling from an ice-based platform is a viable means to acquire critically needed long, continuous rock cores.

Seismic reflection surveys of offshore regions complement the advances in rock coring, extending the record of paleoclimate events and ice-sheet history beyond the vicinity of the drill sites (Decesari et al., 2007). With new technology the acquisition of seismic data over land-fast sea ice and ice shelves is becoming more efficient and reliable, and multichannel seismic reflection data more readily accessible (Wardell et al., 2007). With new drilling core and seismic surveys, the age and lithology for prominent regional reflectors and units are now better constrained. The new seismic datasets are helping to identify coring and drilling sites and map the extent of past ice-sheet grounding events (Pekar et al., 2007). Land-based geologic studies are providing significant new data that complement the insights emerging from the marine studies. Areas of focused studies are permafrost and Paleogene-Neogene outcrops of paleoclimatic significance. There has been a renewed focus on permafrost and cryosol studies under the auspices of SCAR's Antarctic Permafrost and Soils group. These studies are targeting a broad range of ages from the modern to millions of years, leveraging this sensitive index of climatic change (Kowalewski and Marchant, 2007; Vieira et al., 2007).

FRONTIERS ENCASED IN ICE: A VIEW OF SUBGLACIAL LAKES

Exactly 50 years ago, during the last International Polar Year, the former Soviet Union established a research base in the deep interior of the Antarctic continent to be close to the geomagnetic South Pole. It was not until several decades later that it was realized from radio echograms that the station was actually located on ice floating on a lake of liquid water. Named Lake Vostok, this water body has about the same surface area as Lake Ontario, but is three times deeper, and is capped by almost 4 km of ice. Subsequent explorations have produced an inventory of some 145 subglacial lakes, a number that is likely to grow over the next few years.

Rapid changes in our understanding of subglacial lakes emerged at the symposium. Presentations focused on a number of topics, including inventory, tectonic controls for formation of the lakes (Bell et al., 2007), expanding demonstration of lake interconnectedness (Carter et al., 2007), recent discovery of the association between subglacial lakes and ice streams (Fricker et al., 2007), lake-water discharge into the oceans and its potential impact on climate, as well as International Polar Year activities gearing up to explore various subglacial aquatic environments.

Geophysical evidence indicates that at least some of the large lakes are structurally controlled, often exploiting the zones of weakness separating distinct terranes. A detrital zircon recovered from the Vostok ice core (Leitchenkov et al., 2007) yielded ages that clustered between 0.8 Ga and 1.2 Ga and between 1.6 Ga and 1.8 Ga, and supports the concept that the large lakes form along tectonic boundaries. This is ample evidence for the close juxtaposition of basement rocks with a variety of ages in the interior of the East Antarctic Craton, evidently providing basins in which lakes have formed.

A profound discovery of recent years is the interconnectedness of these lakes with subglacial rivers and wetlands. Liquid water is verifiably on the move in many places beneath the ice sheet, and may have profound influence on ice-sheet stability and overall climate. Motion of water along the axis of ice streams is documented for the Recovery Ice Stream by Fricker et al. (2007).

THE NEXT BREAKTHROUGHS: A VIEW TO THE FUTURE

Today, understanding the changes in the polar regions is imperative for our global society, global economy, and global environment. As study of planetary change is critical to all Earth science, knowledge of Antarctica and Antarctic Earth science has never been more important. Convened at the beginning of the International Polar Year 2007-2008, the 10th ISAES was a successful opportunity for our global scientific community to share results, data, and ideas and to plan future cooperative programs.

The International Polar Year 2007-2008 is motivated by both our changing planet and the quest to explore unknown frontiers, both being central to Antarctic Earth science (Bell, 2008, this volume). The work accomplished in the next two years will define future research directions, and collaborations established during the International Polar Year will serve as the basis for decades of future research programs. Antarctica is a global keystone in the Earth system. Antarctic earth science must be both international and global to remain relevant. At the 11th ISAES meeting in four years time, new breakthroughs will be presented based on the International Polar Year results, to enhance the knowledge gained at the 10th ISAES. Fridtjof Nansen noted that humankind is driven to seek knowledge "till every enigma has been solved" (Nansen and Sverdrup, 1897).

REFERENCES

Armadillo, E., F. Ferraccioli, A. Zunino, and E. Bozzo. 2007. Aeromagnetic anomaly patterns reveal buried faults along the eastern margin of the Wilkes Subglacial Basin (East Antarctica). In *Antarctica: A Keystone in a Changing World—Online Proceedings for the Tenth International Symposium on Antarctic Earth Sciences,* eds. Cooper, A. K., C. R. Raymond et al., USGS Open-File Report 2007-1047. Short Research Paper 091, doi:10.3133/of2007-1047.srp091.

Ashworth, A. C., A. R. Lewis, D. R. Marchant, R. A. Askin, D. J. Cantrill, J. E. Francis, M. J. Leng, A. E. Newton, J. I. Raine, M. Williams, and A. P. Wolfe. 2007. The Neogene biota of the Transantarctic Mountains. In *Antarctica: A Keystone in a Changing World—Online Proceedings for the Tenth International Symposium on Antarctic Earth Sciences,* eds. Cooper, A. K., C. R. Raymond et al., USGS Open-File Report 2007-1047, Extended Abstract 071, http://pubs.usgs.gov/of/2007/1047/.

Bell, R. E. 2008, this volume. Antarctic Earth system science in the International Polar Year 2007-2008. In *Antarctica: A Keystone in a Changing World,* eds. Cooper et al. Washington, D.C.: The National Academies Press.

Bell, R. E., M. Studinger, and C. A. Finn. 2007. Tectonic control of subglacial lakes and ice sheet stability. In *Antarctica: A Keystone in a Changing World—Online Proceedings for the Tenth International Symposium on Antarctic Earth Sciences,* eds. Cooper, A. K., C. R. Raymond et al., USGS Open-File Report 2007-1047. Extended Abstract 040, http://pubs.usgs.gov/of/2007/1047/.

Bialas, R. W., W. R. Buck, M. Studinger, and P. G. Fitzgerald. 2007. Plateau collapse model for the Transantarctic Mountains/West Antarctic rift system: Insights from numerical experiments. In *Program Book for the 10th International Symposium on Antarctic Earth Sciences,* eds. A. K. Cooper et al. Short Summary 2.PS-49, p. 69. In *Antarctica: A Keystone in a Changing World—Online Proceedings for the Tenth International Symposium on Antarctic Earth Sciences,* eds. Cooper, A. K., C. R. Raymond et al., USGS Open-File Report 2007-1047, doi:10.3133/of2007-1047.

Bomfleur, B., J. Schneider, R. Schoner, L. Viereck-Gotte, and H. Kerp. 2007. Exceptionally well-preserved Triassic and Early Jurassic floras from North Victoria Land, Antarctica. In *Antarctica: A Keystone in a Changing World—Online Proceedings for the Tenth International Symposium on Antarctic Earth Sciences,* eds. Cooper, A. K., C. R. Raymond et al., USGS Open-File Report 2007-1047, Extended Abstract 034, http://pubs.usgs.gov/of/2007/1047/.

Bradshaw, J. D. 2007. The Ross orogen and Lachlan Fold belt in Marie Byrd Land, Northern Victoria Land and New Zealand: Implication for the tectonic setting of the Lachlan Fold belt in Antarctica. In *Antarctica: A Keystone in a Changing World—Online Proceedings for the Tenth International Symposium on Antarctic Earth Sciences,* eds. Cooper, A. K., C. R. Raymond et al., USGS Open-File Report 2007-1047. Short Research Paper 059, doi:10.3133/of2007-1047.srp059.

Bradshaw, M. A., and F. J. Harmsen. 2007. The paleoenvironmental significance of trace fossils in Devonian sediments (Taylor Group), Darwin Mountains to the Dry Valleys, southern Victoria Land. In *Antarctica: A Keystone in a Changing World—Online Proceedings for the Tenth International Symposium on Antarctic Earth Sciences,* eds. Cooper, A. K., C. R. Raymond et al., USGS Open-File Report 2007-1047, Extended Abstract 133, http://pubs.usgs.gov/of/2007/1047/.

Carter, S. P., D. D. Blankenship, D. A. Young, and J. W. Holt. 2007. Ice surface anomalies, hydraulic potential and subglacial lake chains in East Antarctica. In *Antarctica: A Keystone in a Changing World—Online Proceedings for the Tenth International Symposium on Antarctic Earth Sciences,* eds. Cooper, A. K., C. R. Raymond et al., USGS Open-File Report 2007-1047, Extended Abstract 142, http://pubs.usgs.gov/of/2007/1047/.

Case, J. A. 2007. Opening of the Drake Passage: Does this event correlate to climate change and biotic events from the Eocene La Meseta Formation, Seymour Island, Antarctic Peninsula? In *Antarctica: A Keystone in a Changing World—Online Proceedings for the Tenth International Symposium on Antarctic Earth Sciences,* eds. Cooper, A. K., C. R. Raymond et al., USGS Open-File Report 2007-1047, Extended Abstract 117, http://pubs.usgs.gov/of/2007/1047/.

Collinson, J. W., and W. R. Hammer. 2007. Migration of Triassic tetrapods to Antarctica. In *Antarctica: A Keystone in a Changing World—Online Proceedings for the Tenth International Symposium on Antarctic Earth Sciences,* eds. Cooper, A. K., C. R. Raymond et al., USGS Open-File Report 2007-1047, Extended Abstract 047, http://pubs.usgs.gov/of/2007/1047/.

Convey, P., J. A. E. Gibson, D. A. Hodgson, P. J. A. Pugh, and M. I. Stevens. 2007. New terrestrial biological constraints for Antarctic glaciation. In *Antarctica: A Keystone in a Changing World—Online Proceedings for the Tenth International Symposium on Antarctic Earth Sciences,* eds. Cooper, A.K., C. R. Raymond et al., USGS Open-File Report 2007-1047, Extended Abstract 053, http://pubs.usgs.gov/of/2007/1047/.

Cooper, A. K., C. R. Raymond, and the ISAES Editorial Team. 2007. *Antarctica: A Keystone in a Changing World—Online Proceedings for the Tenth International Symposium on Antarctic Earth Sciences,* USGS Open-File Report 2007-1047, doi:10.3133/of2007-1047.

Decesari, R. C., C. C. Sorlien, B. P. Luyendyk, D. S. Wilson, L. Bartek, J. Diebold, and S. E. Hopkins. 2007. Regional seismic stratigraphic correlations of the Ross Sea: Implications for the tectonic history of the West Antarctic Rift System. In *Antarctica: A Keystone in a Changing World—Online Proceedings for the Tenth International Symposium on Antarctic Earth Sciences,* eds. Cooper, A.K., C. R. Raymond et al., USGS Open-File Report 2007-1047, Short Research Paper 052, doi:10.3133/of2007-1047.srp052.

Finn, C. A., and S. Pisarevsky. 2007. New airborne magnetic data evaluate SWEAT reconstruction. In *Antarctica: A Keystone in a Changing World—Online Proceedings for the Tenth International Symposium on Antarctic Earth Sciences,* eds. Cooper, A.K., C. R. Raymond et al., USGS Open-File Report 2007-1047, Extended Abstract 170, http://pubs.usgs.gov/of/2007/1047/.

Fitzgerald, P. G., R. W. Bialas, W. R. Buck, and M. Studinger. 2007. A plateau collapse model for the formation of the West Antarctic rift system/Transantarctic Mountains. In *Antarctica: A Keystone in a Changing World—Online Proceedings for the Tenth International Symposium on Antarctic Earth Sciences,* eds. Cooper, A. K., C. R. Raymond et al., USGS Open-File Report 2007-1047, Extended Abstract 087, http://pubs.usgs.gov/of/2007/1047/.

Francis, J. E., A. Ashworth, D. J. Cantrill, J. A. Crame, J. Howe, R. Stephens, A.-M. Tosolini, and V. Thorn. 2008, this volume. 100 million years of Antarctic climate evolution: Evidence from fossil plants. In *Antarctica: A Keystone in a Changing World,* eds. A. K. Cooper et al. Washington, D.C.: The National Academies Press.

Fricker, H. A., R. E. Bell, and T. A. Scambos. 2007. Water budget through a series of interconnected subglacial lakes on Recovery Ice Stream, East Antarctica. In *Program Book for the 10th International Symposium on Antarctic Earth Sciences,* eds. A. K. Cooper et al. Short Summary 1.P2.B-3, p. 38. In *Antarctica: A Keystone in a Changing World—Online Proceedings for the Tenth International Symposium on Antarctic Earth Sciences,* eds. Cooper, A. K., C. R. Raymond et al., USGS Open-File Report 2007-1047, doi:10.3133/of2007-1047.

Goodge, J. W., D. M. Brecke, C. M. Fanning, J. D. Vervoort, I. S. Williams, and P. Myrow. 2007. Pieces of Laurentia in East Antarctica. In *Antarctica: A Keystone in a Changing World—Online Proceedings for the Tenth International Symposium on Antarctic Earth Sciences,* eds. Cooper, A. K., C. R. Raymond et al., USGS Open-File Report 2007-1047, Extended Abstract 055, http://pubs.usgs.gov/of/2007/1047/.

Grube, R., and B. Mohr. 2007. Deterioration and/or cyclicity? The development of vegetation and climate during the Eocene and Oligocene in Antarctica. In *Antarctica: A Keystone in a Changing World—Online Proceedings for the Tenth International Symposium on Antarctic Earth Sciences,* eds. Cooper, A. K., C. R. Raymond et al., USGS Open-File Report 2007-1047, Extended Abstract 075, http://pubs.usgs.gov/of/2007/1047/.

Huerta, A. D. 2007. Byrd drainage system: Evidence of a Mesozoic West Antarctic Plateau. In *Antarctica: A Keystone in a Changing World—Online Proceedings for the Tenth International Symposium on Antarctic Earth Sciences,* eds. Cooper, A. K., C. R. Raymond et al., USGS Open-File Report 2007-1047, Extended Abstract 091, http://pubs.usgs.gov/of/2007/1047/.

Kowalewski, D. E., and D. R. Marchant. 2007. Quantifying sublimation of buried glacier ice in Beacon Valley. In *Antarctica: A Keystone in a Changing World—Online Proceedings for the Tenth International Symposium on Antarctic Earth Sciences,* eds. Cooper, A. K., C. R. Raymond et al., USGS Open-File Report 2007-1047, Extended Abstract 115, http://pubs.usgs.gov/of/2007/1047/.

Leitchenkov, G. L., B. V. Belyatsky, N. V. Rodionov, and S. A. Sergeev. 2007. Insight into the geology of the East Antarctic hinterland: Study of sediment inclusions from ice cores of the Lake Vostok borehole. In *Antarctica: A Keystone in a Changing World—Online Proceedings for the Tenth International Symposium on Antarctic Earth Sciences,* eds. Cooper, A. K., C. R. Raymond et al., USGS Open-File Report 2007-1047, Short Research Paper 014, doi:10.3133/of2007-1047.srp014.

Maldonado, A., F. Bohoyo, J. Galindo-Zaldivar, F. J. Hernandez-Molina, F. J. Lobo, A. A. Shreyder, and E. Surinach. 2007. Early opening of Drake Passage: Regional seismic stratigraphy and paleoceanographic implications. In *Antarctica: A Keystone in a Changing World—Online Proceedings for the Tenth International Symposium on Antarctic Earth Sciences,* eds. Cooper, A. K., C. R. Raymond et al., USGS Open-File Report 2007-1047, Extended Abstract 057, http://pubs.usgs.gov/of/2007/1047/.

Martin, A. K. 2007. Double-saloon-door seafloor spreading: A new theory for the breakup of Gondwana. In *Antarctica: A Keystone in a Changing World—Online Proceedings for the Tenth International Symposium on Antarctic Earth Sciences,* eds. Cooper, A. K., C. R. Raymond et al., USGS Open-File Report 2007-1047, Extended Abstract 112, http://pubs.usgs.gov/of/2007/1047/.

Martin, J. E., J. F. Sawyer, M. Reguero, and J. A. Case. 2007. Occurrence of a young elasmosaurid plesiosaur skeleton from the Late Cretaceous (Maastrichtian) of Antarctica. In *Antarctica: A Keystone in a Changing World—Online Proceedings for the Tenth International Symposium on Antarctic Earth Sciences,* eds. Cooper, A. K., C. R. Raymond et al., USGS Open-File Report 2007-1047, Short Research Paper 066, doi:10.3133/of2007-1047.srp066.

Miller, M., and J. Isbell. 2007. Abrupt (how abrupt?) Permian-Triassic changes in southern polar ecosystems. In *Program Book for the 10th International Symposium on Antarctic Earth Sciences* eds. A. K. Cooper et al. Short Summary 2.P1.B-3, p. 57. In *Antarctica: A Keystone in a Changing World—Online Proceedings for the Tenth International Symposium on Antarctic Earth Sciences,* eds. Cooper, A. K., C. R. Raymond et al., USGS Open-File Report 2007-1047, doi:10.3133/of2007-1047.

Naish, T. R., R. D. Powell, P. J. Barrett, R. H. Levy, S. Henrys, G. S. Wilson, L. A. Krisse, F. Niessen, M. Pompilio, J. Ross, R. Scherer, F. Talarico, A. Pyne, and the ANDRILL-MIS Science team. 2008, this volume. Late Cenozoic climate history of the Ross Embayment from the AND-1B drill hole: Culmination of three decades of Antarctic margin drilling. In *Antarctica: A Keystone in a Changing World,* eds. A. K. Cooper et al. Washington, D.C.: The National Academies Press.

Nansen, F., and O. N. Sverdrup. 1897. *Farthest North: Being the Record of a Voyage of Exploration of the Ship "Fram" 1893-96, and of a Fifteen Months' Sleigh Journey by Dr. Nansen and Lieut. Johansen.* New York: Harper.

Pekar, S. F., M. A. Speece, D. M. Harwood, F. Florindo, and G. Wilson. 2007. Using new tools to explore undiscovered country. In *Antarctica: A Keystone in a Changing World—Online Proceedings for the Tenth International Symposium on Antarctic Earth Sciences,* eds. Cooper, A. K., C. R. Raymond et al., USGS Open-File Report 2007-1047, Extended Abstract 169, http://pubs.usgs.gov/of/2007/1047/.

Raub, T. D., and J. L. Kirschvinck. 2008, this volume. A Pan-Precambrian link between deglaciation and environmental oxidation. In *Antarctica: A Keystone in a Changing World,* eds. A. K. Cooper et al. Washington, D.C.: The National Academies Press.

Ryberg, P. E., and E. L. Taylor. 2007. Silicified wood from the Permian and Triassic of Antarctica: Tree rings from polar paleolatitudes. In *Antarctica: A Keystone in a Changing World—Online Proceedings for the Tenth International Symposium on Antarctic Earth Sciences,* eds. Cooper, A. K., C. R. Raymond et al., USGS Open-File Report 2007-1047, Short Research Paper 080, doi:10.3133/of2007-1047.srp080.

Siddoway, C. S. 2008, this volume. Tectonics of the West Antarctic rift system: New light on the history and dynamics of distributed intracontinental extension. In *Antarctica: A Keystone in a Changing World,* eds. A. K. Cooper et al. Washington, D.C.: The National Academies Press.

Smith, N. D., P. J. Makovicky, D. Pol, W. R. Hammer, and P. J. Currie. 2007. The dinosaurs of the Early Jurassic Hanson Formation of the Central Transantarctic Mountains: Phylogenetic review and synthesis. In *Antarctica: A Keystone in a Changing World—Online Proceedings for the Tenth International Symposium on Antarctic Earth Sciences,* eds. Cooper, A. K., C. R. Raymond et al., USGS Open-File Report 2007-1047, Short Research Paper 003, doi:10.3133/of2007-1047.srp003.

Sutherland, R. 2008, this volume. The significance of Antarctica for studies of global geodynamics. In *Antarctica: A Keystone in a Changing World,* eds. A. K. Cooper et al. Washington, D.C.: The National Academies Press.

Vieira, G., M. Ramos, S. Gruber, C. Hauck, and J. Blanco. 2007. The permafrost environment of northwest Hurd Peninsula (Livingston Island, Maritime Antarctic): Preliminary results. In *Antarctica: A Keystone in a Changing World—Online Proceedings for the Tenth International Symposium on Antarctic Earth Sciences,* eds. Cooper, A. K., C. R. Raymond et al., USGS Open-File Report 2007-1047, Extended Abstract 206, http://pubs.usgs.gov/of/2007/1047/.

Wardell, N., J. R. Childs, and A. K. Cooper. 2007. Advances through collaboration: Sharing seismic reflection data via the Antarctic Seismic Data Library System for Cooperative Research (SDLS). *Antarctica: A Keystone in a Changing World—Online Proceedings for the Tenth International Symposium on Antarctic Earth Sciences,* eds. Cooper, A. K., C. R. Raymond et al., USGS Open-File Report 2007-1047, Short Research Paper 001, doi:10.3133/of2007-1047.srp001.

Cooper, A. K., P. J. Barrett, H. Stagg, B. Storey, E. Stump, W. Wise, and the 10th ISAES editorial team, eds. (2008). *Antarctica: A Keystone in a Changing World.* Proceedings of the 10th International Symposium on Antarctic Earth Sciences. Washington, DC: The National Academies Press.

Antarctic Earth System Science in the International Polar Year 2007-2008

R. E. Bell[1]

ABSTRACT

The International Polar Year (IPY) 2007-2008 is the largest coordinated effort to understand the polar regions in our lifetime. This international program of science, discovery, and education involves more than 50,000 scientists from 62 nations. The IPY 2007-2008 Antarctic Earth System Science themes are to determine the polar regions' present environmental status, quantify and understand past and present polar change, advance our understanding of the links between polar regions and the globe, and investigate the polar frontiers of science. There are several key IPY 2007-2008 Earth System Science projects in the Antarctic:

• POLENET will capture the status of the polar lithosphere through new instrument arrays.
• ANDRILL will uncover past change using novel drilling technologies.
• "Plates and Gates" will advance our understanding of the teleconnections between Antarctica and the global climate system.
• AGAP and SALE-UNITED will study hitherto unsampled subglacial mountains and lakes.

A new era of international collaboration will emerge along with a new generation of Antarctic scientists and a legacy of data and enhanced observing systems.

INTRODUCTION TO IPY 2007-2008

Earth science revolves around the study of our changing planet. IPY 2007-2008 is motivated both by the need to improve our understanding of this changing planet and by a quest to explore still unknown frontiers, especially those beneath the vast Antarctic ice sheet. While intuitively we all appreciate that planetary change happens, as the concept of an international polar year emerged the evidence for rapid and dynamic planetary change was becoming omnipresent. Early in 2002 an ice shelf that had been stable for at least 20,000 years (Domack et al., 2005), the Larsen B Ice Shelf, collapsed in a matter of weeks (Scambos et al., 2004). This ice shelf collapse began to bring together the timescale of planetary change with the more familiar human timescale of days, week, and months.

By 2002 change in the polar regions was undeniable. The surface melt in Greenland was increasing in extent (Steffen et al., 2004), sea ice cover in the Arctic was beginning to decrease notably (Johannessen et al., 1999), the glaciers feeding the Weddell Sea accelerated after the Larsen Ice Shelf collapsed (Rignot et al., 2004; Scambos et al., 2004), and the Amundsen Sea sector of the West Antarctic ice sheet was thinning and accelerating (Joughin et al., 2003; Shepherd et al., 2004). The polar environments were clearly changing, and doing so rapidly. While change is being observed globally, the environmental change at the poles is taking place faster than environmental change anywhere else on the planet. The ramifications of these polar changes reach far beyond the Antarctic and Arctic, because as ice melts, sea levels rise. The need to understand the changes in the polar regions—past, present, and future—is imperative for our global society, the global economy, and the global environment, as well as being a fascinating scientific challenge in its

[1]Lamont-Doherty Earth Observatory of Columbia University, Palisades, New York, 10964-8000, USA (robinb@ldeo.columbia.edu).

own right. This overwhelming sense of rapid and increasing change, and what we can learn from it, especially in the polar regions where the change is fastest, is one of the major motivations for the development of the IPY (Albert, 2004).

Beyond the sense of planetary change, discovery is a keen motivator. In the 1800s exploration was conducted by teams of men pressing for the poles, led by Nansen, Amundsen, Scott, and Shackleton. Nansen captured the essence of geographically motivated exploration, noting that humankind's spirit "will never rest till every spot of these regions has been trod upon" (Nansen and Sverdrup, 1897). Today we seek to push the frontiers of our knowledge of processes, rather than geographic frontiers. The community remains intensely motivated by Nansen's second restless quest, which is to seek knowledge "till every enigma has been solved." While Nansen sought to understand the circulation in the Arctic Ocean, today we see major opportunities for scientific discovery in Antarctic Earth science. Discovery is the second motivation for this IPY. The frontiers are no longer the geographic poles but the regions and processes hidden by kilometers of ice and water.

Propelled forward by the sense of planetary change and a sense of discovery, IPY 2007-2008 will be the largest internationally coordinated research program in 50 years, actively engaging over 50,000 scientists from 62 nations. The result of a five-year community-wide planning process, the IPY 2007-2008 will be an intensive period of interdisciplinary science focused on both polar regions—the Antarctic and the Arctic. The projects of the IPY 2007-2008 focus on deciphering these processes of change in the polar regions and their linkages with the rest of the globe while also exploring some of the final frontiers.

FRAMEWORK

Today in our global interconnected world, "year" events occur at a frenetic pace. For example, 2002 was the U.N. International Year of Mountains and 2005 was the U.N. Year of Physics. These U.N.-sponsored events tend to focus on celebration, awareness, and education. For instance, in parallel with the IPY 2007-2008, Earth scientists have developed the U.N. International Year of Planet Earth, a celebration of the role of Earth science in society. Awareness, celebration, and education have consistently been a key facet of all IPYs since 1882-1883, but the central framework for the polar years has always been to facilitate collaborative science at a level impossible for any individual nation.

The concept of collaborative international polar science focused on a specific period was developed by Lt. Karl Weyprecht, an Austrian naval officer. Weyprecht was a scientist and co-commander of the Austro-Hungarian North Pole Expedition that set off in 1872 in a three-masted schooner, the *Admiral Tegetthoff*. The expedition returned two years later without the schooner. The expedition had abandoned the ship frozen into the pack ice and hauled sledges over the pack

ice for 90 days to the relative safety of open water. Weyprecht was acutely aware that the systematic observations necessary to advance the understanding of the fundamental problems of meteorology and geophysics were impossible for men hauling sledges across the ice and struggling to survive. His frustration with the inability to understand polar phenomena with the data from a single national expedition is captured in his "Fundamental Principles of Arctic Research" (Weyprecht, 1875). He noted that "whatever interest all these observations may possess, they do not possess that scientific value, even supported by a long column of figures, which under other circumstances might have been the case. They only furnish us with a picture of the extreme effects of the forces of Nature in the Arctic regions, but leave us completely in the dark with respect to their causes."

Weyprecht believed that the systematic successful study of the polar regions and large-scale polar phenomena by a single nation was impossible. He argued that fixed stations where coordinated observations could be made were necessary for consistency of measurements. Weyprecht's insights were central to the planning and execution of the first IPY.

A realization of Weyprecht's vision, the first IPY (1882-1883) involved 12 countries launching 15 expeditions to the poles: 13 to the Arctic and 2 to the Antarctic, using coal- and steam-powered vessels (Figure 1). At each station a series of

FIGURE 1 First International Polar Year. *Top:* Norwegian ship in the ice, Kara Sea (from Steen [1887]). *Bottom:* Observer making temperature observations at Fort Conger, 1882 (from Greely [1886]).
SOURCE: See http://www.arctic.noaa.gov/aro/ipy-1/US-LFB-P4.htm.

FIGURE 2 Second International Polar Year. *Left:* Aircraft in Antarctica (from the Ohio State University Archives, Papers of Admiral Richard E. Byrd, #7842_18). *Right:* Launching a weather balloon in northern Canada (from the University of Saskatchewan Archives, 1931).

regular observations were recorded with pencil and paper, from meteorology to Earth's magnetic field variations (Figure 1). Beyond the advances to science and geographical exploration, a principal legacy of the first IPY was setting a precedent for international science cooperation. The clear gap between Weyprecht's vision and the outcomes was that the data were never fully integrated and analyzed together (Wood and Overland, 2006).

During the first IPY (1882-1883) the Brooklyn Bridge opened in New York City, five years before the Eiffel Tower opened in Paris. Fifty years later the second IPY (1932-1933) began, and this was the year the Empire State Building opened in New York City. Routine flights by aircraft and wireless communication were both now possible. During the second IPY, 40 nations conducted Arctic research focusing on meteorology, magnetism, atmospheric science, and ionospheric physics. Forty permanent observation stations were established in the Arctic. The U.S. contribution to the second IPY was the second Byrd Antarctic expedition. The Byrd expedition established the first inland research station, a winter-long meteorological station on the Ross Ice Shelf at the southern end of Roosevelt Island. Scientists employed aircraft to extend the range of their observations and for the first time received data transmitted back from balloons as they drift upward, allowing the first vertical sampling of the polar atmosphere (Figure 2).

The third IPY expanded beyond the polar regions, quickly becoming global. This polar year was renamed the International Geophysical Year (IGY) and ran from July 1, 1957, to December 31, 1958 (see Figure 3). Coincident with the groundbreaking for the Sydney Opera House, the IGY

was the brainchild of a small number of eminent physicists, including Sydney Chapman, James Van Allen, and Lloyd Berkner. These physicists realized that the technology developed during World War II, such as rockets and radar, could be deployed to advance science. Sixty-seven nations participated in the IGY. The IGY's research, discoveries, and vast array of synoptic observations set the stage for decades of geophysical investigations. Data collected from ships were used subsequently to advance the theory of plate tectonics, while satellites detected the Van Allen Radiation Belt. Seismic measurements collected along geophysical traverses measured the thickness of the Antarctic ice sheet, enabling the first estimates of Antarctica's ice mass. Emerging from the IGY was the Scientific Committee on Antarctic Research (SCAR) in 1958 and the Antarctic Treaty in 1961. Permanent stations were established for the first time in Antarctica as a direct result of the IGY.

At the end of the IGY, Hugh Odishaw, executive director of the U.S. National Committee, noted, "We have only scratched the surface of our ignorance with respect to Antarctic. . . . There is at hand an unparalleled situation for stimulating the best in man" (Odishaw, 1959). Having scratched the surface of Antarctica, scientists and engineers of 1958 handed the baton to our generation to bring together scientists and engineers to understand the role the poles play in our rapidly changing world. It is for us to explore the remaining frontiers using the cutting-edge technologies available to us today: jet aircraft, ships, satellites, lasers, the Global Positioning System (GPS), advanced communications, computers, numerical modeling, passive seismics, autonomous observatories, and novel coring technologies.

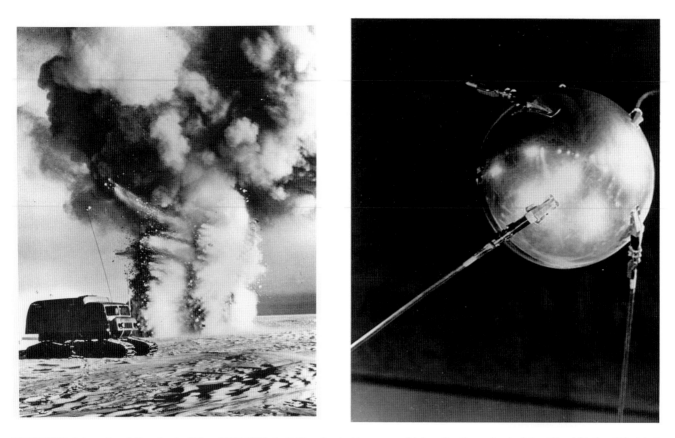

FIGURE 3 International Geophysical Year 1957-1958. *Left:* Geophysical traverse vehicle collecting seismic data at Byrd Station, West Antarctica (from http://www.nas.edu/history/igy/seismology.html). *Right: Sputnik,* the first satellite launched into space (image from NASA).

IPY 2007-2008 PLANNING PROCESS

The development of the IPY 2007-2008 began with an extended period of community education with scientists studying the accounts of the previous IPYs and discussing the results. The concept of another polar year emerged through discussions among communities and with funding agencies. Websites documenting the contributions of the first three IPYs were launched. These discussions were first documented in SCAR meeting reports in 2000. At the 2000 SCAR meeting in Tokyo, K. Erb, president of the Council of Managers of National Antarctic Programs, reported on discussions to "prepare for recognition of the 50th Anniversary of the International Polar Year." One year later the discussions were still on the margins, with a short note in the report from the SCAR executive meeting, under the title "Any Other Business." Again the focus was on celebration, not action. The Neumayer International Symposium in 2001 provided a further opportunity for the community to discuss the impending 50th anniversary of the IGY and the 125th anniversary of the first IPY, but the focus had not yet moved beyond celebration plans.

The first concrete plan for IPY 2007-2008 science was presented by Heinz Miller at the 2002 SCAR meeting in Shanghai (SCAR, 2002). Miller proposed a series of mul-

tidisciplinary traverses along all the major ice divides of East Antarctica. The SCAR delegates supported a proposal to develop a "celebration." In late 2002 the U.S. National Academy of Sciences Polar Research Board convened a one-day international workshop addressing the question of whether an IPY was an applicable framework for science in 2007. Workshop participants strongly supported the IPY concept, and within the next six months the U.S. National Academies began the planning process in the United States (Albert, 2004). Simultaneously Chris Rapley and Robin Bell submitted a proposal to the International Council for Science (ICSU) Executive Committee to form a planning committee for the IPY. ICSU approved the proposal and the Planning Committee met for the first time in August 2003. The planning had moved past a celebration to setting a science agenda for a major international interdisciplinary effort (Figure 4).

Independently a group at the World Meteorological Organization (WMO), a sponsor of the earlier IPYs, had begun considering involvement in the proposed IPY 2007-2008. WMO joined the ICSU planning process in an advisory role. With the completion of the framework document (Rapley and Bell, 2004), the ICSU Planning Committee's work was complete. Together ICSU and WMO formed a new joint steering group called the Joint Committee, with

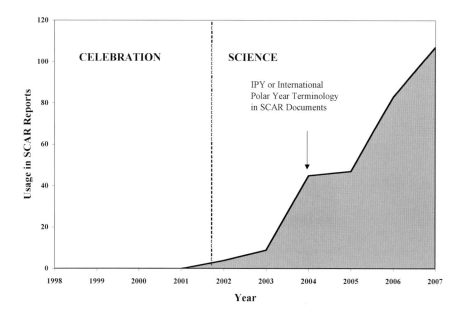

FIGURE 4 Emergence of International Polar Year 2007-2008 as documented in SCAR reports. Usage of the terms "IPY" and "International Polar Year" in SCAR reports from 1998 to 2007. Discussion focused on a celebration until 2002, when the emphasis became science.

two co-chairs with representation from ICSU and WMO. A competition for an IPY International Program Office (IPO) was announced, and the successful bid was submitted by the British Antarctic Survey. An executive director for the IPY was hired, and the IPY 2007-2008 moved from the design phase to the implementation phase.

The implementation phase required that the Joint Committee develop a process to review the proposed work for the IPY 2007-2008 and develop a science plan. In due course the Joint Committee reviewed over 300 proposals that form the core of the IPY Science Program. The committee worked with the authors of proposals to ensure that many smaller projects could be accommodated under larger umbrella projects, like CASO (Climate of Antarctica and the Southern Ocean), which includes several smaller projects that started as Letters of Intent, but all contribute to the overall CASO goal. The aim was to encourage development of a relatively small number of large projects that would "make a difference."

Development of the project proposals was not top-down but a direct result of the community input. Taken together the 300 or so projects can be seen to make up a comprehensive and integrated science plan, which was published by the Joint Committee in 2007 (Allison et al., 2007). In March 2007 the polar year opened around the globe with events in over 30 nations signaling the beginning of two years of intensive polar observation and analysis. This current year is just the beginning.

THEMES

This four-year grassroots planning process defined six scientific themes that are the essential framework for IPY 2007-2008:

1. **Status:** Determine the present environmental status of the polar regions;

2. **Change:** Quantify and understand past and present natural, environmental, and social change in the polar regions and improve projections of future change;

3. **Global Linkages:** Advance understanding on all scales of the links and interactions between the polar regions and the rest of the globe and of the controlling processes;

4. **New Frontiers:** Investigate the frontiers of science in the polar regions;

5. **Vantage Point:** Use the unique vantage point of the polar regions to develop and enhance observatories from the interior of Earth to the sun and the cosmos beyond; and

6. **The Human Dimension:** Investigate the cultural, historical, and social processes that shape the sustainability of circumpolar human societies and identify their unique contributions to global cultural diversity and citizenship.

The first three themes—Status, Change, and Global Linkages—capture the changing planet, with an emphasis on changing climate at all times and scales (Figure 5 as the red triangle, or delta symbol). The majority of IPY 2007-2008 projects address these themes. The fourth—New Frontiers, or the exploration and discovery theme—has its largest projects in Antarctic Earth sciences as predicted by Odishaw at the close of the IGY. These frontiers in Earth science are primarily beneath the ice sheet or beneath the ocean floor (blue box at the base of Figure 5). The Vantage Point theme encompasses everything from observatories to examine the inner core to telescopes monitoring distant galaxies. The Human Dimension theme of this IPY 2007-2008 has a strong presence in the north but is not developed in the south due to the absence of a large permanent human population there.

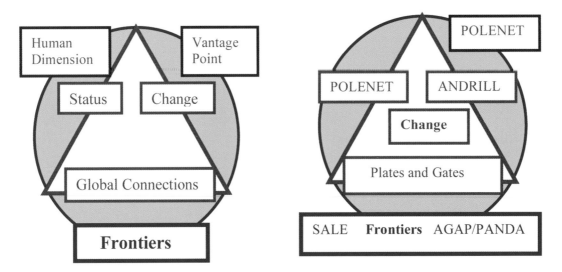

FIGURE 5 *Left:* Schematic of major IPY 2007-2008 themes. *Right:* Mapping major Antarctic Earth science projects.

MAJOR INTERNATIONAL POLAR YEAR EARTH SCIENCE PROJECTS

The principal IPY 2007-2008 Antarctic Earth science projects address change and all its components and exploration of new frontiers. The Antarctic Earth science components of the IPY 2007-2008 are primarily found within the Earth, land, and ice sector of the framework (Allison et al., 2007). Twelve projects fall into the category of Antarctic Earth sciences (Table 1).

Many of these projects are inherently interdisciplinary. This paper highlights five large projects that capture the breadth of the IPY 2007-2008 Earth science programs, both in terms of the time periods they address and in terms of their geographic locations. POLENET (Polar Earth Observing Network) will be implemented in both polar regions and will capture the tectonic and isostatic status of the Antarctic plate. POLENET is the major Antarctic Earth science project under the Status theme.

ANDRILL (Antarctic Geologic Drilling), which in late 2006 recovered the longest rock core in Antarctica, is a highly visible element of the Antarctic Climate Evolution (ACE) program, which integrates modeling studies and observational data from the Antarctic margin to resolve the continent's paleoenvironmental history. ANDRILL together with ACE clearly illustrate both the Change theme and the Global Linkages theme, as they examine the teleconnections between Northern and Southern Hemisphere climate change.

"Plates and Gates," by focusing on the tectonic and sedimentary formation of ocean gateways that are critical for controlling major water masses and global change, is a clear example of an investigation of the global linkages between polar processes and global climate. SALE-UNITED and the AGAP/PANDA projects, which target the origin, evolution, and setting of subglacial lakes and the Gamburtsev subglacial highlands and the structure of the Dome A region, are both clear examples of the New Frontiers theme. As the POLENET observatories will include components capable of examining structure deep within Earth's interior, that project is also emblematic of the Vantage Point theme.

POLENET

POLENET is a consortium of scientists from 24 nations that will deploy a diverse suite of geophysical instruments aiming at interactions of the atmosphere, oceans, polar ice sheets, and Earth's crust and mantle (Figure 6). The POLENET teams will be deploying new GPS instruments, seismic stations, magnetometers, tide gauges, ocean-floor sensors, and meteorological recorders. The science program of the POLENET consortium will investigate polar geodynamics; Earth's magnetic field, crust, mantle, and core structure and dynamics; and systems-scale interactions of the solid Earth, cryosphere, oceans, and atmosphere. Activities will focus on the deployment of autonomous observatories at remote sites on the continents and offshore, coordinated with measurements made at permanent stations and by satellite campaigns. Geophysical observations made by POLENET will contribute to many branches of geoscience and glaciology. For example, sea-level and ice-sheet monitoring can be fully modeled only when measurements of solid Earth motions are incorporated. Both plate tectonic and paleoclimate studies benefit from crustal deformation results. POLENET's approach to install autonomous observatories collecting coordinated measurements captures Weyprecht's vision of coordinated observations from fixed stations.

TABLE 1 Antarctic Earth Science IPY 2007-2008 Proposals from Scope of Science Report

Proposal No. and Short Title		Long Title
67	AGAP	Antarctica's Gamburtsev Province; origin, evolution, and setting of the Gamburtsev subglacial highlands; exploring an unknown territory
256	ANDRILL	Antarctic Geologic Drilling: Antarctic continental margin drilling to investigate Antarctica's role in global environmental change
109	Rift System Dynamics	Geodynamics of the West Antarctic Rift System and its implications for the stability of the West Antarctic ice sheet
77	Plates and Gates	Plate tectonics and polar ocean gateways
152	IDEA	Ice Divide of East Antarctica: Trans-Antarctic Scientific Traverses expeditions
185	POLENET	Polar Earth Observing Network
886	USGS	U.S. Geological Survey: Integrated Research
54	ACE	Antarctic Climate Evolution
97	ICECAP	Investigating the Cryospheric Evolution of the Antarctic Plate
33	ANTPAS	Antarctic Permafrost and Soils
313	PANDA	The Prydz Bay, Amery Ice Shelf, and Dome-A Observatories
42	SALE-UNITED	Subglacial Antarctic Lake Environments-Unified International Team for Exploration and Discovery

FIGURE 6 *Left:* Location of planned POLENET sites in Antarctica. Stars = New stations. *Center:* Deploying remote GPS station. *Right:* Deploying remote seismic station. SOURCE: www.polenet.org.

ANDRILL

ANDRILL is the latest of a series of floating-ice-based drilling projects on the Antarctic margin, complementing ship-based projects that date back to Deep Sea Drilling Project Leg 28 in 1972 (Figure 7). ANDRILL is a multinational collaboration of more than 200 scientists, students, and educators from Germany, Italy, New Zealand, and the United States; it is also the largest Antarctic Earth science IPY 2007-2008 project targeting the Change and global linkages themes.

ANDRILL is an example of a major international project that developed in parallel with the IPY 2007-2008 planning and became an integral part of the IPY 2007-2008 program. ANDRILL is a scientific drilling project investigating Antarctica's role in global climate change over the last 60 million years. Employing new drilling technology designed specifically for ice-shelf conditions as well as state-of-the-art core analysis and ice-sheet modeling, ANDRILL addresses four scientific issues: (1) the history of Antarctica's climate and ice sheets; (2) the evolution of polar biota and ecosystems; (3) the timing and nature of major tectonic and volcanic episodes; and (4) the role of Antarctica in Earth's ocean-climate system. ANDRILL's goal is to drill a series of holes in the McMurdo Sound area in regions that have previously been inaccessible to ship-based drilling technologies. The strati-

FIGURE 7 *Top left:* ANDRILL drill rig in the McMurdo Sound. *Top right:* Conducting core analysis on site. *Bottom:* Core in core trays. SOURCE: Images are from www.andrill.org.

graphic records retrieved from these boreholes will cover critical time periods in the development of Antarctica's major ice sheets. The sediment cores will be used to construct an overall glacial and interglacial history for the region, including documentation of sea-ice coverage, sea level, terrestrial vegetation, and meltwater discharge events. The cores will also provide a general chronostratigraphic framework for regional seismic studies to help unravel the area's complex tectonic history. The first borehole was drilled in 2006-2007, and drilling the second borehole will begin in the 2007-2008 field season.

Plates and Gates

"Plates and Gates" is a multidisciplinary project aimed at understanding key polar oceanographic gateways for major water masses. Ocean gateways are the deepwater passageways that form as continents rift apart and that are destroyed when ocean basins close (Figure 8). As part of the global ocean circulation, water masses move through these passageways, and major shifts in the ocean gateways produce changes in the transport of heat, salt, and nutrients. These changes in global ocean circulation may trigger changes in global climate. Plates and Gates will establish detailed tectonic, geodynamic, sedimentary, and paleotopographic histories of major oceanic gateways, providing basic constraints for global climate modeling. In Antarctica, key areas are the Drake Passage and the former Tasman Gateway, the last barriers to the establishment of the Antarctic Circumpolar Current. The Antarctic Circumpolar Current is the fundamental vehicle for mass flux between the Pacific, Atlantic, and Indian oceans and has the largest flux of all the globe's ocean

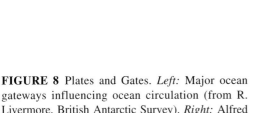

FIGURE 8 Plates and Gates. *Left:* Major ocean gateways influencing ocean circulation (from R. Livermore, British Antarctic Survey). *Right:* Alfred Wegner Institute's R/V *Polarstern* (from K. Gohl, Alfred Wegner Institute).

currents. The establishment of this large global ocean current has been suggested as one of the major triggers for the onset of Antarctic glaciation (Kennett, 1977). Field projects and modeling will be linked from the beginning to yield a model of the present status of water mass exchange in these regions, development of "gateway flow dynamics models" as well as a new range of high-resolution paleooceanographic and paleoclimate models. Plates and Gates will employ a coupled ocean atmosphere general circulation model to examine the impact of opening and closing of high- and low-latitude gateways on Eocene-Oligocene and Pliocene-Pleistocene climate changes. This project started in January 2007 with an expedition onboard the R/V *Polarstern.*

AGAP and PANDA

AGAP and PANDA together are a multinational, multidisciplinary aerogeophysical, traverse, and passive seismic instrumentation effort to explore East Antarctic ice-sheet history and the lithospheric structure of the Gamburtsev Subglacial Mountains (see Figure 9). AGAP and PANDA together assembled scientists from six nations to launch the IPY 2007-2008 Gamburtsev Subglacial Mountains Expedition. The Gamburtsev Subglacial Mountains are a major mountain range, larger than the Alps but virtually unexplored since they were discovered during the IGY in 1958. The Gamburtsev Subglacial Mountains are a 400-km-wide elevated massif rising 2000-3000 m above the regional topography and resting beneath the ice divide at Dome A, the

highest plateau of the Antarctic ice sheet. While both the age and origin of the Gamburtsev Mountains in the framework of Antarctic and Gondwana tectonics has been a matter of considerable speculation, this unusual mountain range has been advanced as the nucleation point for two continental-scale glaciations, one in the late Paleozoic and one in the Cenozoic. The Gamburtsev Subglacial Mountains today are encased beneath 1 to 4 km of continental ice, but their role in ice-sheet dynamics, specifically as the nucleation point for continental-wide glaciation, remains speculative given our lack of knowledge about their age and origin.

AGAP will use aircraft and surface instrumentation to collect major new datasets, including gravity, magnetics, ice radar, and other geological observations. The airborne team (GAMBIT) will acquire an extensive new airborne dataset, including gravity, magnetics, ice thickness, synthetic aperture radar images of the ice-bed interface, near-surface and deep internal layers, and ice surface elevation. The nested survey design will include a dense high-resolution survey over Dome A augmented by long regional lines to map the tectonic structures. The interpretation of these datasets will advance our understanding of ice-sheet dynamics, subglacial lakes, and Antarctic tectonics. The seismic team (GAMSEIS) will deploy portable broadband seismographs to examine the seismic structure of the crust and upper mantle. The PANDA transect from Prydz Bay-Amery Ice Shelf-Lambert Glacier Basin-Dome A covers an interconnected ocean, ice-shelf, and ice-sheet system, which plays a very important role in East Antarctica mass balance, sea level, and climate change.

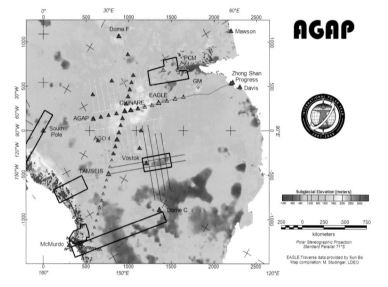

FIGURE 9 *Top:* Aerogeophysical Twin Otter in East Antarctica. *Bottom:* AGAP survey design over ice surface (yellow grid).
SOURCE: Images are from M. Studinger, Lamont-Doherty Earth Observatory.

SALE-UNITED

SALE-UNITED is a federation of interdisciplinary researchers who will conduct expeditions to explore all facets of subglacial lake environments under the New Frontiers theme during the IPY 2007-2008 (Figure 10). The SALE-UNITED research and exploration project will investigate subglacial lake environments of differing ages, evolutionary histories, and physical settings. These comparative studies will provide a holistic view of subglacial environments over millions of years and under differing climatic conditions. Included in the SALE-UNITED IPY 2007-2008 projects will be the drilling into Lake Vostok in 2008-2009, acquisition of surface geophysics over Lake Ellsworth in West Antarctica and Lake Concordia near Dome Concordia, and acquisition of both airborne and surface geophysics over the Recovery Lakes in Queen Maud Land. Each one of these projects represents a major advance in the study of subglacial lakes. The successful recovery of water from drilling into Lake Vostok will provide key new insights into the fundamental questions as to the basic nature of the water in the lakes and whether life can be supported in the water column. Lakes Ellsworth and Concordia are also targeted for sampling. The surface geophysics acquired during IPY 2007-2008 will provide seismic constraints on the volume of water and the processes within

these lakes. The Recovery Lakes appear to be triggering the onset of rapid ice flow into the Recovery Ice Stream (Bell et al., 2007). Seismic data collected by the U.S.–Norwegian traverse team will provide insights into the volume of subglacial water, while the airborne geophysics will constrain the flux of ice and water through the system. Additionally, the SALE-UNITED group will conduct genomic studies of accreted ice and modeling of lake stability. Together these projects will advance our understanding of the subglacial lake environment and the role lakes play in ice-sheet stability.

AN INTEGRATED VISION FOR ANTARCTIC EARTH SCIENCE IN THE IPY 2007-2008

The five major projects described above capture the major IPY 2007-2008 themes that were defined in the planning process. Together they illustrate the breadth of the Antarctic Earth science programs for the IPY 2007-2008. While at first glance these projects may appear to be narrowly focused on climate, they capture the entire scope of major Antarctic geophysical events (Table 2). For example, AGAP and PANDA will be targeting the basic cratonic structure of East Antarctica, structures that likely formed during the assembly of Rodinia and other key events early in the history of the Antarctic continent. Plates and Gates focuses on the

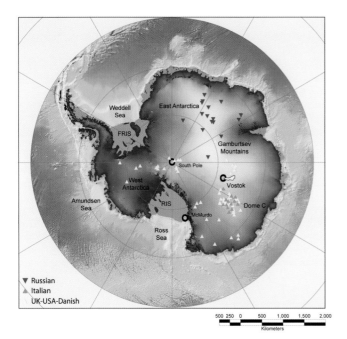

FIGURE 10 *Top:* SALE-UNITED. Location of subglacial lakes (from M. Studinger, http://www.ldeo.columbia.edu/~mstuding/vostok.html) *Bottom:* Traverse vehicles for U.S.–Norwegian traverse of the Recovery Lakes (from Norwegian Polar Institute).

break-up of Gondwana and the subsequent formation of the ocean gateways between continental fragments. Both AGAP and POLENET will contribute to resolving the history of extension in the Ross Sea, while controls on Transantarctic Mountains uplift will be elucidated by both Plates and Gates and ANDRILL. The origin, history, and stability of the Antarctic ice sheets will be addressed by all these projects.

OUTSTANDING CHALLENGES

As of this writing we are still at an early stage of the IPY 2007-2008, less than six months after the official opening in Paris and around the globe on March 1, 2007. Since the identification of funding for national components of IPY 2007-2008–approved projects is still ongoing in many countries, it remains difficult to fathom the full breadth of the IPY 2007-2008. But several hundred million dollars of new money for polar science is already available, and more is likely.

Even at this early stage it is useful to consider the pitfalls that other IPYs have encountered that prevented them from realizing their full potential. The results of the first IPY were never fully realized because each nation worked to publish their data individually. The data were not openly shared, and long-term collaboration between nations did not materialize. In our more electronically connected world there is little excuse for data not to be openly shared and deposited in the appropriate data repository so that future generations can make use of this precious resource.

Building collaborations within a discipline is simple. The challenge for this IPY 2007-2008 is to establish long-lasting, effective working relationships across disciplines. Many of the projects contain the seeds of these difficult multidisciplinary relationships, whether between modelers and field scientists or between biologists and geophysicists. These interactions must be fostered and developed. Collaborations built on shared data and shared passions in Earth systems are essential. Interdisciplinary science is difficult, but it will be the only way Earth science will remain relevant to society, and it is the only way to gain the full return from our investment. True, open interdisciplinary collaboration will serve to advance our understanding of the poles and the role they play in our planetary system.

TABLE 2 Five Major Antarctic Geophysical Events Targeted by IPY 2007-2008 Programs

Major Event	Formation of East Antarctic Craton	Breakup of Gondwana	Extension of the Ross Sea	Uplift of the Transantarctic Mountains	Origin, History, Status, Stability of Antarctic Ice Sheets
IPY 2007-2008 Programs	AGAP, PANDA	Plates and Gates	ANDRILL, POLENET	ANDRILL, Plates and Gates	AGAP, PANDA, ANDRILL, Plates and Gates, POLENET, SALE-UNITED

ANTICIPATED OUTCOMES

This IPY 2007-2008 marks the beginning of a new era in polar science that will routinely involve cooperation between a wide range of research disciplines from geophysics to ecology. While much of our science today is international, the IPY 2007-2008 is truly the largest international scientific endeavor of any kind that most of us will ever witness. Just as many of today's leaders in science and engineering entered these fields because of the IGY, the next generation will be captivated by the powerful science projects of the IPY 2007-2008. The IPY 2007-2008 will open new frontiers, specifically East Antarctica and the subglacial environment, to the international scientific community. From the IPY 2007-2008 will emerge new collaborative frameworks for science that will continue to enable each of us to accomplish more than we could have accomplished as individual scientists or as individual nations.

REFERENCES

Albert, M. R. 2004. The International Polar Year. *Science* 303(5663): 1437.

Allison, I. B. M. et al. 2007. The scope of science for the International Polar Year 2007-2008. *World Meterologic Organization Technical Document* No. 1364, 81 pp. Geneva: World Meterologic Organization.

Bell, R. E., M. Studinger, C. A. Shuman, M. A. Fahnestock, and I. Joughin. 2007. Large subglacial lakes in East Antarctica at the onset of fast-flowing ice streams. *Nature* 445(7130):904-907.

Domack, E., D. Duran, A. Leventer, S. Ishman, S. Doane, S. McCallum, D. Amblas, J. Ring, R. Gilbert, and M. Prentice. 2005. Stability of the Larsen B Ice Shelf on the Antarctic Peninsula during the Holocene epoch. *Nature* 436(7051):681-686.

Greely, A. W. 1886. Report on the Proceedings of the United States expedition to Lady Franklin Bay, Grinnell Land. Washington, D.C.: U.S. Government Printing Office.

Johannessen, O. M., E. V. Shalina, V. Elena, and M. W. Miles. 1999. Satellite evidence for an Arctic sea ice cover in transformation. *Science* 286(5446):1937-1939.

Joughin, I., E. Rignot, C. E. Rosanova, B. K. Lucchitta, and J. Bohlander. 2003. Timing of recent accelerations of Pine Island Glacier, Antarctica. *Geophysical Research Letters* 30(13):1706.

Kennett, J. P. 1977. Deep-sea drilling contributions to studies of evolution of Southern Ocean and Antarctic glaciation. *Antarctic Journal of the United States* 12(4):72-75.

Nansen, F., and O. N. Sverdrup. 1897. *Farthest North: Being the Record of a Voyage of Exploration of the Ship "Fram" 1893-96, and of a Fifteen Months' Sleigh Journey by Dr. Nansen and Lieut. Johansen.* New York: Harper.

Odishaw, H. 1959. "The Meaning of the International Geophysical Year," U.S. President's Committee on Information Activities Abroad (Sprague Committee) Records, 1959-1961, Box 6, A83-10, Dwight D. Eisenhower Library, Abilene, Kansas.

Rapley, C., and R. E. Bell. 2004. *A Framework for the International Polar Year 2007-2008.* Paris: ICSU.

Rignot, E., G. Casassa, P. Gogineni, W. Krabill, A. Rivera, and R. Thomas. 2004. Accelerated ice discharge from the Antarctic Peninsula following the collapse of Larsen B Ice Shelf. *Geophysical Research Letters* 31(18).

Scambos, T. A., J. A. Bohlander, C.A. Shuman, and P. Skvarca. 2004. Glacier acceleration and thinning after ice shelf collapse in the Larsen B embayment, Antarctica. *Geophysical Research Letters* 31(18).

SCAR. 2002. *Scientific Committee on Antarctic Research Bulletin,* Cambridge, Scientific Committee on Antarctic Research 149:1-16.

Shepherd, A., D. Wingham, A. Payne, and P. Skvarca. 2004. Warm ocean is eroding West Antarctic ice sheet. *Geophysical Research Letters* 31(23).

Steen, A. S. 1887. Die internationale Polarforschung, 1882-1883. Beobachtungs-Ergebnisse der Norwegischen Polarstation Bossekop in Alten. Christiania: Grödahl & Sons. 2 vols.

Steffen, K., S. V. Nghiem, R. Huff, and G. Neumann. 2004. The melt anomaly of 2002 on the Greenland ice sheet from active and passive microwave satellite observations. *Geophysical Research Letters* 31(20).

Weyprecht, L. K. 1875. Scientific work of the Second Austro-Hungarian Polar Expedition, 1872-4. *Journal of the Royal Geographical Society of London* 45:19-33.

Wood, K. R., and J. E. Overland. 2006. Climate lessons from the first International Polar Year. *Bulletin of the American Meteorological Society* 87(12):1685-1697.

Cooper, A. K., P. J. Barrett, H. Stagg, B. Storey, E. Stump, W. Wise, and the 10th ISAES editorial team, eds. (2008). *Antarctica: A Keystone in a Changing World.* Proceedings of the 10th International Symposium on Antarctic Earth Sciences. Washington, DC: The National Academies Press.

100 Million Years of Antarctic Climate Evolution: Evidence from Fossil Plants

J. E. Francis,[1] A. Ashworth,[2] D. J. Cantrill,[3] J. A. Crame,[4] J. Howe,[1] R. Stephens,[1] A.-M. Tosolini,[1] V. Thorn[1]

ABSTRACT

The evolution of Antarctic climate from a Cretaceous greenhouse into the Neogene icehouse is captured within a rich record of fossil leaves, wood, pollen, and flowers from the Antarctic Peninsula and the Transantarctic Mountains. About 85 million years ago, during the mid-Late Cretaceous, flowering plants thrived in subtropical climates in Antarctica. Analysis of their leaves and flowers, many of which were ancestors of plants that live in the tropics today, indicates that summer temperatures averaged 20°C during this global thermal maximum. After the Paleocene (~60 Ma) warmth-loving plants gradually lost their place in the vegetation and were replaced by floras dominated by araucarian conifers (monkey puzzles) and the southern beech *Nothofagus*, which tolerated freezing winters. Plants hung on tenaciously in high latitudes, even after ice sheets covered the land, and during periods of interglacial warmth in the Neogene small dwarf plants survived in tundra-like conditions within 500 km of the South Pole.

INTRODUCTION

The Antarctic Paradox is that, despite the continent being the most inhospitable continent on Earth with its freezing climate and 4-km-thick ice cap, some of the most common fossils preserved in its rock record are those of ancient plants. These fossils testify to a different world of globally warm and ice-free climates, where dense vegetation was able to survive very close to the poles. The fossil plants are an important source of information about terrestrial climates in high latitudes, the regions on Earth most sensitive to climate change.

Although plants of Permian and Triassic age provide a signal of terrestrial climates for the Transantarctic region for times beyond 100 million years (Taylor and Taylor, 1990), it is during the Cretaceous that the Antarctic continent reached the approximate position that it is in today over the South Pole (Lawver et al., 1992). Without the cover of ice Cretaceous vegetation flourished at high latitudes. Even when ice built up on the continent during the Cenozoic, fossil plants indicate that vegetation was still able to grow in relatively inhospitable conditions. The plant record thus provides us with a window into past climates of Antarctica, particularly during the critical transition from greenhouse to icehouse (Francis et al., forthcoming).

Reviews of the paleobotany that focus on floral evolution, rather than paleoclimate, can be found in the work of Askin (1992) and Cantrill and Poole (2002, 2005) and references therein. This paper is not intended to be an exhaustive review of all paleoclimate or paleobotanical studies but presents new information about Antarctic climate evolution deduced from some new and some selected published studies of fossil plants.

WARMTH IN THE CRETACEOUS GREENHOUSE

The most detailed record of Cretaceous terrestrial climates from plants comes from Alexander Island on the Antarctic

---

[1] School of Earth and Environment, University of Leeds, UK.

[2] Department of Geosciences, North Dakota State University, Fargo, ND 58105-5517, USA.

[3] National Herbarium of Victoria, Royal Botanic Gardens Melbourne, Private Bag 2000, South Yarra, Victoria 3141, Australia.

[4] British Antarctic Survey, High Cross, Madingley Road, Cambridge, UK.

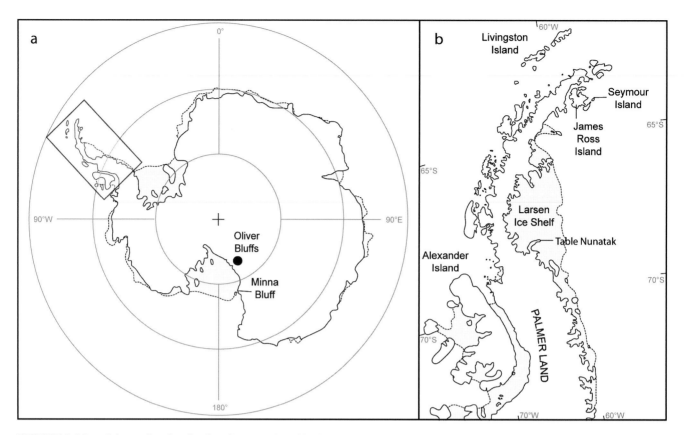

FIGURE 1 Map of Antarctica showing locations mentioned in the text.

Peninsula (Figure 1). Here the mid-Cretaceous (Albian) Fossil Bluff Group sediments include a series of marine, submarine fan, fluviatile, and deltaic sediments formed as the infill of a subsiding fore-arc basin. Floodplains and midchannel bars of braided and meandering river systems (Triton Point Formation, Nichols and Cantrill, 2002) were extensively forested by large trees with a rich undergrowth.

Many of the plants are preserved in their original positions, having been encased by sheet flood and crevasse splay deposits during catastrophic flooding events. In addition, finely laminated flood deposits draped small plants, preserving them in place.

The *in situ* preservation of the plants has allowed detailed reconstructions of the forest environments by Howe (2003). Three vegetation assemblages were identified through statistical analysis of field data: (1) a conifer and fern assemblage with mature conifers of mainly araucarian type and an understory of *Sphenopteris* ferns; (2) a mixed conifer, fern, and cycad assemblage with araucarian conifers and ginkgo trees; and (3) a disturbance flora of liverworts, *Taeniopteris* shrubs, ferns, and angiosperms. Reconstructions of these assemblages and their depositional setting are presented in Figure 2.

An artist's reconstruction is presented in Figure 3. This

reconstruction portrays spacing of the large trees, such as ginkgos, podocarps, and araucarian conifers, as accurately as possible from field data. The undergrowth consists of ferns, cycadophytes, liverworts, mosses, and small shrubby angiosperms. The extinct plant *Taeniopteris* formed thickets in the disturbed clearings in the forest (Cantrill and Howe, 2001). The picture also shows the volcanoes on the adjacent arc that were the source of volcaniclastic sediment, much of which was deposited as catastrophic deposits that engulfed standing trees (in the distance in this picture).

Paleoclimatic interpretation using fossil plants on Alexander Island is based mainly on comparison with the ecological tolerances of similar living Southern Hemisphere taxa (Falcon-Lang et al., 2001). This indicates that the climate was generally warm and humid to allow the growth of large conifers, with mosses and ferns in the undergrowth. According to Nichols and Cantrill (2002), river flow is considered to have been perennial, based on sedimentary evidence, but with periodic floods indicative of high rainfall.

Evidence from fossil soils, however, provides a somewhat different climate story. The paleosols, in which trees are rooted, have structures such as blocky peds, clay cutans, and mottling that are typical of modern soils that form under seasonally dry climates (Howe and Francis, 2005). Although

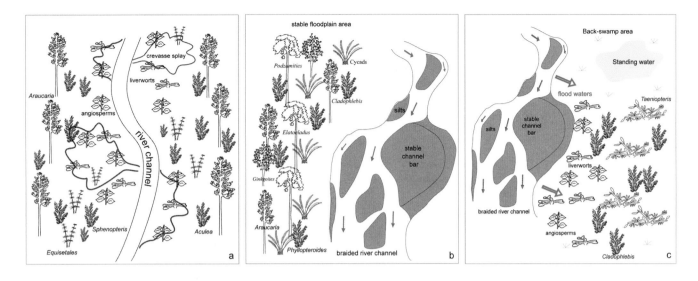

FIGURE 2 Reconstruction of paleoenvironment and community structure of (a) an open woodland community that grew on areas of the floodplain distal to a meandering river channel but subjected to frequent, low-energy floods; (b) a patch forest community of mature conifers and ferns, growing on floodplain areas affected by catastrophic floods; and (c) disturbance vegetation growing in back-swamp areas of a braided river floodplain, subjected to frequent low-energy flood events and ponding (Howe, 2003).

evidence of intermittent flooding is apparent from the fluvial sandstones and mudstones that volumetrically dominate the rock sequence, the paleosols indicate that this high-latitude mid-Cretaceous environment was predominantly seasonally dry. Whereas the flood sediments are likely to have been deposited relatively rapidly (days or weeks), the paleosols likely represent hundreds to thousands of years for soil development and forest growth, suggesting a predominantly dry climate for this high-latitude site.

Climate models for the mid-Cretaceous predicted warm, humid climates for this region. Valdes et al. (1996) predicted high summer temperatures of 20-24°C and low winter temperatures just above freezing. There is no evidence of freezing in the paleosols. However, it is likely that the more significant climate parameter that influenced the soils was rainfall and soil moisture. The models of Valdes et al. (1996) of seasonal mean surface soil moisture distribution (the balance between precipitation and evaporation) predict a seasonal moisture regime with dry conditions in summer but wet in winter, supporting the seasonal signature seen in the paleosols.

Younger Cretaceous strata preserved within the James Ross Island back-arc basin, and which crop out on James Ross, Seymour, and adjacent islands, contain a series of fossil floras that are providing new information about biodiversity and climate. During the Late Cretaceous angiosperms (flowering plants) became an important component of the vegetation. The Coniacian Hidden Lake Formation (Gustav Group) and the Santonian-early Campanian Santa Marta Formation (Marambio Group) in particular contain fossil angiosperm leaves that have yielded information about paleoclimate. The leaves represent the remains of vegetation that grew at approximately 65°S on the emergent volcanic arc that is now represented by the Antarctic Peninsula. They were subsequently transported and buried in marine sediments in the adjacent back-arc basin, and are now preserved as impressions and compressions in volcaniclastic siltstones and carbonate concretions (Hayes et al., 2006).

The angiosperm leaf morphotypes have been tentatively compared with those of living families such as Sterculiaceae, Lauraceae, Winteraceae, Cunoniaceae, and Myrtaceae (Hayes et al., 2006) (Figure 4). Most of these families today can be found in warm temperate or subtropical zones of the Southern Hemisphere. Sterculiaceae is a tropical and subtropical family found in Australia, South Asia, Africa, and northern South and Central America. The Laurales today live in tropical or warm temperate regions with a moist equable climate. Both the Cunoniaceae and Elaeocarpaceae are tropical and subtropical trees and shrubs found in equatorial tropical regions, and the Winteraceae are concentrated in wet tropical montane to cool temperate rainforests of the southwest Pacific and South America.

A more quantitative analysis of the leaves based on physiognomic aspects of the leaves, using leaf margin analysis and simple and multiple linear regression models, provided data on paleotemperatures and precipitation. Estimates of mean annual temperatures (MATs) range from 15.2 (±2) to 18.6 (±1.9)°C for the Coniacian, and 17.1 (±2)°C to 21.2 (±1.9)°C for the late Coniacian-early Santonian. Averaging all data from all methods gives a MAT for the Coniacian of 16.9°C and 19.1°C for the late Coniacian-early Santonian (Hayes et al., 2006). Estimates of annual precipitation range

FIGURE 3 Reconstruction of the forest environment on Alexander Island during the Cretaceous, based on the work of Howe (2003) and British Antarctic Survey geologists (see text and references).
SOURCE: Image is a painting by Robert Nicholls of Paleocreations.com, and is housed at the British Antarctic Survey.

from 594 to 2142 (±580) mm for the Coniacian and 673 to 1991 (±580) mm, for the late Coniacian-early Santonian, with very high rainfalls of 2630 (±482) and 2450 (±482) mm for the growing seasons, respectively, comparable with those of equatorial tropical rainforests today. However, these data must be considered with caution as some important aspects of leaf morphology related to rainfall are not preserved in the fossil assemblage.

These fossil plants are thus indicative of tropical and subtropical climates at high paleolatitudes during the mid-Late Cretaceous, without extended periods of winter temperatures below freezing and with adequate moisture for growth. In other Antarctic paleoclimate studies evidence for strong warmth at this time was found by Dingle and Lavelle (1998) in their analyses of clay minerals and from analysis of fossil woods (Poole et al., 2005). This warm peak may have also been the trigger for the expansion of the angiosperms in the Antarctic, represented by a marked increased abundance of angiosperm pollen in Turonian sediments (Keating et al., 1992). On a global scale this corresponds to the Cretaceous thermal maximum, from about 100-80 Ma, reported from

many sites (e.g., Clark and Jenkyns, 1999; Huber et al., 2002), and possibly attributed to rising atmospheric CO_2 levels due to a tectonically driven oceanographic event in the opening of the equatorial Atlantic gateway (Poulsen et al., 2003).

The high levels of Cretaceous climate warmth are also recorded in an unusual assemblage of late Santonian fossil plants from Table Nunatak on the east side of the Antarctic Peninsula. A small isolated outcrop has yielded layers of charcoal, produced by wildfires, within a sequence of sandstones, siltstones, and mudstones, deposited in shallow marine conditions at the mouth of an estuary or in a delta distributary channel. The charcoal layer contains the remains of burnt plants, including megaspores; fern rachids; conifer leaves, wood, shoots, seeds and pollen cones; angiosperms leaves, fruits and seeds (Eklund et al., 2004).

The most unusual of these fossils are tiny fossil flowers, the oldest recorded from Antarctica. Eklund (2003) identified several types of angiosperm flowers, some of which are most likely related to the living families Siparunaceae, Wintera-

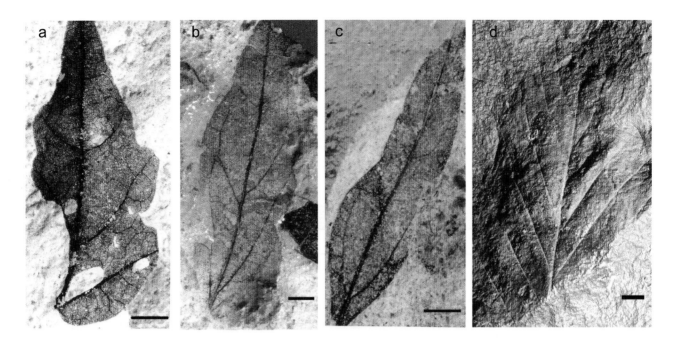

FIGURE 4 A selection of fossil leaves from the Hidden Lake Formation: (a) D.8754.1a, Morphotype 2 (Sterculiaceae); (b) D.8754.8.54a, Morphotype 11 (Laurales); (c) D.8754.8.57a, Morphotype 11 (Laurales); (d) D.8621.27a, Morphotype 10 (unknown affinity). Botanical names in parentheses refer to most similar modern family (Hayes et al., 2006). Scale bar 5 mm for all leaves. Numbers refer to the British Antarctic Survey numbering system.

ceae, and Myrtaceae, as well as several unidentified types. The Siparunaceae family is now confined to the tropics, and the Winteraceae (e.g., mountain pepper tree) and Myrtaceae (e.g., eucalypts) are warm-cool temperate types from the Southern Hemisphere. Interestingly, the *Siparuna*-like fossil flowers have a flat-roofed structure with a central pore, which in the living plants acts as a landing platform for gall midges that lay their eggs inside the flower head. Since the developing gall midge larvae usually destroy the flowers as they grow (Eklund, 2003), the preservation of such flowers as fossils would normally be extremely unlikely. The fortuitous occurrence of wildfires during the Cretaceous has thus helped preserve such flowers as rare fossils.

FROM GREENHOUSE TO ICEHOUSE

After the peak warmth of the mid-Late Cretaceous, climate appears to have cooled globally during the latest part of the Cretaceous, as seen also in the Antarctic fossil wood record (Francis and Poole, 2002; Poole et al., 2005). However, warm climates returned to the high latitudes during the Paleocene and early Eocene, as is reflected in fossil plants.

The Paleogene sedimentary sequence on Seymour Island (Figure 1) contains plant-rich horizons that hold signals to ancient climates. An unusual new flora from La Meseta Formation has been studied by Stephens (2008). The flora is dominated by permineralized branches of conifers and compressions of angiosperm leaves, and found within car-

bonate concretions within the shallow marine Campamento allomember of Marenssi et al. (1998) (equivalent to Telm 3 of Sadler [1988]). This unit has been dated as 51.5-49.5 Ma through links with global eustatic lowstands (Marenssi, 2006) and by Sr dating by Stephens, which yielded a latest early Eocene (late Ypresian) age.

The flora is unusual in that it is dominated by leaves, cone scales, and leafy branches of araucarian conifers, very similar in all respects to living *Araucaria araucana* (monkey puzzle) from Chile. Study of three-dimensional arrangement of leaves using the technique of neutron tomography illustrates that leaf and branch arrangement is characteristic of living *Araucaria araucana* (Figure 5). A limited array of angiosperm leaves are preserved in the nodules and include morphotypes that are similar to leaves of living Lauraceae, Myricaceae, Myrtaceae, and Proteaceae. It is notable that many of the fossil leaves appear to have particularly thick carbonaceous compressions and, indeed, the living equivalents are evergreen Southern Hemisphere trees that have thick waxy cuticles. This may signal taphonomic sorting of some kind rather than a climate signal, as the more fragile leaves, including *Nothofagus*, are unusually absent from this flora. Climate interpretation of this flora, based on the nearest living types, especially the araucarian conifers, suggests that the latest Eocene climate was cool and moist. Oxygen isotope analyses of marine molluscs in the same concretions yielded cool marine paleotemperatures of 8.3-12.5°C, comparable to other estimates for Telm 3 from marine oxygen isotopes

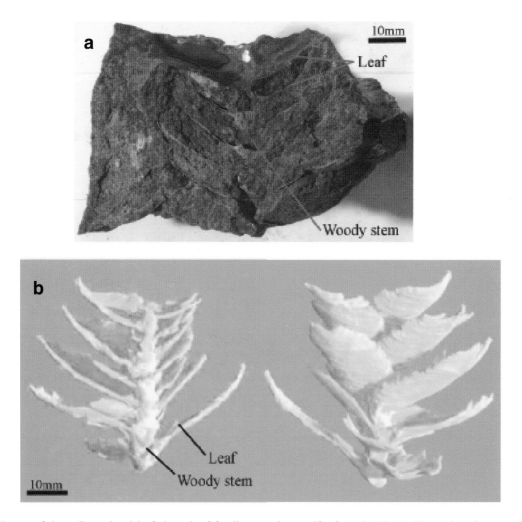

FIGURE 5 Image of three-dimensional leafy branch of fossil araucarian conifer from La Meseta Formation, Seymour Island (sample DJ.1103.189). (a) The branch as it appears in hand specimen. (b) Reconstruction from tomographic image showing the leafy branch in three dimensions. Neutron tomography to create the figure was undertaken at the Institut Laue Langevin facility, Grenoble, France.

of 8-13.2°C (Pirrie et al., 1998) and 10.5-19.6°C (Dutton et al., 2002).

New paleoclimate data have also been derived from new collections of fossil angiosperm leaves from the Nordenskjold flora in the Cross Valley Formation of Late Paleocene age (Elliot and Trautman, 1982) and from the Middle Eocene Cucullaea 1 flora (Telm 5 of Sadler [1988], Cucullaea 1 allomember of Marenssi et al. [1998], dated as 47-44 Ma by Dutton et al. [2002]). In the Late Paleocene flora 36 angiosperm leaf morphotypes were identified, along with two pteridophytes (ferns), and podocarp and araucarian conifers. The angiosperm leaf component is dominated by types with possibly modern affinities to the Nothofagacae, Lauraceae, and Proteaceae, along with other types, such as Myrtaceae, Elaeocarpaceae, Winteraceae, Moraceae, Cunoniaceae, and Monimiaceae, some of which have no identifiable modern affinities as yet. The fossil assemblage contains a mix of families of subtropical and cool-temperate nature and represents cool-warm temperate mixed conifer-broad-leaved evergreen and deciduous forest, dominated by large trees of *Nothofagus* (southern beech) and araucarian and podocarp conifers, with other angiosperms as mid-canopy trees and understorey shrubs. In comparison the Middle Eocene flora shows a 47 percent decrease in diversity with only 19 leaf morphotypes. The flora is dominated by *Nothofagus*, most similar to modern cool-temperate types, and has much rarer subtropical and warm-temperate types, such as Proteaceae. The decline in diversity indicates a substantial cooling of climate over this interval.

Paleoclimate data were derived using CLAMP (Climate Leaf Analysis Multivariate Program) (Wolfe and Spicer, 1999) and several leaf margin analysis techniques based on physiognomic properties of the leaves. CLAMP results are given below. A MAT of 13.5 ± 0.7°C was determined for the

Late Paleocene. A strongly seasonal climate was implied, with a warm month mean of 25.7 ± 2.7°C and a cold month mean of 2.2 ± 2.7°C, with 2110 mm annual rainfall. By the Middle Eocene the climate had cooled considerably; the MAT was 10.8 ± 1.1°C with a warm month mean of 24 ± 2.7°C, and a cold month mean of −1.17 ± 2.7°C, with 1534 mm annual rainfall. This suggests that the climate was still markedly seasonal but that winter temperatures were often below freezing.

The Late Paleocene floras thus represent warm Antarctic climates probably without ice, even in winter. However, Seymour Island floras suggest that by the Middle Eocene (47-44 Ma) climates had cooled considerably and ice may have been present on the Antarctic continent, at least during winter months. Several Eocene macrofloras dominated by *Nothofagus* from King George Island, South Shetland Islands, suggest mean annual temperatures of about 10-11°C (seasonality unknown) (e.g., Birkenmajer and Zastawniak, 1989; Hunt and Poole, 2003), and the highest latitude Eocene floras, from Minna Bluff in the McMurdo Sound region, are suggestive of cool temperate climates (Francis, 2000). Other geological evidence also points to significant cooling by the latter part of the Eocene; fluctuating volumes of ice on Antarctica from about 42 Ma have been implied by Tripati et al. (2005) from isotope measurements of marine sediments and, from the Antarctic region, Birkenmajer et al. (2005) proposed a glacial period from 45-41 Ma, based on valley-type tillites of Eocene-Oligocene age from King

George Island. Ivany et al. (2006) have also reported the glacial deposits of possible latest Eocene or earliest Oligocene age from Seymour Island.

PLANTS IN THE FREEZER

Despite the onset of glaciation and the growth of ice sheets on Antarctica during the latest Eocene or earliest Oligocene, vegetation was not instantly wiped out but clung on tenaciously in hostile environments. Even though many warmth-loving plant taxa disappeared during the mid-Eocene, floras dominated by *Nothofagus* remained for many millions of years. Single leaves of *Nothofagus* of Oligocene and Miocene age have been found in CRP-3 and CIROS-1 drill cores, respectively (Hill, 1989; Cantrill, 2001), but one of the most remarkable Antarctic floras is preserved within glacial diamictites of the Meyer Desert Formation, Sirius Group, at Oliver Bluffs (85°S) in the Transantarctic Mountains.

In a meter-thick layer of sandstone, siltstone, and mudstone sandwiched between tillites, small twigs of fossil *Nothofagus* wood are preserved entwined around cobbles (Figure 6). The twigs are the branches and rootlets of small dwarf shrubs preserved within their positions of growth. Fossil leaf mats of *Nothofagus* leaves (*Nothofagus beardmorensis*) (Hill et al., 1996) are also preserved in the same horizon. Several other fossil plants have also been recovered, including moss cushions, liverworts, and fruits, stems, and seeds of several

FIGURE 6 Sirius Group sediments and fossil plants, Oliver Bluffs, Transantarctic Mountains. (a) General view of Oliver Bluffs. (b) View of bluff composed of glacial diamictites. Arrow points to meter-thick horizon of glacial sandstone, equivalent to plant-bearing horizon. (c) *Nothofagus* leaves from upper part of plant-bearing horizon. (d) Small branch of fossil wood in growth position within paleosol horizon. (e) Small branch in situ in paleosol horizon, with delicate bark still attached. Scale bars represent 1 cm.

vascular plants, as well as remains of beetles, a gastropod, bivalves, a fish, and a fly (Ashworth and Cantrill, 2004).

All the evidence for paleoclimate from these remarkable fossils (Ashworth and Cantrill, 2004), from growth rings in the twigs (Francis and Hill, 1996), and from paleosols (Retallack et al., 2001) points toward a tundra environment with a mean annual temperature of about –12°C, with short summers up to +5°C but with long cold winters below freezing. Dated as Pliocene in age but disputed (see Ashworth and Cantrill, 2004), this periglacial or interglacial environment existed at a time when glaciers retreated briefly from the Oliver Buffs region, allowing dwarf shrubs to colonize the exposed tundra surface only about 500 km from the South Pole.

As Antarctica's climate cooled further into the Pleistocene deep-freeze, vascular plants were lost from the continent. However, its rich fossil plant record retains its legacy of past warmth.

ACKNOWLEDGMENTS

New information presented here was obtained during postgraduate research projects of Howe (Alexander Island) and Stephens (Seymour Island), funded by Natural Environment Research Council and the British Antarctic Survey (BAS). Stephens wishes to thank Martin Dawson (Leeds) and Andreas Hillenbach and Hendrik Ballhausen (Institut Laue Langevin, Grenoble) for help with neutron tomography. New data on fossil plants from Seymour Island were collected as part of a NERC-Antarctic Funding Initiative project AFI1/01 to Francis, Cantrill, and Tosolini. BAS is thanked for field support in all these projects. Work in the Transantarctic Mountains was funded by NSF/OPP grants 9615252 and 0230696 to Ashworth. The Trans-Antarctic Association and Geological Society of London are thanked for additional support. This work has been aided by discussions with Hunt, Falcon-Lang, Hayes, Poole, Eklund (Leeds/BAS), and Roof.

REFERENCES

Ashworth, A. C., and D. J. Cantrill. 2004. Neogene vegetation of the Meyer Desert Formation (Sirius Group), Transantarctic Mountains, Antarctica. *Palaeogeography, Palaeoclimatology, Palaeoecology* 213:65-82.

Askin, R. A. 1992. Late Cretaceous-early Tertiary Antarctic outcrop evidence for past vegetation and climate. In *The Antarctic Paleoenvironment: A Perspective on Global Change*, eds. J. P. Kennett and D. A. Warnke, *Antarctic Research Series* 56:61-75. Washington, D.C.: American Geophysical Union.

Birkenmajer, K., and E. Zastawniak. 1989. Late Cretaceous-early Tertiary floras of King George Island, West Antarctica: Their stratigraphic distribution and palaeoclimatic significance, origins and evolution of the Antarctic biota. *Geological Society of London Special Publication* 147:227-240.

Birkenmajer, K., A. Gazdzicki, K. P. Krajewski, A. Przybycin, A. Solecki, A. Tatur, and H. I. Yoon. 2005. First Cenozoic glaciers in West Antarctica. *Polish Polar Research* 26:3-12.

Cantrill, D. J. 2001. Early Oligocene *Nothofagus* from CRP-3, Antarctica: Implications for the vegetation history. *Terra Antarctica* 8:401-406.

Cantrill, D. J., and J. Howe. 2001. Palaeoecology and taxonomy of Pentoxylales from the Albian of Antarctica. *Cretaceous Research* 22:779-793.

Cantrill, D. J., and I. Poole. 2002. Cretaceous to Tertiary patterns of diversity changes in the Antarctic Peninsula. In *Palaeobiogeography and Biodiversity Change: A Comparison of the Ordovician and Mesozoic-Cenozoic Radiations*, eds. A. W. Owen and J. A. Crame, *Geological Society of London Special Publication* 194:141-152.

Cantrill, D. J., and I. Poole. 2005. Taxonomic turnover and abundance in Cretaceous to Tertiary wood floras of Antarctica: Implications for changes in forest ecology. *Palaeogeography, Palaeoclimatology, Palaeoecology* 215:205-219.

Clark, L. J., and H. C. Jenkyns. 1999. New oxygen isotope evidence for long-term Cretaceous climatic change in the Southern Hemisphere. *Geology* 27:699-702.

Dingle, R., and M. Lavelle. 1998. Late Cretaceous-Cenozoic climatic variations of the northern Antarctic Peninsula: New geochemical evidence and review. *Palaeogeography, Palaeoclimatology, Palaeoecology* 141:215-232.

Dutton, A., K. Lohmann, and W. J. Zinsmeister. 2002. Stable isotope and minor element proxies for Eocene climate of Seymour Island, Antarctica. *Paleoceanography* 17(5), doi:10.1029/2000PA000593.

Eklund, H. 2003. First Cretaceous flowers from Antarctica. *Review of Palaeobotany and Palynology* 127:187-217.

Eklund, H., D. J. Cantrill, and J. E. Francis. 2004. Late Cretaceous plant mesofossils from Table Nunatak, Antarctica. *Cretaceous Research* 25:211-228.

Elliot, D. H., and T. A. Trautman. 1982. Lower Tertiary strata on Seymour Island, Antarctic Peninsula. In *Antarctic Geoscience*, ed. C. Craddock, pp. 287-297. Madison: University of Wisconsin Press.

Falcon-Lang, H. J., D. J. Cantrill, and G. J. Nichols. 2001. Biodiversity and terrestrial ecology of a mid-Cretaceous, high-latitude floodplain, Alexander Island, Antarctica. *Journal of the Geological Society of London* 158:709-724.

Francis, J. E. 2000. Fossil wood from Eocene high latitude forests, McMurdo Sound, Antarctic. In *Paleobiology and Palaeoenvironments of Eocene Rocks, McMurdo Sound, East Antarctica*, eds. J. D. Stilwell and R. M. Feldmann, *Antarctic Research Series* 76:253-260, Washington, D.C.: American Geophysical Union.

Francis, J. E., and R. S. Hill. 1996. Fossil plants from the Pliocene Sirius group, Transantarctic Mountains: Evidence for climate from growth rings and fossil leaves. *Palaios* 11:389-396.

Francis, J. E., and I. Poole. 2002. Cretaceous and early Tertiary climates of Antarctica: Evidence from fossil wood. *Palaeogeography, Palaeoclimatology, Palaeoecology* 182:47-64.

Francis, J. E., S. Marenssi, R. Levy, M. Hambrey, V. T. Thorn, B. Mohr, H. Brinkhuis, J. Warnaar, J. Zachos, S. Bohaty, and R. DeConto. forthcoming. From Greenhouse to Icehouse—The Eocene/Oligocene in Antarctica, In *Antarctic Climate Evolution*, chap. 6, eds. F. Florindo and M. J. Siegert. Amsterdam: Elsevier.

Hayes, P. A., J. E. Francis, and D. J. Cantrill. 2006. Palaeoclimate of Late Cretaceous Angiosperm leaf floras, James Ross Island, Antarctic. In *Cretaceous-Tertiary High Latitude Palaeoenvironments, James Ross Basin.*, eds. J. E. Francis, D. Pirrie, and J. A. Crame, *Geological Society London Special Publication* 258:49-62.

Hill, R. S. 1989. Palaeontology—fossil leaf. In *Antarctic Cenozoic History from the CIROS-1 Drillhole, McMurdo Sound*, ed. P. J. Barrett. *New Zealand DSIR Bulletin* 245:143-144.

Hill, R. S., D. M. Harwood, and P. N. Webb. 1996. *Nothofagus beardmorensis* (Nothofagaceae), a new species based on leaves from the Pliocene Sirius Group, Transantarctic Mountains, Antarctica. *Review of Palaeobotany and Palynology* 94:11-24.

Howe, J. 2003. Mid Cretaceous fossil forests of Alexander Island, Antarctic, Ph.D. thesis. University of Leeds, Leeds, U.K.

Howe, J., and J. E. Francis. 2005. Metamorphosed palaeosols from the mid-Cretaceous fossil forests of Alexander Island, Antarctica. *Journal of the Geological Society of London* 162:951-957.

Huber, B. T., R. D. Norris, and K. G. MacLeod. 2002. Deep-sea paleo-temperature record of extreme warmth during the Cretaceous. *Geology* 30:123-126.

Hunt, R. J., and I. Poole. 2003. Paleogene West Antarctic climate and vegetation history in light of new data from King George Island. In *Causes and Consequences of Globally Warm Climates in the early Paleogene,* eds. S. L. Wing, P. D. Gingerich, B. Schmitz, and E. Thomas. *Geological Society of America* Special paper 369:395-412, Boulder, Colorado: Geological Society of America.

Ivany, L. C., S. Van Simaeys, E. W. Domack, and S. D. Samson. 2006. Evidence for an earliest Oligocene ice sheet on the Antarctic Peninsula. *Geology* 34:377-380.

Keating, J. M., M. Spencer-Jones, and S. Newham. 1992. The stratigraphical palynology of the Kotick Point and Whisky Bay formations, Gustav Group (Cretaceous), James Ross Island. *Antarctic Science* 4:279-292.

Lawver, L. A., L. M. Gahagan, and M. E. Coffin. 1992. The development of paleoseaways around Antarctica. In *The Antarctic Paleoenvironment: A Perspective on Global Change,* eds. J. P. Kennett and D. A. Warnke, *Antarctic Research Series* 56:7-30, Washington, D.C.: American Geophysical Union.

Marenssi, S. A. 2006. Eustatically controlled sedimentation recorded by Eocene strata of the James Ross Basin, Antarctic. In *Cretaceous–Tertiary High-Latitude Palaeoenvironments, James Ross Basin, Antarctica,* eds. J. E. Francis, D. Pirrie, and J. A. Crame, pp. 125-133. *Geological Society of London Special Publication* 258.

Marenssi, S. A., S. Santillana, and C. A. Rinaldi. 1998. Stratigraphy of the La Meseta Formation (Eocene), Marambio (Seymour) Island, Antarctic. In *Paleógeno de América del Sur y de la Península Antártica.* ed. S. Casadío, *Asociación Paleontológica Argentina Publicación Especial* 5:137-146. Buenos Aires: Asociación Paleontológica Argentina.

Nichols, G. J., and D. J. Cantrill. 2002. Tectonic and climatic controls on a Mesozoic fore-arc basin succession, Alexander Island, Antarctic. *Geology Magazine* 139:313-330.

Pirrie, D., J. D. Marshall, and J. A. Crame. 1998. Marine high-Mg calcite cements in Teredolites-bored fossil wood; evidence for cool paleoclimates in the Eocene La Meseta Formation, Seymour Island, Antarctic. *Palaios* 13:276-286.

Poole, I., D. Cantrill, and T. Utescher. 2005. A multi-proxy approach to determine Antarctic terrestrial palaeoclimate during the Late Cretaceous and early Tertiary. *Palaeogeography, Palaeoclimatology, Palaeoecology* 222:95-121.

Poulsen, C. J., A. S. Gendaszek, and R. L. Jacob. 2003. Did the rifting of the Atlantic Ocean cause the Cretaceous thermal maximum? *Geology* 31:115-118.

Retallack, G. J., E. S. Krull, and G. Bockheim. 2001. New grounds for reassessing the palaeoclimate of the Sirius Group, Antarctica. *Journal of the Geological Society of London* 158:925-935.

Sadler, P. M. 1988. Geometry and stratification of uppermost Cretaceous and Paleogene units on Seymour Island, northern Antarctic Peninsula. In *Geology and Paleontology of Seymour Island, Antarctic Peninsula,* eds. R. M. Feldmann and M. O. Woodburne, *Geological Society of America Memoir* 169:303-320. Boulder, Colorado: Geological Society of America.

Stephens, R. 2008. Palaeoenvironmental and climatic significance of an Araucaria-dominated Eocene flora from Seymour Island, Antarctic. Ph.D. thesis. University of Leeds, Leeds, U.K.

Taylor, E. L., and T. N. Taylor. 1990. Structurally preserved Permian and Triassic floras from Antarctic. In *Antarctic Paleobiology and Its Role in the Reconstruction of Gondwana,* eds. T. N. Taylor and E. L. Taylor, pp. 149-163. New York: Springer-Verlag.

Tripati, A., J. Backman, H. Elderfield, and P. Ferretti. 2005. Eocene bipolar glaciation associated with global carbon cycle changes. *Nature* 436:341-346.

Valdes, P. J., B. W. Sellwood, and G. D. Price. 1996. Evaluating concepts of Cretaceous equability. *Palaeoclimates, Data and Modelling* 2:139-158.

Wolfe, J. A., and R. A. Spicer. 1999. Fossil leaf character states: Multivariate analysis. In *Fossil Plants and Spores: Modern Techniques,* eds. T. P. Jones and N. P. Rowe, pp. 233-239. London: Geological Society of London.

Antarctica's Continent-Ocean Transitions: Consequences for Tectonic Reconstructions

K. Gohl[1]

ABSTRACT

Antarctica was the centerpiece of the Gondwana supercontinent. About 13,900 km of Antarctica's 15,900-km-long continental margins (87 percent) are of rifted divergent type, 1600 km (10 percent) were converted from a subduction type to a passive margin after ridge-trench collision along the Pacific side of the Antarctic Peninsula, and 400 km (3 percent) are of active convergent type. In recent years the volume of geophysical data along the continental margin of Antarctica has increased substantially, which allows differentiation of the crustal characteristics of its continent-ocean boundaries and transitions (COB/COT). These data and geodynamic modeling indicate that the cause, style, and process of breakup and separation were quite different along the Antarctic margins. A circum-Antarctic map shows the crustal styles of the margins and the location and geophysical characteristics of the COT. The data indicate that only a quarter of the rifted margins are of volcanic type. About 70 percent of the rifted passive margins contain extended continental crust stretching between 50 and 300 km oceanward of the shelf edge. Definitions of the COT and an understanding of its process of formation has consequences for plate-kinematic reconstructions and geodynamic syntheses.

INTRODUCTION

About 13,900 km of the 15,900-km-long continental margins of the Antarctic plate are of rifted divergent type, 1600 km were converted from a subduction-type to a passive margin after ridge-trench collision along the Pacific side of the Ant-

arctic Peninsula, and 400 km are of active convergent type. The structure and composition of continental margins, in particular those of rifted margins, can be used to elucidate the geodynamic processes of continental dispersion and accretion. The margins of Antarctica have mostly been subject to regional studies mainly in areas near research stations in the Ross Sea, Prydz Bay, Weddell Sea, and along the Antarctic Peninsula near national research facilities. In recent years—mainly motivated by the United Nations Convention on the Law of the Sea—large volumes of new offshore geophysical data have been collected, primarily along the East Antarctic margin. For the first time this provides the opportunity to make a comprehensive analysis of the development of these continental margins over large tracts of extended continental crust that were previously unknown. The coverage of circum-Antarctic multichannel seismic lines from the Antarctic Seismic Data Library System for Cooperative Research (SDLS) of the Scientific Committee on Antarctic Research (SCAR) (Wardell et al., 2007) (Figure 1) is, with the exception of some areas in the central Weddell Sea, off western Marie Byrd Land and off the Ross Sea shelf, dense enough for quantifying basement types and volcanic and nonvolcanic characteristics of the margins. The track map (Figure 1) shows that deep crustal seismic data, necessary for a complete and accurate characterization of the marginal crust to upper mantle level, are still absent over most margins.

In this paper I first present a compilation of the structural types of the circum-Antarctic continental margins based on a review of relevant published data of diverse types together with new data. Then I contemplate implications of the knowledge of margin crustal types and properties for plate-kinematic and paleobathymetric reconstructions and for isostatic response models.

[1]Alfred Wegener Institute for Polar and Marine Research, Postbox 120161, 27515 Bremerhaven, Germany (karsten.gohl@awi.de).

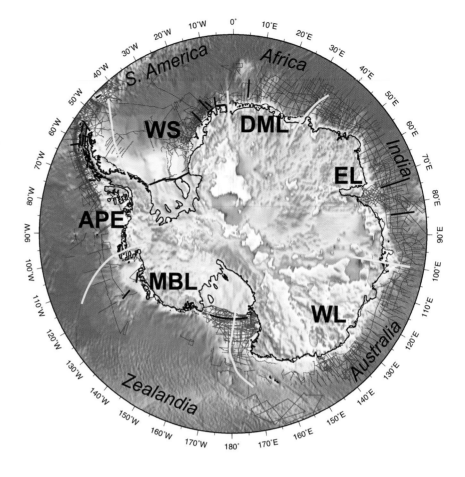

FIGURE 1 Overview map of Antarctica and surrounding ocean floor using a combined BEDMAP (bedrock topography with ice sheet removed) (Lythe et al., 2001) and satellite-derived bathymetry grid (McAdoo and Laxon, 1997). White areas on the continent are below sea level if the ice sheet is removed (without isostatic compensation). Yellow lines mark boundaries between the six crustal breakup sectors of respective conjugate continents in this paper. Red lines show the tracks of offshore multichannel seismic profiles of the SCAR Seismic Data Library System (SDLS) (compiled by M. Breitzke, AWI). Black lines mark the locations of offshore deep crustal seismic refraction profiles. WS = Weddell Sea sector; DML = Dronning Maud Land sector; EL = Ellsworth-Lambert Rift sector; WL = Wilkes Land sector; MBL = Marie Byrd Land sector; APE = Antarctic Peninsula-Ellsworth Land sector.

STRETCHING AND BREAKING: PASSIVE MARGIN TYPES

Weddell Sea (WS) Sector Conjugate to South America

The complex tectonic development of the Weddell Sea sector (Figure 2) has more recently been reconstructed by Hübscher et al. (1996a), Jokat et al. (1996, 1997, 2003, 2004), Leitchenkov et al. (1996), Ghidella and LaBrecque (1997), Golynsky and Aleshkova (2000), Ghidella et al. (2002), Rogenhagen and Jokat (2002), and König and Jokat (2006). Deciphering of the crustal types in the central Weddell Sea is still hampered by the lack of deep crustal seismic data. In the southern Weddell Sea seismic refraction data reveal a thinned continental crust of about 20 km thickness beneath the northern edge of the Filchner-Ronne ice shelf (Hübscher et al., 1996a). It can be assumed that this thinned crust extends northward to a boundary marked by the northern limit of a large positive gravity anomaly (Figure 2). König and Jokat (2006) associate an east-west rifting of this crust (stretching factor of 2.5) with the motion of the Antarctic Peninsula from East Antarctica as the earliest event in the Weddell Sea plate circuit at about 167 Ma prior to the early Weddell Sea opening in a north-south direction

at about 147 Ma. It is not clear, however, whether some of the crustal extension is also associated with this early Weddell Sea opening. The crust between the northern end of the large positive gravity anomaly and the magnetic Orion Anomaly and Andenes Anomaly (Figure 2) is interpreted as a COT with the Orion Anomaly suggested to represent an extensive zone of volcanics that erupted during the final breakup between South America and Antarctica (König and Jokat, 2006). Deep crustal seismic refraction data across the Orion and Andenes anomalies and the assumed COT south of it are needed in order to constrain their crustal composition and type. Although the Orion Anomaly may provide a hint toward a volcanic-type margin, the few seismic data do not allow a complete characterization of the COT in the southern Weddell Sea. Identified magnetic spreading anomalies (oldest is M17) show evidence that oceanic crust exists north of the Orion and Andenes anomalies with the prominent T-Anomaly marking supposedly the changeover from slow to ultraslow spreading-type crust (König and Jokat, 2006).

The Weddell Sea margin along the east coast of the Antarctic Peninsula is still rather enigmatic due to missing data. König and Jokat (2006) show that it rifted from the western Patagonian margin as part of the earliest plate motion in the Weddell Sea region at about 167 Ma. They follow Ghidella

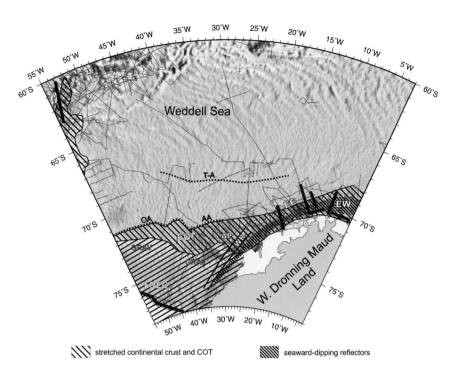

FIGURE 2 Weddell Sea sector with satellite-derived gravity field (McAdoo and Laxon, 1997) and continental margin features. Red lines show the tracks of offshore multichannel seismic profiles of SDLS. Dark blue lines mark the locations of offshore deep crustal seismic refraction profiles. COT = continent-ocean boundary; GRAV = positive gravity anomaly marking northern limit of thinned continental crust; FREC = Filchner-Ronne extended crust; EE = Explora Escarpment; EW = Explora Wedge; AP = Andenes Plateau; WR = Weddell Sea Rift; T-A = T-Anomaly; OA = Orion Anomaly; AA = Andenes Anomaly.

and LaBrecque's (1997) argument for a nonvolcanic margin based on low-amplitude magnetic anomalies and a characteristic bathymetry.

The margin of the eastern Weddell Sea along the western Dronning Maud Land coast is more clearly characterized by the prominent bathymetric expression of the Explora Escarpment and the massive volcanic flows along the Explora Wedge (Figure 2), identified by the abundance of seaward dipping reflectors (SDRs) in the seismic reflection data. A number of deep crustal seismic refraction profiles cross the Explora Wedge and the Explora Escarpment and allow models showing a 70-km to 90-km-wide transitional crust thinning from about 20 km thickness to 10 km thickness toward the north (e.g., Jokat et al., 2004). Relatively high P-wave velocities in the lower crust and in the upper crustal section of the SDRs (Jokat et al., 2004) are evidence for a volcanic-type continental margin.

Dronning Maud Land (DML) Sector Conjugate to Africa

Data and syntheses of the central and eastern Dronning Maud Land (DML) margin stem from work by Hinz and Krause (1982), Hübscher et al. (1996b), Roeser et al. (1996), Jokat et al. (2003, 2004), Hinz et al. (2004), and König and Jokat (2006). The recent plate-kinematic reconstructions by Jokat et al. (2003) and König and Jokat (2006) show that southeast Africa was conjugate to the DML margin (Figure 3) from just east of the Explora Escarpment to the Gunnerus Ridge. However, this DML margin has two distinct parts, separated by the Astrid Ridge. The eastern

Lazarev Sea margin is characterized by a COT consisting of a broad stretched continental crust and up to 6-km-thick volcanic wedges clearly identified by SDR sequences. Deep crustal seismic data show that the crust thins in two steps from 23 km to about 10 km thickness over a distance of 180 km (König and Jokat, 2006). Their velocity-depth model reveals high seismic P-wave velocities in the lower crust of this COT, suggesting voluminous underplating and intrusion of magmatic material. A coast-parallel strong positive and negative magnetic anomaly pair marks the northern limit of the COT and is interpreted as the onset of the first oceanic crust generated by spreading processes at chron M12 (136 Ma). The volcanic characteristics of the eastern Lazarev Sea segment is very likely to be related to the same magmatic events leading to the Early Cretaceous crustal accretion of a Large Igneous Province (LIP) consisting of the separated oceanic plateaus Maud Rise, Agulhas Plateau, and Northeast Georgia Rise (Gohl and Uenzelmann-Neben, 2001) and to which also parts of the Mozambique Ridge may have belonged.

At the Riiser-Larsen Sea margin east of the Astrid Ridge, the outer limit of the COB is constrained by densely spaced aeromagnetic data revealing spreading anomalies up to M24 (155 Ma) (Jokat et al., 2003). This is so far the oldest magnetic seafloor spreading anomaly observed along any of the circum-Antarctic margins. Seismic reflection data (Hinz et al., 2004) indicate a COT of stretched continental crust that is with 50 km width much narrower compared with the COT of the Lazarev Sea margin. However, the data do not seem to indicate a strong magmatic influence of the COT as major

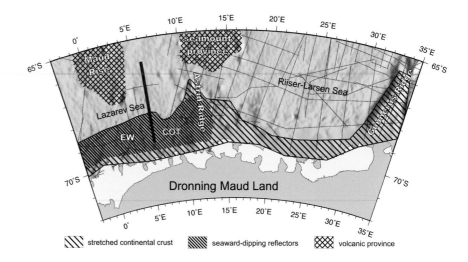

FIGURE 3 Dronning Maud Land sector with satellite-derived gravity field (McAdoo and Laxon, 1997) and continental margin features. Red lines show the tracks of offshore multichannel seismic profiles of SDLS. The dark blue line marks the location of an offshore deep crustal seismic refraction profile. COT = continent-ocean boundary; EW = Explora Wedge.

SDRs are missing (Hinz et al., 2004). Deep crustal seismic refraction data do not exist to better characterize this part of the DML margin and its COT or COB.

Enderby Land to Lambert Rift (EL) Sector Conjugate to India

Most of the crustal and sedimentary structures of the continental margins off Enderby Land (east of Gunnerus Ridge), between Prydz Bay and the Kerguelen Plateau, off Wilhelm II Land and Queen Mary Land (Figure 4) have been revealed by large seismic datasets acquired by Russian and Australian surveys (e.g., Stagg et al., 2004, 2005; Guseva et al., 2007; Leitchenkov et al., 2007a; Solli et al., 2007). Gaina et al. (2007) developed a breakup model between India and this East Antarctic sector based on compiled magnetic data of the southernmost Indian Ocean and the structures of the continental margin interpreted from the seismic data. Despite the large amount of high-quality seismic reflection data, the definition of the COB or COT is equivocal due to the lack of deep crustal seismic refraction data with the exception of nonreversed sonobuoy data. Stagg et al. (2004, 2005) defined the COB as the boundary to a zone by which the first purely oceanic crust was accreted and which shows a changeover from faulted basement geometry of stretched continental crust to a relatively smoother basement of ocean crust. This boundary is often accompanied by a basement ridge or trough.

The Enderby margin can be divided into two zones of distinct character. West of about 58°W the ocean fractures zone terminates in an oblique sense at the margin, thus giving it a mixed rift-transform setting (Stagg et al., 2004). Their defined COB lies between 100 km and 170 km oceanward of the shelf edge. Gaina et al. (2007) identified spreading anomalies from M0 to M9 east of Gunnerus Ridge from relatively sparse shipborne magnetic data. Most of both magnetic and seismic profiles do not parallel the spreading flow lines

and may be biased by the structure and signal of crossing fracture zones. The eastern Enderby margin zone has more of a normal rifted margin setting with a COB up to 300 km north of the shelf edge (Stagg et al., 2004). The prominent magnetic Mac Robertson Coastal Anomaly (MCA) correlates with the northern limit of the COT in this eastern zone. Ocean-bottom seismograph data along two seismic refraction profiles were recently acquired in the eastern Enderby Basin and across the Princess Elizabeth Trough (PET) between the Kerguelen Plateau and Princess Elizabeth Land as part of a German-Russian cooperation project (Gohl et al., 2007a) (Figure 4). The western profile confirms an extremely stretched crystalline continental crust, which thins to 7 km thickness (plus 4 km sediments on top), from the shelf edge to the location of the MCA. It is interesting to note that apart from a few scattered observations close to the marked COBs, major SDR sequences do not seem to exist on the Enderby Land margin (Stagg et al., 2004, 2005), suggesting the lack of a mantle plume at the time of breakup at about 130 Ma (Gaina et al., 2007).

The characteristics of the margin off central and eastern Prydz Bay is affected by both the inherited structure of the Paleozoic-Mesozoic Lambert Rift system as well as by magmatic events of the Kerguelen Plateau LIP, probably postdating the initial India-Antarctica breakup by about 10 million to 15 million years (Gaina et al., 2007). Guseva et al. (2007) proposed a direct connection between the volcanic Southern Kerguelen Plateau crust and stretched continental crust of Wilhelm II Land. The recent deep crustal seismic refraction data and a helicopter-magnetic survey, however, provides constraints that the central part of the PET consists of oceanic crust, possibly affected by the LIP event (Gohl et al., 2007a). This result is used to draw a narrow zone of stretched continental crust that widens eastward along the margin of the Davis Sea (Figure 4), where it reaches the width of the COT in the area of Bruce Rise as suggested by Guseva et

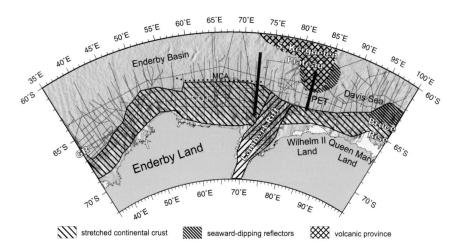

FIGURE 4 Enderby Land-Lambert Rift sector with satellite-derived gravity field (McAdoo and Laxon, 1997) and continental margin features. Red lines show the tracks of offshore multichannel seismic profiles of SDLS. Dark blue lines mark the locations of offshore deep crustal seismic refraction profiles. COT = continent-ocean boundary; PET = Princess Elizabeth Trough; GR = Gunnerus Ridge; MCA = Mac Robertson Coastal Anomaly.

al. (2007). Similar to observations along the Enderby Land margin, major SDR sequences are not observed on the Antarctic margin of the Davis Sea but only around the margins of the Southern Kerguelen Plateau. However, SDRs appear as strongly reflecting sequences at Bruce Rise (Guseva et al., 2007).

Wilkes Land (WL) Sector Conjugate to Australia

A vast amount of seismic reflection, gravity, and magnetic data as well as nonreversed sonobuoy refraction data collected by Australian and Russian scientists in the last few years allows a characterization of the Wilkes Land and Terre Adélie Coast margin east of Bruce Rise (Figure 5) (Stagg et al., 2005; Colwell et al., 2006; Leitchenkov et al., 2007b). Along most of this margin the shelf slope is underlain by a marginal rift zone of stretched and faulted basement. From the northern limit of this marginal rift zone to 90-180 km farther oceanward, Eittreim et al. (1985), Eittreim and Smith (1987), Colwell et al. (2006), and Leitchenkov et

(2007b) identified a COT consisting primarily of strongly stretched and faulted continental crust with embedded magmatic segments following linear trends parallel to the margin. The interpretation of the COB along parts of the margin is debatable, as mainly basement characteristics were used for the differentiation of crustal types. It cannot be completely excluded that the magnetic anomalies interpreted as magmatic components are actually true seafloor spreading anomalies C33y and C34y, which would move the COB farther south. However, this seems unlikely based on the results of Colwell et al. (2006), Direen et al. (2007), and Sayers et al. (2001) in a comparison of the conjugate magnetic anomalies as well as new theoretical and analogue examples published in Sibuet et al. (2007). Reversed deep crustal refraction data would provide better constraints on the composition of the crustal units but are missing anywhere along this margin. However, in the absence of better data I adopted the interpretation by Colwell et al. (2006) and Leitchenkov et al. (2007b) of an up-to-300-km-wide zone of stretched continental crust (Figure 5).

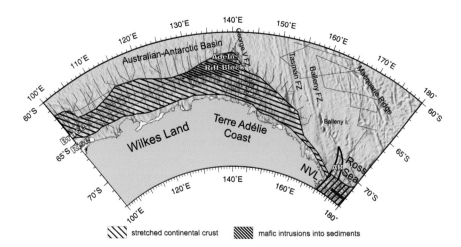

FIGURE 5 Wilkes Land sector with satellite-derived gravity field (McAdoo and Laxon, 1997) and continental margin features. Red lines show the tracks of offshore multichannel seismic profiles of SDLS. The dark blue line marks the location of an offshore deep crustal seismic refraction profile. NVR = Northern Victoria Land; AR = Adare Rift.

Off the Terre Adélie Coast margin the Adélie Rift Block, which is interpreted as a continental crustal block, is part of the stretched marginal crust. Major sequences of SDRs are not observed along the margin of the Wilkes Land sector, although Eittreim et al. (1985), Eittreim and Smith (1987), and Colwell et al. (2006) describe significant volumes of mafic intrusions within the sedimentary sequences of the landward edge of the Adélie Rift Block. The margin east of about 155°E toward the Ross Sea is characterized by prominent oblique fracture zones (e.g., Balleny FZ) reaching close to the shelf edge. This is similar to the rift-transform setting as observed in the western Enderby Basin but in an opposite directional sense. Stock and Cande (2002) and Damaske et al. (2007) suggest that a broad zone of distributed deformation was active at the margin, affecting the continental crustal blocks of Northern Victoria Land even after the initiation of ocean spreading.

The continental margin in the western Ross Sea underwent a rather complex development that makes a clear delineation of the COB difficult. A major proportion of the Ross Sea crustal extension is associated with a 180 km plate separation between East and West Antarctica when the Adare Trough was formed in Eocene and Oligocene time (Cande et al., 2000; Stock and Cande, 2002). The southward extension of this rift dissects the continental shelf region of the western Ross Sea (Davey et al., 2006).

Marie Byrd Land (MBL) Sector Conjugate to Zealandia

The eastern Ross Sea margin (Figure 6) is conjugate to the southwesternmost margin of the Campbell Plateau of New Zealand. Unlike the western Ross Sea margin, this margin can be considered a typical rifted continental margin with the oldest identified shelf-edge parallel spreading anomaly being C33y just east of the Iselin Rift (Stock and Cande, 2002).

Seismic refraction data from a profile across the Ross Sea shelf by Cooper et al. (1997) and Trey et al. (1997) show a crust of 17-km to 24-km thickness with the thinnest parts beneath two troughs in the eastern and western region. It can be assumed that this type of stretched continental crust extends to the shelf edge at least. Whether crustal thinning is a result of the New Zealand-Antarctic breakup or extensional processes of the West Antarctic Rift System or both cannot be answered due to insufficient data in this area. It is also not known whether magmatic events affected the structure of the marginal crust.

Structural models of the continental margin of western Marie Byrd Land also suffer from the lack of seismic and other geophysical data. The only assumption for a sharp breakup structure and a very narrow transitional crust is derived from the close fit of the steep southeastern Campbell Plateau margin to the western MBL margin and coastal gravity anomaly (e.g., Mayes et al., 1990; Sutherland, 1999; Larter et al., 2002; Eagles et al., 2004). The same approach was applied for the closest fit between Chatham Rise and eastern MBL. However, new deep crustal seismic data from the Amundsen Sea indicate that the inner to middle shelf of the Amundsen Sea Embayment consists of crust thinned to about 21- to 23-km thickness with a pre- or syn-breakup failed rift structure (Gohl et al., 2007b). Here the continental margin structure is rather complex due to the propagating and rotating rifting processes between the breakup of Chatham Rise from the western Thurston Island block at about 90 Ma (Eagles et al., 2004), the opening of the Bounty Trough between Chatham Rise and Campbell Plateau (Eagles et al., 2004; Grobys et al., 2007), and the initiation of the breakup of Campbell Plateau from MBL at about 84-83 Ma (e.g., Larter et al., 2002; Eagles et al., 2004). The analysis of a crustal seismic refraction profile (Gohl et al., 2007b) and a seismic reflection dataset (Gohl et al., 1997b) between the

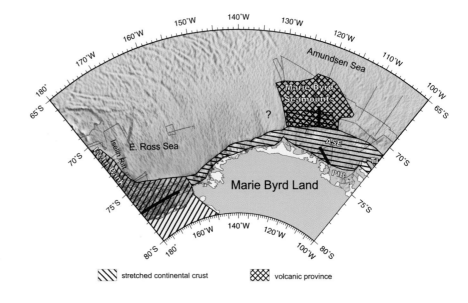

FIGURE 6 Marie Byrd Land sector with satellite-derived gravity field (McAdoo and Laxon, 1997) and continental margin features. Red lines show the tracks of offshore multichannel seismic profiles of SDLS. Dark blue lines mark the locations of offshore deep crustal seismic refraction profiles. ASE = Amundsen Sea Embayment; PIB = Pine Island Bay; TI = Thurston Island.

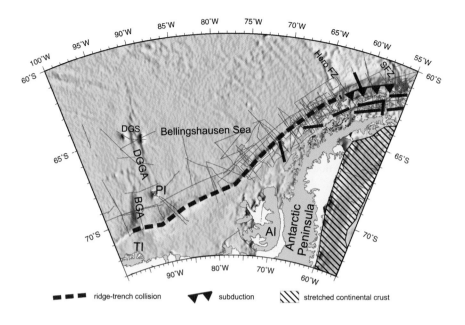

FIGURE 7 Antarctic Peninsula-Ellsworth Land sector with satellite-derived gravity field (McAdoo and Laxon, 1997) and continental margin features. Red lines show the tracks of offshore multichannel seismic profiles of SDLS. Dark blue lines mark the locations of offshore deep crustal seismic refraction profiles. AI = Alexander Island; TI = Thurston Island; PI = Peter Island; DGS = De Gerlache Seamounts; DGGA = De Gerlache Gravity Anomaly; BGA = Bellingshausen Gravity Anomaly; SFZ = Shackleton Fracture Zone.

shelf edge and the Marie Byrd Seamount province reveals that the crust in this corridor is 12-18 km thick, highly fractured, and volcanically overprinted. Lower crustal velocities suggest a continental affinity, which implies highly stretched continental crust or continental fragments, possibly even into the area of the seamount province. SDRs are not observed in the few existing seismic reflection profiles crossing the shelf and margin (Gohl et al., 1997a,b, 2007b; Nitsche et al., 2000; Cunningham et al., 2002; Uenzelmann-Neben et al., 2007).

SUBDUCTION TO RIDGE COLLISIONS: THE CONVERTED ACTIVE-TO-PASSIVE MARGIN

Antarctic Peninsula to Ellsworth Land (APE) Sector— Proto-Pacific Margin

Subduction of the Phoenix plate and subsequent collision of the Phoenix-Pacific spreading ridge resulted in a converted active to passive nonrifted margin along the Ellsworth Land and western Antarctic Peninsula margin between Thurston Island and the Hero Fracture Zone (e.g., Barker, 1982; Larter and Barker, 1991) (see Figure 7). The age for the oldest ridge-trench collision in the western Bellingshausen Sea can be roughly estimated to be about 50 Ma, using a few identified spreading anomalies (Eagles et al., 2004; Scheuer et al., 2006). Oceanic crustal age along the margin becomes progressively younger toward the northeast (Larter and Barker, 1991). Ultraslow subduction is still active in the segment of the remaining Phoenix plate between the Hero FZ and the Shackleton FZ along the South Shetland Trench. The only deep crustal seismic data along this converted margin south of the Hero FZ are from two profiles by Grad et al. (2002), showing that the crust thins oceanward from about 37 km thickness beneath the Antarctic Peninsula to a thickness of 25

to 30 km beneath the middle and outer continental shelf. The remains of the collided ridge segments cannot be resolved by the models. Data are also not sufficient to estimate any possible crustal extension of the western Antarctic Peninsula due to stress release after subduction ceded.

GEODYNAMIC AND PLATE-TECTONIC IMPLICATIONS AND COMPLICATIONS

The compilation of circum-Antarctic margin characteristics indicates that only a quarter (3400 km) of the rifted margins is of volcanic type, although uncertainties on volcanic and nonvolcanic affinity exist for about 4900 km (35 percent) of the rifted margins due to the lack of data in the Weddell Sea and along the MBL and Ross Sea margins. Only the eastern WL and western DML sectors as well as some isolated areas such as Bruce Rise of the EL sector and the Adélie Rift Block in the WL sector show well-observed magmatic characteristics that are explained by syn-rift mantle plumes or other smaller-scale magmatic events. Most other margins seem to have been formed by processes similar to those proposed for other nonvolcanic margins, such as the Iberian-Newfoundland conjugate margins (e.g., Sibuet et al., 2007) and the Nova Scotia margin (e.g., Funck et al., 2003), which exhibit widely extended, thinned, and block-faulted continental crust. Whether any updoming of lower continental crust is a possible process, as Gaina et al. (2007) suggest for the Enderby Land margin, is difficult to assess because of a lack of drill information.

Approximately 70 percent of the rifted passive margins contain continental crust stretched over more than 50 km oceanward of the shelf edge. In about two-thirds of these extended margins the COT has a width of more than 100 km, in many cases up to 300 km. The total area of extended continental crust on the continental shelf and beyond the shelf edge, including COTs with substantial syn-rift magmatic-volcanic accretion, is estimated to be about 2.9×10^6 km^2. Crustal stretching factors still remain uncertain for a good proportion of the extended margins due to a lack of crustal thickness measurements in most sectors. However, assuming stretching factors between 1.5 close to the shelf edge and in marginal rifts and increasing to 2 or 3 outboards in the deep sea, the continental crust, which has to be added to the original continent of normal crustal thickness, makes up about 1.5×10^6 km^2. The quantification of extended marginal crust has implications for plate-kinematic reconstructions, paleo-bathymetric models, and possibly for isostatic balancing of the Antarctic continent in glacial-interglacial cycles.

In almost all large-scale plate-kinematic reconstructions in which Antarctica is a key component, plate motions are calculated by applying rotation parameters derived from spreading anomalies and fracture zone directions, and continent-ocean boundaries are fixed single-order discontinuities, in most cases identified by the shelf edge or the associated margin-parallel gravity anomaly gradient (e.g., Lawver and Gahagan, 2003; Cande and Stock, 2004). This has caused misfits in terms of substantial overlaps or gaps when plates are reconstructed to close fit. For instance, a large misfit occurs when fitting the southeastern Australian margin to the eastern Wilkes Land margin in the area between 142°E and 160°E where extended continental crust reaches up to 300 km off the Terre Adélie Coast (Stock and Cande, 2002; Cande and Stock, 2004). In their reconstruction of the breakup processes of the Weddell Sea region, König and Jokat (2006) accounted for extended continental crust in the southern Weddell Sea and derived a reasonable fit of the conjugate margins of South America, Antarctica, and Africa. An appropriate approach for reconstructing best fits of continents and continental fragments is to apply a crustal balancing technique by restoring crustal thickness in rift zones and plate margins (Grobys et al., 2008). This, however, requires detailed knowledge of pre- and postrift crustal thickness, crustal composition, and magmatic accretion.

Detailed delineations of the COT and COB are important ingredients for paleobathymetric reconstructions. In the reconstruction of the circum-Antarctic ocean gateways (e.g., Lawver and Gahagan, 2003; Brown et al., 2006; Eagles et al., 2006), but also of the bathymetric features along nongateway continental margins, the differentiation between continental crust that was stretched and faulted and possibly intruded by magmatic material on one side and oceanic crust generated from spreading processes on the other side may make a significant difference in estimating the widths and depths along and across pathways for paleo-ocean currents.

Parameters of crustal and lithospheric extensions, depths, and viscoelastic properties are key boundary conditions for accurate calculation of the isostatic response from a varying ice sheet in glacial-interglacial cycles. Ice-sheet

modelers still use relatively rudimentary crustal and lithospheric models of the Antarctic continent. Although the tomographic inversion of seismological data has improved the knowledge of the lithospheric structure beneath Antarctica and surrounding ocean basins (e.g., Morelli and Danesi, 2004), its spatial resolution of the upper 60-70 km of the relevant elastic lithosphere (Ivins and James, 2005), and in particular of the boundary between continental and oceanic lithosphere, is still extremely crude. Considering that ice sheets advanced to the shelf breaks of most Antarctic continental margins during glacial maxima, the isostatic response must be directly related to the width and depths of any extended continental crust and lithosphere oceanward of the shelf, which would probably have an effect on estimates of sea-level change. To quantify this effect, good-quality deep crustal and lithospheric data are needed to derive the geometries and rheologies of the extended crust and lithospheric mantle.

CONCLUSIONS

In a comprehensive compilation of circum-Antarctic continental margin types, about 70 percent of the rifted passive margins contain extended continental crust stretching more than 50 km oceanward of the shelf edge. Most of these extended margins have a continent-ocean transition with a width of more than 100 km—in many cases up to 300 km. Only a quarter of the rifted margins seem to be of volcanic type. The total area of extended continental crust on the shelf and oceanward of the shelf edge, including COTs with substantial syn-rift magmatic-volcanic accretion, is estimated to be about 2.9×10^6 km^2. This has implications for improved plate-kinematic and paleobathymetric reconstructions and provides new constraints for accurate calculations of isostatic responses along the Antarctic margin.

ACKNOWLEDGMENTS

I gratefully acknowledge the masters, crews, and scientists of the many marine expeditions during which the geophysical data used in this paper were acquired. Many thanks go to Nick Direen and an anonymous reviewer whose review comments substantially helped improve the paper. I also thank the co-editor Howard Stagg for useful comments and the editorial work.

REFERENCES

Barker, P. F. 1982. The Cenozoic subduction history of the Pacific margin of the Antarctic Peninsula: Ridge crest-trench interactions. *Journal of the Geological Society of London* 139:787-801.

Brown, B., C. Gaina, and R. D. Müller. 2006. Circum-Antarctic palaeo-bathymetry: Illustrated examples from Cenozoic to recent times. *Palaeogeography, Palaeoclimatology, Palaeoecology* 231:158-168.

Cande, S. C., and J. M. Stock. 2004. Cenozoic reconstructions of the Australia-New Zealand-South Pacific Sector of Antarctica. In *The Cenozoic Southern Ocean: Tectonics, Sedimentation and Climate Change Between Australia and Antarctica*, eds. N. Exon, J. K. Kennett, and M. Malone. Geophysical Monograph 151:5-17. Washington, D.C.: American Geophysical Union.

Cande, S. C., J. M. Stock, D. Müller, and T. Ishihara. 2000. Cenozoic motion between East and West Antarctica. *Nature* 404:145-150.

Colwell, J. B., H. M. J. Stagg, N. G. Direen, G. Bernardel, and I. Borissova. 2006. The structure of the continental margin off Wilkes Land and Terre Adélie Coast, East Antarctica. In *Antarctica: Contributions to Global Earth Sciences*, eds. D. K. Fütterer, D. Damaske, G. Kleinschmidt, H. Miller, and F. Tessensohn, pp. 327-340. New York: Springer-Verlag.

Cooper, A. K., H. Trey, G. Pellis, G. Cochrane, F. Egloff, M. Busetti, and ACRUP Working Group. 1997. Crustal structure of the southern Central Trough, western Ross Sea. In *The Antarctic Region: Geological Evolution and Processes*, ed. C. A. Ricci, pp. 637-642. Siena: *Terra Antartica* Publication.

Cunningham, A. P., R. D. Larter, P. F. Barker, K. Gohl, and F. O. Nitsche. 2002. Tectonic evolution of the Pacific margin of Antarctica. 2. Structure of Late Cretaceous-early Tertiary plate boundaries in the Bellingshausen Sea from seismic reflection and gravity data. *Journal of Geophysical Research* 107(B12):2346, doi:10.1029/2002JB001897.

Damaske, D., A. L. Läufer, F. Goldmann, H.-D. Möller, and F. Lisker. 2007. Magnetic anomalies northeast of Cape Adare, northern Victoria Land (Antarctica), and their relation to onshore structures. In *Antarctica: A Keystone in a Changing World—Online Proceedings for the Tenth International Symposium on Antarctic Earth Sciences*, eds. Cooper, A. K., C. R. Raymond et al., USGS Open-File Report 2007-1047, Short Research Paper 016, doi:10.3133/of2007-1047.srp016.

Davey, F. J., S. C. Cande, and J. M. Stock. 2006. Extension in the western Ross Sea region—links between Adare Basin and Victoria Basin. *Geophysical Research Letters* 33, L20315, doi:10.1029/2006GL027383.

Direen, N. G., J. Borissova, H. M. J. Stagg, J. B. Colwell, and P. A. Symonds. 2007. Nature of the continent-ocean transition zone along the southern Australian continental margin: A comparison of the Naturaliste Plateau, south-western Australia, and the central Great Australian Bight sectors. In *Imaging, Mapping and Modelling Continental Lithosphere Extension and Breakup*, eds. G. Karner, G. Manatschal, and L. M. Pinheiro. *Geological Society Special Publication* 282:239-263.

Eagles, G., K. Gohl, and R. B. Larter. 2004. High resolution animated tectonic reconstruction of the South Pacific and West Antarctic margin. *Geochemistry, Geophysics, Geosystems (G³)* 5, doi:10.1029/2003GC000657.

Eagles, G., R. Livermore, and P. Morris. 2006. Small basins in the Scotia Sea: The Eocene Drake Passage gateway. *Earth and Planetary Science Letters* 242:343-353.

Eittreim, S. L., and G. L. Smith. 1987. Seismic sequences and their distribution on the Wilkes Land margin. In *The Antarctic Continental Margin: Geology and Geophysics of Offshore Wilkes Land*, eds. S. L. Eittreim and M. A. Hampton. *Earth Sciences Series* 5A:15-43. Tulsa, OK: Circum-Pacific Council for Energy and Mineral Resources.

Eittreim, S. L., M. A. Hampton, and J. R. Childs. 1985. Seismic-reflection signature of Cretaceous continental breakup on the Wilkes Land margin, Antarctica. *Science* 229:1082-1084.

Funck, T., J. R. Hopper, H. C. Larsen, K. E. Louden, B. E. Tucholke, and W. S. Holbrook. 2003. Crustal structure of the ocean-continent transition at Flemish Cap: Seismic refraction results. *Journal of Geophysical Research* 108(B11), 2531, doi:10.1029/2003JB002434.

Gaina, C., R. D. Müller, B. Brown, T. Ishihara, and S. Ivanov. 2007. Breakup and early seafloor spreading between India and Antarctica. *Geophysical Journal International* 170:151-169, doi:10.1111/j.1365-246X.2007.03450.

Ghidella, M. E., and J. L. LaBrecque. 1997. The Jurassic conjugate margins of the Weddell Sea: Considerations based on magnetic, gravity and paleobathymetric data. In *The Antarctic Region: Geological Evolution and Processes*, ed. C. A. Ricci, pp. 441-451. Siena: *Terra Antartica* Publication.

Ghidella, M. E., G. Yáñez, and J. L. LaBrecque. 2002. Revised tectonic implications for the magnetic anomalies of the western Weddell Sea. *Tectonophysics* 347:65-86.

Gohl, K., and G. Uenzelmann-Neben. 2001. The crustal role of the Agulhas Plateau, southwest Indian Ocean: Evidence from seismic profiling. *Geophysical Journal International* 144:632-646.

Gohl, K., F. Nitsche, and H. Miller. 1997a. Seismic and gravity data reveal Tertiary interplate subduction in the Bellingshausen Sea, southeast Pacific. *Geology* 25:371-374.

Gohl, K., F. O. Nitsche, K. Vanneste, H. Miller, N. Fechner, L. Oszko, C. Hübscher, E. Weigelt, and A. Lambrecht. 1997b. Tectonic and sedimentary architecture of the Bellingshausen and Amundsen Sea Basins, SE Pacific, by seismic profiling. In *The Antarctic Region: Geological Evolution and Processes*, ed. C. A. Ricci, pp. 719-723. Siena: *Terra Antartica* Publication.

Gohl, K., G. L. Leitchenkov, N. Parsiegla, B.-M. Ehlers, C. Kopsch, D. Damaske, Y. B. Guseva, and V. V. Gandyukhin. 2007a. Crustal types and continent-ocean boundaries between the Kerguelen Plateau and Prydz Bay, East Antarctica. In *Antarctica: A Keystone in a Changing World—Online Proceedings for the Tenth International Symposium on Antarctic Earth Sciences*, eds. Cooper, A. K., C. R. Raymond et al., USGS Open-File Report 2007-1047, Extended Abstract 038, http://pubs.usgs.gov/of/2007/1047/.

Gohl, K., D. Teterin, G. Eagles, G. Netzeband, J. W. G. Grobys, N. Parsiegla, P. Schlüter, V. Leinweber, R. D. Larter, G. Uenzelmann-Neben, and G. B. Udintsev. 2007b. Geophysical survey reveals tectonic structures in the Amundsen Sea embayment, West Antarctica. In *Antarctica: A Keystone in a Changing World—Online Proceedings for the Tenth International Symposium on Antarctic Earth Sciences*, eds. Cooper, A. K., C. R. Raymond et al., USGS Open-File Report 2007-1047, Short Research Paper 047, doi:10.3133/Of2007-1047.srp047.

Golynsky, A. V., and N. D. Aleshkova. 2000. New aspects of crustal structure in the Weddell Sea region from aeromagnetic studies. *Polarforschung* 67:133-141.

Grad, M., A. Guterch, T. Janik, and P. Sroda. 2002. Seismic characteristics of the crust in the transition zone from the Pacific Ocean to the northern Antarctic Peninsula, West Antarctica. In *Antarctica at the Close of a Millennium*, eds. J. A. Gamble, D. N. B. Skinner, and S. Henrys. *Royal Society of New Zealand Bulletin* 35:493-498.

Grobys, J. W. G., K. Gohl, B. Davy, G. Uenzelmann-Neben, T. Deen, and D. Barker. 2007. Is the Bounty Trough, off eastern New Zealand, an aborted rift? *Journal of Geophysical Research* 112, B03103, doi:10.1029/2005JB004229.

Grobys, J. W. G., K. Gohl, and G. Eagles. 2008. Quantitative tectonic reconstructions of Zealandia based on crustal thickness estimates. *Geochemistry, Geophysics, Geosystems (G³)* 9:Q01005, doi:10.1029/2007GC001691.

Guseva, Y. B., G. L. Leitchenkov, and V. V. Gandyukhin. 2007. Basement and crustal structure of the Davis Sea region (East Antarctica): Implications for tectonic setting and COB definition. In *Antarctica: A Keystone in a Changing World—Online Proceedings for the Tenth International Symposium on Antarctic Earth Sciences*, eds. Cooper, A. K., C. R. Raymond et al., USGS Open-File Report 2007-1047, Short Research Paper 025, doi:10.3133/of2007-1047.srp025.

Hinz, K., and W. Krause. 1982. The continental margin of Queen Maud Land/Antarctica: Seismic sequences, structural elements and geological development. *Geologisches Jahrbuch E* 23:17-41.

Hinz, K., S. Neben, Y. B. Gouseva, and G. A. Kudryavtsev. 2004. A compilation of geophysical data from the Lazarev Sea and the Rijser-Larsen Sea, Antarctica. *Marine Geophysical Researches* 25:233-245, doi:10.1007/s11001-005-1319-y.

Hübscher, C., W. Jokat, and H. Miller. 1996a. Structure and origin of southern Weddell Sea crust: Results and implications. In *Weddell Sea Tectonics and Gondwana Break-up*, eds. B. C. Storey, E. C. King, and R. A. Livermore. *Geological Society Special Publication* 108:201-211.

Hübscher, C., W. Jokat, and H. Miller. 1996b. Crustal structure of the Antarctic continental margin in the eastern Weddell Sea. In *Weddell Sea Tectonics and Gondwana Break-up*, eds. B. C. Storey, E. C. King, and R. A. Livermore. *Geological Society Special Publication* 108:165-174.

Ivins, E. R., and T. S. James. 2005. Antarctic glacial isostatic adjustment: A new assessment. *Antarctic Science* 17:537-549, doi:10.1017/S0954102004.

Jokat, W., C. Hübscher, U. Meyer, L. Oszko, T. Schöne, W. Versteeg, and H. Miller. 1996. The continental margin off East Antarctica between 10°W and 30°W. In *Weddell Sea Tectonics and Gondwana Break-up*, eds. B. C. Storey, E. C. King, and R. A. Livermore. *Geological Society Special Publication* 108:129-141.

Jokat, W., N. Fechner, and M. Studinger. 1997. Geodynamic models of the Weddell Sea Embayment in view of new geophysical data. In *The Antarctic Region: Geological Evolution and Processes*, ed. C. A. Ricci, pp. 453-459. Siena: *Terra Antartica* Publication.

Jokat, W., T. Boebel, M. König, and U. Meyer. 2003. Timing and geometry of early Gondwana breakup. *Journal of Geophysical Research* 108(B9), doi:10.1029/2002JB001802.

Jokat, W., O. Ritzmann, C. Reichert, and K. Hinz. 2004. Deep crustal structure of the continental margin off the Explora Escarpment and in the Lazarev Sea, East Antarctica. *Marine Geophysical Researches* 25:283-304, doi:10.1007/s11001-005-1337-9.

König, M., and W. Jokat. 2006. The Mesozoic breakup of the Weddell Sea. *Journal of Geophysical Research* 111, B12102, doi:10.1029/2005JB004035.

Larter, R. D., and P. F. Barker. 1991. Effects of ridge crest-trench interaction on Phoenix-Antarctic spreading: Forces on a young subducting plate. *Journal of Geophysical Research* 96(B12):19586-19607.

Larter, R. D., A. P. Cunningham, P. F. Barker, K. Gohl, and F. O. Nitsche. 2002. Tectonic evolution of the Pacific margin of Antarctica. 1. Late Cretaceous tectonic reconstructions. *Journal of Geophysical Research* 107, B12:2345, doi:10.1029/2000JB000052.

Lawver, L. A., and L. M. Gahagan. 2003. Evolution of Cenozoic seaways in the circum-Antarctic region. *Palaeogeography, Palaeoclimatology, Palaeoecology* 3115:1-27.

Leitchenkov, G. L., H. Miller, and E. N. Zatzepin. 1996. Structure and Mesozoic evolution of the eastern Weddell Sea, Antarctica: History of early Gondwana break-up. In *Weddell Sea Tectonics and Gondwana Break-up*, eds. B. C. Storey, E. C. King, and R. A. Livermore. *Geological Society Special Publication* 108:175-190.

Leitchenkov, G. L., Y. B. Guseva, and V. V. Gandyukhin. 2007a. Cenozoic environmental changes along the East Antarctic continental margin inferred from regional seismic stratigraphy. In *Antarctica: A Keystone in a Changing World—Online Proceedings for the Tenth International Symposium on Antarctic Earth Sciences*, eds. Cooper, A. K., C. R. Raymond et al., USGS Open-File Report 2007-1047, Short Research Paper 005, doi:10.3133/of2007-1047.srp005.

Leitchenkov, G. L., V. V. Gandyukhin, Y. B. Guseva, and A. Y. Kazankov. 2007b. Crustal structure and evolution of the Mawson Sea, western Wilkes Land margin, East Antarctica. In *Antarctica: A Keystone in a Changing World—Online Proceedings for the Tenth International Symposium on Antarctic Earth Sciences*, eds. Cooper, A. K., C. R. Raymond et al., USGS Open-File Report 2007-1047, Short Research Paper 028, doi:10.3133/of2007-1047.srp028.

Lythe, M. B., D. G. Vaughan, and the BEDMAP Consortium. 2001. BEDMAP: A new ice thickness and subglacial topographic model of Antarctica. *Journal of Geophysical Research* 106(B6):11335-11351.

Mayes, C. L., L. A. Lawver, and D. T. Sandwell. 1990. Tectonic history and new isochron chart of the South Pacific. *Journal of Geophysical Research* 95(B6):8543-8567.

McAdoo, D. C., and S. Laxon. 1997. Antarctic tectonics: Constraints from an ERS-1 satellite marine gravity field. *Science* 276:556-560.

Morelli, A., and S. Danesi. 2004. Seismological imaging of the Antarctic continental lithosphere: A review. *Global Planetary Change* 42:155-165.

Nitsche, F. O., A. P. Cunningham, R. D. Larter, and K. Gohl. 2000. Geometry and development of glacial continental margin depositional systems in the Bellingshausen Sea. *Marine Geology* 162:277-302.

Roeser, H. A., J. Fritsch, and K. Hinz. 1996. The development of the crust off Donning Maud Land, East Antarctica. In *Weddell Sea Tectonics and Gondwana Break-up*, eds. B. C. Storey, E. C. King, and R. A. Livermore. *Geological Society Special Publication* 108:243-264.

Rogenhagen, J., and W. Jokat. 2002. Origin of the gravity ridges and Anomaly-T in the southern Weddell Sea. In *Antarctica at the Close of a Millennium*, eds. J. A. Gamble, D. N. B. Skinner, and S. Henrys. *Royal Society of New Zealand Bulletin* 35:227-231.

Sayers, J., P. A. Symonds, N. G. Direen, and G. Bernardel. 2001. Nature of the continent-ocean transition on the non-volcanic rifted margin of the central Great Australian Bight. In *Non-volcanic Rifting of Continental Margins: A Comparison of Evidence from Land and Sea*, eds. R. C. L. Wilson, R. B. Whitmarsh, B. Taylor, and N. Froitzheim. *Geological Society Special Publication* 187:51-76.

Scheuer, C., K. Gohl, R. D. Larter, M. Rebesco, and G. Udintsev. 2006. Variability in Cenozoic sedimentation along the continental rise of the Bellingshausen Sea, West Antarctica. *Marine Geology* 277:279-298.

Sibuet, J.-C., S. Srivastava, and G. Manatschal. 2007. Exhumed mantle-forming transitional crust in the Newfoundland-Iberia rift and associated magnetic anomalies. *Journal of Geophysical Research* 112, B06105, doi:10.1029/2005JB003856.

Solli, K., B. Kuvaas, Y. Kristoffersen, G. Leitchenkov, J. Guseva, and V. Gandyukhin. 2007. The Cosmonaut Sea Wedge. In *Antarctica: A Keystone in a Changing World—Online Proceedings for the Tenth International Symposium on Antarctic Earth Sciences*, eds. Cooper, A. K., C. R. Raymond et al., USGS Open-File Report 2007-1047, Short Research Paper 009, doi:10.3133/of2007-1047.srp009.

Stagg, H. M. J., J. B. Colwell, N. G. Direen, P. E. O'Brien, G. Bernardel, I. Borissova, B. J. Brown, and T. Ishihara. 2004. Geology of the continental margin of Enderby and Mac Robertson Lands, East Antarctica: Insights from a regional data set. *Marine Geophysical Researches* 25:183-218.

Stagg, H. M. J., J. B. Colwell, N. G. Direen, P. E. O'Brien, B. J. Brown, G. Bernadel, I. Borissova, L. Carson, and D. B. Close. 2005. Geological framework of the continental margin in the region of the Australian Antarctic Territory. *Geoscience Australia Record* 2004/25.

Stock, J. M., and S. C. Cande. 2002. Tectonic history of Antarctic seafloor in the Australia-New Zealand-South Pacific sector: Implications for Antarctic continental tectonics. In *Antarctica at the Close of a Millennium*, eds. J. A. Gamble, D. N. B. Skinner, and S. Henrys. *Royal Society of New Zealand Bulletin* 35:251-259.

Sutherland, R. 1999. Basement geology and tectonic development of the greater New Zealand region: An interpretation from regional magnetic data. *Tectonophysics* 308:341-362.

Trey, H., J. Makris, G. Brancolini, A. K. Cooper, G. Cochrane, B. Della Vedova, and ACRUP Working Group. 1997. The Eastern Basin crustal model from wide-angle reflection data, Ross Sea, Antarctica. In *The Antarctic Region: Geological Evolution and Processes*, ed. C. A. Ricci, pp. 637-642. Siena: *Terra Antartica* Publication.

Uenzelmann-Neben, G., K. Gohl, R. D. Larter, and P. Schlüter. 2007. Differences in ice retreat across Pine Island Bay, West Antarctica, since the Last Glacial Maximum: Indications from multichannel seismic reflection data. USGS Open-File Report 2007-1047, Short Research Paper 001, doi:10.3133/of2007-1047.srp084.

Wardell, N., J. R. Childs, and A. K. Cooper. 2007. Advances through collaboration: Sharing seismic reflection data via the Antarctic Seismic Data Library System for Cooperative Research (SDLS). In *Antarctica: A Keystone in a Changing World—Online Proceedings for the Tenth International Symposium on Antarctic Earth Sciences*, eds. Cooper, A. K., C. R. Raymond et al., USGS Open-File Report 2007-1047, Short Research Paper 001, doi:10.3133/of2007-1047.srp001.

Cooper, A. K., P. J. Barrett, H. Stagg, B. Storey, E. Stump, W. Wise, and the 10th ISAES editorial team, eds. (2008). *Antarctica: A Keystone in a Changing World.* Proceedings of the 10th International Symposium on Antarctic Earth Sciences. Washington, DC: The National Academies Press.

Landscape Evolution of Antarctica

S. S. R. Jamieson and D. E. Sugden[1]

ABSTRACT

The relative roles of fluvial versus glacial processes in shaping the landscape of Antarctica have been debated since the expeditions of Robert Scott and Ernest Shackleton in the early years of the 20th century. Here we build a synthesis of Antarctic landscape evolution based on the geomorphology of passive continental margins and former northern mid-latitude Pleistocene ice sheets. What makes Antarctica so interesting is that the terrestrial landscape retains elements of a record of change that extends back to the Oligocene. Thus there is the potential to link conditions on land with those in the oceans and atmosphere as the world switched from a greenhouse to a glacial world and the Antarctic ice sheet evolved to its present state. In common with other continental fragments of Gondwana there is a fluvial signature to the landscape in the form of the coastal erosion surfaces and escarpments, incised river valleys, and a continent-wide network of river basins. A selective superimposed glacial signature reflects the presence or absence of ice at the pressure melting point. Earliest continental-scale ice sheets formed around 34 Ma, growing from local ice caps centered on mountain massifs, and featured phases of ice-sheet expansion and contraction. These ice masses were most likely cold-based over uplands and warm-based across lowlands and near their margins. For 20 million years ice sheets fluctuated on Croll-Milankovitch frequencies. At ~14 Ma the ice sheet expanded to its maximum and deepened a preexisting radial array of troughs selectively through the coastal mountains and eroded the continental

shelf before retreating to its present dimensions at ~13.5 Ma. Subsequent changes in ice extent have been forced mainly by sea-level change. Weathering rates of exposed bedrock have been remarkably slow at high elevations around the margin of East Antarctica under the hyperarid polar climate of the last ~13.5 Ma, offering potential for a long quantitative record of ice-sheet evolution with techniques such as cosmogenic isotope analysis.

INTRODUCTION

Our aim is to synthesize ideas about the evolution of the terrestrial landscapes of Antarctica. There are advantages to such a study. First, the landscape evidence appears to extend back beyond earliest Oligocene times when the first ice sheets formed. Thus events on land may potentially be linked with atmospheric and oceanic change as the world switched from a greenhouse to a glacial world and saw the development of the Antarctic ice sheet. This helps in establishing correlations or leads and lags between different components of the Earth system as a way of establishing cause and effect in global environmental change. Second, the evidence of landscape evolution can be used to refine models of Earth surface processes and their interaction with the wider global system (e.g., by linking conditions in the terrestrial source areas with the marine record of deposition). Third, there are analogies in the Northern Hemisphere of similar-size former Pleistocene ice sheets where the bed is exposed for study. The glaciological body of evidence and theory built on such a base over 150 years is helpful in assessing the nature of the inaccessible topography beneath the Antarctic ice sheet. As such it can illuminate interpretations of subglacial conditions and the dynamics of the present ice sheet.

The crux of any reconstruction of landscape evolution

---

[1]School of GeoSciences, University of Edinburgh, Edinburgh, EH8 9XP, Scotland, UK (Stewart.Jamieson@ed.ac.uk, David.Sugden@ed.ac.uk).

in a glaciated area is the extent to which ice sheets have transformed a preexisting fluvial topography. For 150 years there has been debate between those highlighting the erosive power of ice and those indicating its preservative capacity. In 1848 Charles Lyell, on his way over the formerly glaciated eastern Grampians of Scotland to receive a knighthood from Queen Victoria, wrote in his diary, "Here as on Mt Washington and in the White Mountains the decomposing granite boulders and the bare surfaces of disintegrating granite are not scored with glacial furrows or polished." Such observations subsequently led to the idea of unglaciated enclaves that escaped glaciation completely (e.g., in Britain: Linton, 1949; and in North America: Ives, 1966). In Antarctica, scientists on the early 20th-century expeditions of Scott and Shackleton debated the issue, with Taylor (1922) arguing that the landscape was essentially glacial in origin but cut under earlier warmer glacial conditions, and Priestley (1909) arguing that glaciers had modified an existing fluvial landscape. The debate has continued; some argue for dissection of the Transantarctic Mountains by glaciers since the Pliocene (van der Wateren et al., 1999) while others point to the important role of fluvial erosion at an earlier time (Sugden and Denton, 2004).

This paper contributes to this debate by outlining the main variables influencing landscape evolution in Antarctica and then developing hypotheses about what might be expected from both a fluvial and a glacial perspective and the interaction of the two.

WIDER CONTEXT

East Antarctica consists of a central fragment of Gondwana and is surrounded by rifted margins (Figure 1). The initial breakup of Gondwana around most of East Antarctica took place between 160 Ma and 118 Ma. The separation of India and Antarctica took place first, while the separation of Australia from Antarctica took place in earnest after 55 Ma. Tasmania and New Zealand separated from the Ross Sea margins at around 70 Ma. West Antarctica comprises four separate mini-continental blocks: Antarctic Peninsula, Thurston, Marie Byrd Land, and Ellsworth-Whitmore, thought to be associated with extensional rifting. The drift of the continents opened up seaways around Antarctica and changed ocean circulation and productivity. A long-held view is that this permitted the development of the Antarctic Circumpolar Current, which introduced conditions favorable for glaciation (i.e., cooler temperatures and increased precipitation) (Kennett, 1977). In addition, recent climate modeling studies have suggested that a reduction in atmospheric greenhouse gases may have played an important role in the triggering of Antarctic glaciation (DeConto and Pollard, 2003; Huber and Nof, 2006). Critical dates for the development of ocean gateways are ~33 Ma, when a significant seaway opened up between Antarctica and Australia (Stickley et al., 2004), and ~31 Ma, when Drake Strait between the Antarctic Peninsula and South America became a significant seaway (Lawver and Gahagan, 2003).

The stepwise glacial history of Antarctica has been pieced together from marine records. Ice sheets first built up at ~34 Ma and their growth was marked by a sudden rise in benthic $\delta^{18}O$ values (Zachos et al., 1992). On land on the Ross Sea margin of Antarctica, beech forest similar to that in Patagonia today gave way to scrub forest and this change was accompanied by a progressive shift in clay minerals from smectite, typical of forest soils, to chlorite and illite characteristic of polar environments (Raine and Askin, 2001; Ehrmann et al., 2005; Barrett, 2007). Glaciation for the next ~20 million years was marked by ice volume fluctuations similar in scale to those of the Pleistocene ice sheets of the Northern Hemisphere. These fluctuations are demonstrated

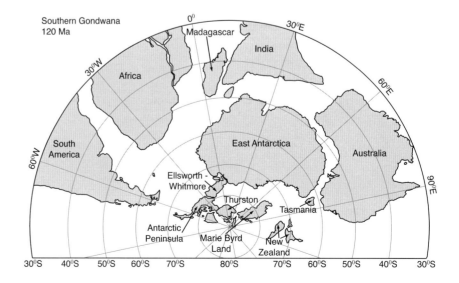

FIGURE 1 The location of Antarctica within Gondwana. The reconstruction shows the fragmentation of the supercontinent at 120 Ma (modified from Lawver et al., 1992).

by strata from 34 Ma to 17 Ma cored off the Victoria Land coast near Cape Roberts (Naish et al., 2001; Dunbar et al., forthcoming), high-resolution isotopic records from near Antarctica (Pekar and DeConto, 2006), and by ice-sheet modeling forced by reduced atmospheric CO_2 levels and contemporary orbital insolation changes (DeConto and Pollard, 2003). During the same period, the Cape Roberts record shows a progressive decline in meltwater sediments accumulating offshore and a vegetation decline to moss tundra, both indicating progressive cooling. A second stepwise cooling of Pacific surface waters of 6-7°C accompanied by a glacial expansion occurred in the mid-Miocene at ~14 Ma and is indicated in the marine isotope record (Shevenell et al., 2004; Holbourn et al., 2005). This event is also recorded in the transition from ash-bearing temperate proglacial deposits to diamict from cold ice in the Olympus Range of South Victoria Land at the edge of the South Polar Plateau (Lewis et al., 2007). Geomorphic evidence indicates that the present hyperarid polar climate of interior Antarctica was established at this time, along with the present structure of polar ocean circulation. Subsequent increases in ice-sheet volume, such as occurred in the Quaternary, involved ice thickening at the coast in response to a lowering of global sea level (Denton et al., 1989).

LANDSCAPE EVOLUTION: THE FLUVIAL SIGNAL

Hypothesis

One can predict in qualitative terms the landscape that would accompany the separation of a long-lived continent such as Gondwana. There would be integrated continental-scale river networks similar in scale to that of the Orange River in South Africa and the Murray River in Australia. Presumably these Antarctic rivers would have developed in a semiarid, seasonally wet climate, especially those in the interior of the supercontinent far from the sea. The rifted margins would have been subjected to fluvial processes related to the evolution of passive continental margins. This has been the focus of much research in recent years and is well summarized by Summerfield (2000). Beaumont et al. (2000) developed a coupled surface process-tectonic model of passive continental margin evolution at a classic diverging rift margin (Figure 2A). The main feature is the uplift of the rifted margin inland of a bounding fault running parallel to the coast. The uplift is a result of thermal buoyancy and crustal flexure in response to denudation near the coast and deposition offshore. Erosion is stimulated by the steep surface gradients created by the lower base level as the rift opens up. Mass wasting and rivers with steep gradients carve a coastal lowland and valleys into the upland rim. A seaward-facing escarpment forms at the drainage divide. Preexisting rivers may keep pace with the uplift and traverse the escarpment and allow dissection to spread behind the escarpment. Subsidiary erosion surfaces may form in response to lithological contrasts in rock type

and in response to secondary faults that develop parallel to the coast. Observations on other Gondwanan margins suggest that most erosion occurs within 10-20 million years of the rift opening up (Persano et al., 2002). Subsequently, cooling and crustal flexure may cause subsidence and the seaward ends of the valleys are flooded by the sea.

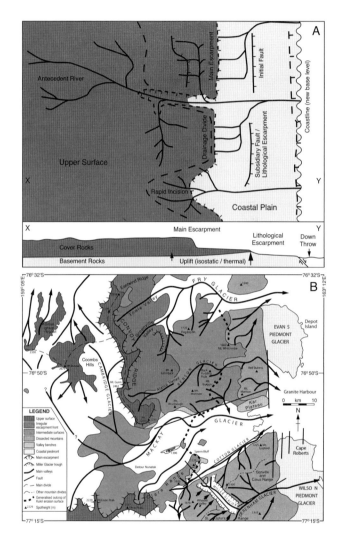

FIGURE 2 (A) Typical landscapes of a passive continental margin as modeled by Beaumont et al. (2000), using a numerical coupled tectonic-earth surface process model of landscape evolution. The new base level creates a pulse of erosion and associated rock uplift that leads to a staircase of coastal erosion surfaces separated by escarpments and incised river valleys, some of which may maintain their valleys across the escarpment crest. (B) Geomorphic map of the Convoy Range and Mackay Glacier area, showing the staircase of seaward-facing escarpments, erosion surfaces, dissected mountain landscapes near the coast, and the dendritic valley patterns radiating from a high point in the Coombs Hills. Principal faults parallel to and at right angles to the mountain front are taken from Fitzgerald (1992).
SOURCE: Sugden and Denton (2004). The model in (A) simulates the field evidence in remarkable detail.

The model applies to idealized young rift margins, and there are often more complex relationships affecting other continental margins of Gondwana (Bishop and Goldrick, 2000). Factors such as time elapsed since rifting, proximity to the rift, inherited altitude of the margin, tectonic down-warping, and crustal flexure can all affect the amplitude of the topography and the evolution of any escarpment.

Antarctic Evidence

Figure 3 shows a reconstruction of the continental river patterns inferred to exist if the Antarctic ice sheet is removed and the land compensated for isostatic depression by full flexural rebound. The map is based on the BEDMAP reconstruction of subglacial topography with a nominal resolution of 5 km (Figure 4) (Lythe et al., 2001). Sea level is assumed to lie at –100 m to represent the subsidence associated with the crustal cooling and flexure of mature passive continental margins. The reconstruction is based on hydrological modeling methods for cell-based digital elevation models whereby water is deemed to flow to the lowest adjacent cell. There are many uncertainties, not the least of which is that large areas have little data, but a sensitivity analysis suggests that models forced by a range of different assumptions yield a network that is essentially similar from run to run (Jamieson et al., 2005). The reconstruction shows that there are integrated networks leading radially to major depressions at the coast,

such as the western Weddell Sea, Lambert, Wilkes Land, and Oates Land basins. The dendritic pattern of the network, the centripetal pattern of flow from the subglacial Gamburtsev Mountains, and the topological coherence of the tributaries are demonstrated by bifurcation ratios that are representative of other river basins, such as the Orange River in South Africa (Jamieson et al., 2005). These observations imply a fluvial signature in the subglacial landscape on a continental scale.

Such a conclusion is reinforced by investigation of the valley patterns on mountain areas rising above the present ice-sheet surface. Good examples of former fluvial valley systems now occupied by local glaciers exist in the Transantarctic Mountains. The valley networks of several basins in northern Victoria Land have been ordered according to fluvially based Horton-Strahler rules (Horton, 1945; Strahler, 1958) and the dendritic pattern and hierarchical relationships between valley segments, and variables such as cumulative mean length are typical of river networks (Baroni et al., 2005). The Royal Society Range shows a similarly dendritic network radiating from one of the highest summits in Antarctica at over 4000 m (Sugden et al., 1999).

The landforms near the coast of the Antarctic passive continental margin are well displayed in the relatively glacier-free area of the Dry Valleys and adjacent Convoy Range (Figure 2B). Many features expected of a fluvial landscape are present. These include erosion surfaces rising inland from

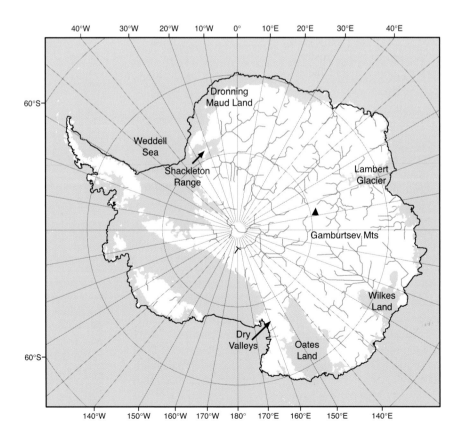

FIGURE 3 Reconstruction of the continental-scale river patterns beneath the present ice sheet. We assume that the land is isostatically compensated and that sea level is 100 m lower than today. The model uses the BEDMAP subglacial topography at a nominal cellular scale of 5 km (Figure 4). BEDMAP data are from Lythe et al. (2001).

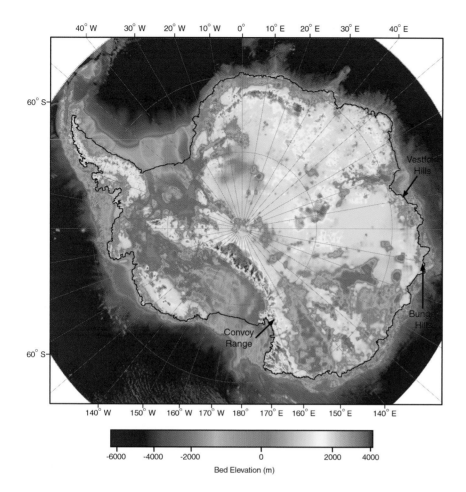

FIGURE 4 The present-day subglacial topography of Antarctica. BEDMAP data are from Lythe et al. (2001).

the coast and separated by escarpments, a coastal piedmont, an undulating upper erosion surface dotted with 100-450 m inselbergs, a seaward-facing escarpment 1000-1500 m in height, and intermediate erosion surfaces delimited by lithological variations, notably near horizontal dolerite sills. Sinuous valleys with a dendritic pattern can also be found. Most run from the escarpment to the sea, but others, such as that occupied by Mackay Glacier, breach the escarpment and drain the interior through tributaries bounded by additional escarpments. In detail the valleys have a sinuous planform, rectilinear valley sides with angles of 26-36°, and often lower-angle pediment slopes in the valley floors (Figure 5). The floor of the southernmost Dry Valley, Taylor Valley, is below sea level toward the coast, where it is filled with over 300 m of sediments of late-Miocene to recent age (Webb and Wrenn, 1982).

The erosion of the coastal margin has been accompanied by the denudation of a wedge of rock thickest at the coast and declining inland. This is displayed in the coastal upwarping of basement rocks and in the denudation history indicated by apatite fission track analysis. Both lines of evidence agree and point to the denudation since rifting of 4.5-5 km of rock at the coast falling to ~1 km of rock at locations 100 km inland (Fitzgerald, 1992; Sugden and Denton, 2004). The

fission track analyses suggest that most denudation occurred shortly after 55 Ma.

The evidence presented above is powerful affirmation that the normal fluvial processes of the erosion of a passive continental margin explain the main landscape features of three mountain blocks of the Transantarctic Mountains extending over a distance of 260 km. In these cases it is the fault pattern and the position of the drainage divide that are the main controls on topography and determine the differences in the landscapes of each block. It is not possible to apply the model of passive continental margin evolution to other sectors of East Antarctica without more detailed evidence, but similar landscapes occur in the Shackleton Range (Kerr and Hermichen, 1999) and in Dronning Maud Land (Näslund, 2001). One exception is the Bunger Hills and Vestfold Hills sector of East Antarctica east of the Lambert Glacier, where there is no escarpment. The matching piece of Gondwana is India, which separated early, and a longer and more complex evolution of the rift margin may explain the lack of an escarpment today.

It seems reasonable to argue that the framework of passive continental margin evolution applies to Antarctica and that in many areas a major pulse of fluvial erosion and accompanying uplift was a response to new lower base levels

FIGURE 5 George Denton and David Marchant in Victoria Valley, Dry Valleys, a typically fluvial landscape with rectilinear slopes and a shallow pediment slope leading to the valley axis. The slopes have escaped modification by overriding ice.

following continental breakup. Since breakup took place at different times, the stage of landscape evolution will vary from place to place in Antarctica. Furthermore, where large-scale tectonic features such as the Lambert graben disrupt the continental margin, they are likely to focus the fluvial system in a distinctive way (Jamieson et al., 2005).

LANDSCAPE EVOLUTION: THE GLACIAL SIGNAL

Hypothesis

The beds of former Northern Hemisphere mid-latitude ice sheets form the basis of understanding how glaciers transform preexisting landscapes. There are differences in that the Antarctic ice sheet has existed for tens of millions of years rather than a few million, but there are important similarities in that the North American ice sheet was of similar size and volume as the Antarctic ice sheet, and in that both have fluctuated in size in response to orbital forcing during their evolution, albeit to varying extents.

At the outset it is helpful to distinguish two scales of feature: Those that reflect the integrated radial flow of the ice sheet at a continental scale when it is close to or at its maximum and those that reflect the local and regional flow patterns as ice flows radially from topographic highs.

The key to landscape change by large ice sheets is the

superposition of a continental-scale radial flow pattern on the underlying topography. Such patterns are easily obscured by local signals but one can identify the following:

- Erosion in the center and wedges of deposition beneath and around the peripheries (Sugden, 1977; Boulton, 1996);
- Continental shelves that are deeper near the continent and shallower offshore as a result of erosion near the coast, often at the junction between basement and sedimentary rocks (Holtedahl, 1958);
- Radial pattern of large 10-km-scale troughs that breach and dissect the drainage divides near the coast and may continue offshore (e.g., in Norway, Greenland, and Baffin Island) (Løken and Hodgson, 1971; Holtedahl, 1967);
- Radial pattern of ice streams with beds tens of km wide, streamlined bedforms in bedrock and drift, and sharply defined boundaries (Stokes and Clark, 1999); and
- Radial pattern of meltwater flow crossing regional interfluves, as revealed, for example, by the pattern of eskers in North America (Prest, 1970).

Local and regional patterns also display radial configurations and reflect multiple episodes of reduced glaciation. Distinguishing features of marginal glaciation are corries or cirques, the dominant orientation of which, northeast facing, is determined in the Northern Hemisphere by slopes shaded from the sun and subject to wind drift by prevailing westerly winds (Evans, 1969). Stronger local glaciation typically builds ice caps on mountain massifs with ice flow carving a radial pattern of troughs, so well displayed in the English Lake District and Scotland, for example.

Clearly there will be a complex interaction between local, regional, and continental modes of flow depending on such factors as climate, ice extent, and topographic geometry. We attempt to model this complexity in Figure 6 by showing various stages of evolution of the Antarctic ice sheet using GLIMMER, a three-dimensional thermomechanical ice-sheet model as described by Payne (1999) and Jamieson et al. (forthcoming). The intention is to illustrate the range of different ice-sheet geometries that would be expected at various stages of Croll-Milankovitch glacial-deglacial cycles. The model is run for an arbitrary 1 million years with stepped temperature changes every 50 kyr falling from present-day Patagonian values to present-day Antarctic values. This timescale is designed to allow the ice to achieve approximate equilibrium at all times and to allow the isostatic response of the bedrock to reach a balance with these fluctuations in ice thickness. Patagonian climate statistics are used to simulate the climate inferred from vegetation associated with the initial glaciation of Antarctica (Cape Roberts Science Team, 2000; Raine and Askin, 2001). Modeled precipitation follows the pattern of net surface mass balance derived by Vaughan et al. (1999). At the beginning of the model run,

maximum coastal precipitation at sea level is scaled up to 2 m per year, four times that of the present day. The maximum then falls linearly to 0.5 m per year by the end of the model run. Mean annual temperatures fall from 7°C at sea level to present values of −12°C through the model run. Melt rates under warmer climatic conditions are calculated using a positive degree-day model (Reeh, 1991) whereby ablation is proportional to the number of days where temperature is above the freezing point. Diurnal variability is accounted for by using a normal distribution of temperature with a 5°C standard deviation. The pattern of mass balance used to drive ice growth is shown in Figure 7.

The bed topography (Figure 4) is derived from BEDMAP (Lythe et al., 2001) and is flexurally rebounded to compensate for the lack of an ice sheet. The use of an isostatically compensated present-day topography ignores tectonic movements and means that the results of the modeling become more uncertain as one goes further back in time. However, there is less risk in East Antarctica, where the main topographic features were established by Oligocene times. For example, geological evidence in the form of basement clasts in Oligocene strata cored off the Victoria Land coast (CIROS-1 drillcore) (Barrett et al., 1989) and Cape Roberts (Cape Roberts Science Team, 2000) suggests that the Trans-antarctic Mountains had been eroded deep enough to form a significant feature by Oligocene times. Furthermore, the Gamburtsev Mountains are considered to be a Pan-African feature with an age of 500 Ma (van de Flierdt et al., 2007).

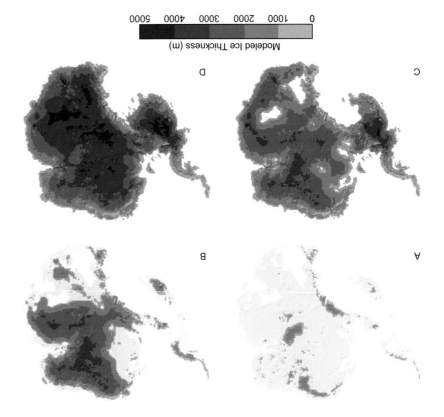

Modeled ice thickness (m)

0 1000 2000 3000 4000 5000

A   B

C   D

FIGURE 6 Model of the Antarctic ice sheet, generated using the GLIMMER 3-D thermomechanical model and the stepped transition from a Patagonian-style climate to the present polar climate. The four stages (A-D) illustrate the range of variability to be expected as the Antarctic ice sheet experienced many Croll-Milankovitch glacial cycles during its early evolution.

The different stages of growth illustrate the principal pattern of glaciation of Antarctica. Initial growth is in coastal mountains, such as in Dronning Maud Land, along the Trans-antarctic Mountains, in the West Antarctic archipelago, and in the high Gamburtsev Mountains in the interior. The ice spreads out from these mountain centers, first linking the main East Antarctic centers and then the West Antarctic centers. The model is deliberately simple but it suffices to show that glaciation starts preferentially in maritime mountains and in interior mountains if they are high enough. It also serves to illustrate the complexity of the changing pattern of flow as different ice centers merge and ice flow evolves from locally radial to continentally radial. The subglacial landscape of Antarctica can be expected to consist of a palimpsest of landforms related to these local, regional, and continental stages, while eroded material will experience a complex history of temporary deposition and changing flow

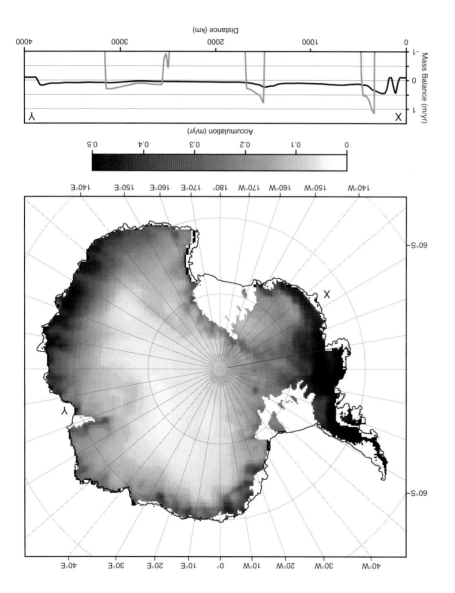

FIGURE 7 Present-day accumulation is used to drive a simulated Antarctic ice sheet (Vaughan et al., 1999). Profile X-Y shows that under a Patagonian-style regime (grey line) there are zones of increased accumulation at high altitudes. Accumulation is discontinuous because of high levels of ablation across much of the continental lowlands. The present-day accumulation is shown by the black line.

paths before being delivered to the coastal margin by the continental ice sheet.

Antarctic Evidence

The evidence of continental radial patterns of erosion is spectacular. The Lambert trough, which is 40-50 km across and 1 km deep, drains 10 percent of the East Antarctic ice sheet. It is coincident with a graben and is comparable to, but deeper and longer than, the North American equivalents, such as Frobisher Bay in Baffin Island. And then there is the spectacular series of troughs cutting through the Transant-arctic Mountain rim. Webb (1994) has previously suggested that the Beardmore trough, 200 km long, 15-45 km wide, and over 1200 m deep, exploited a preexisting river valley. Unloading due to glacial erosion may have contributed to isostatic uplift of the adjacent mountains (Stern et al., 2005). Offshore there are continuations of such troughs incised

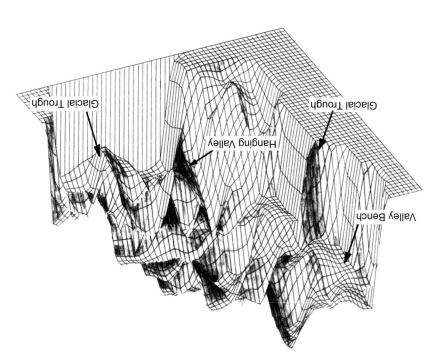

into the continental shelf of both East and West Antarctica (Wellner et al., 2001). Other troughs, such as that running parallel to Adelaide Island on the Antarctic Peninsula, have exploited the junction between basement and sedimentary rocks (Anderson, 1999). A series of ice streams flow into the Ross Sea and Weddell Sea embayments. They are underlain by deformable till and, in the case of the Rutford ice stream, by streamlined bedforms (Smith et al., 2007). There is also growing evidence of radial outflow of basal meltwater (Evatt et al., 2006). Hundred-meter-deep rock channels and massive staircases of giant potholes represent large-scale outbursts of subglacial meltwater across the Transantarctic Mountains rim in the McMurdo area (Denton and Sugden, 2005; Lewis et al., 2006).

There is also evidence of local and regional glacial landforms. Early studies of the subglacial Gamburtsev Mountains revealed characteristic trough overdeepening and the presence of hanging valleys, pointing to a local glaciation (Figure 8). The glacial landscapes of Dronning Maud Land too were created by local mountain glaciation and not the present ice sheet (Holmlund and Näslund, 1994). Recent geophysical surveying has demonstrated the presence of lakes in overdeepened troughs radiating from the Ellsworth Mountain core (Siegert, pers. comm., 2007). Overdeepening in the topographically constrained part of a fjord, and the rock threshold at the point when the trough opens out, are well-known characteristics of fjords in the Northern Hemisphere (Løken and Hodgson, 1971; Holtedahl, 1967). Similar features are found to radiate from uplands now below sea level in the Ross Sea embayment (De Santis et al., 1995;

Sorlien et al., 2007). In these latter cases the troughs are revealed by seismic survey. In the McMurdo Dry Valleys area of the Transantarctic Mountains a phase of local glaciation is represented by troughs identified on the inland flank of the mountains (Drewry, 1982) and troughs exploiting sinuous river valleys, such as the Mackay Glacier. Finally, the compact wet-based glacial deposits (the Sirius Group deposits) distributed at high elevations along 1000 km of the Transantarctic Mountains in the Ross Sea sector represent glaciation centered on the mountains (Denton et al., 1991, 1993). The deposits are characterized by local lithologies and the till components contain striated stones and a matrix typical of glacial erosion under warm-based ice. Some of these deposits incorporate remains of *Nothofagus* (southern beech) forest representative of a cool temperate environment.

LANDSCAPE EVOLUTION: THE COMBINED SIGNAL

Hypotheses

It is possible to relate subglacial landscapes to the processes by which glaciers modify preexisting topography. A simple classification scheme recognizes landscapes of areal scouring with abundant evidence of glacial scour; those of selective linear erosion where troughs dissect plateaus; those with no sign of glacial erosion and devoid of glacial landforms; and depositional landscapes composed of till and meltwater deposits (Sugden, 1978). The differences are related to whether the basal ice is at the pressure melting point. A key assumption is that ice erodes effectively when the base

FIGURE 8 An early reconstruction of the landscape of the subglacial Gamburtsev Mountains based on the analysis of radio-echo sounding data. The arrows pick out diagnostic glacial features such as overdeepening and hanging valleys presumed to have been formed by local mountain glaciation (modified from Perkins, 1984).

Glacial Trough

Glacial Trough

Hanging Valley

Valley Bench

is at the pressure melting point because sliding takes place between the ice and bedrock, permitting several processes to entrain bedrock and to deposit material. Such a situation explains landscapes of areal scouring, the linear erosion of troughs, and zones of deposition. The converse is that when the basal temperature is below the pressure melting point there is no sliding at the ice and rock interface. In such situations ice can be essentially protective and leave little sign of erosion. There is debate as to how protective the ice may be (Cuffey et al., 2000), but recent work on cosmogenic isotope analysis demonstrating the age of exposure and the time buried beneath ice has shown that the hypothesis holds in many areas of the Northern Hemisphere (Briner et al., 2006; Stroeven et al. 2002).

Armed with these observations it is possible to hypothesize about the landscape beneath the Antarctic ice sheet and suggest local, regional, and continental patterns. The numerical model of ice-sheet growth can be used to predict the changing pattern of glacial erosion during glacial cycles. Figure 9 shows the distribution of basal ice at the pressure melting point during stages of regional and continental glaciation. The regional pattern shows how the peripheries of the regional ice sheets are warm-based near their margins and inland to the vicinity of the equilibrium line where the

ice discharge is highest and generating most internal heat. These will be the areas of glacial erosion. The influence of topography is clear in that warm-based ice is focused on the depressions and major valleys where ice is thicker and flows faster. These low-lying areas at the pressure melting point are also those in which subglacial lakes might accumulate. What is striking is the way this zone of peripheral warm-based ice and erosion is a wave that sweeps across the landscape as the ice sheet grows to its continental maximum. At the macroscale the zone of warm-based ice is more extensive around the continental margins, especially in the vicinity of the main preexisting drainage basin outlets. Under both scenarios the ice over the main uplands remains below the pressure melting point.

From the above we can predict that the landscape in lowland areas of the Antarctic ice sheet will be underlain by a landscape of areal scouring. This relates both to the presence of ice at the pressure melting point and to the progressive erosion of rock debris by radial outflow of ice at different scales. Interaction between local and ice-sheet maximum flow directions and rock structure will determine the roughness and degree of streamlining of landforms. Flow in the same direction under both local and continental modes will favor elongated streamlined bedforms, perhaps with plucked

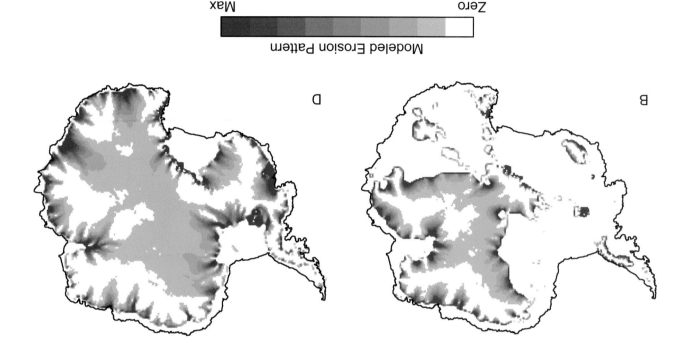

FIGURE 9 Modeled distribution of basal ice at the pressure melting point during intermediate and full stages of Antarctic glaciation (letters correspond to snapshots in Figure 6). Erosion, thought to be associated with sliding under such basal conditions, is concentrated toward the ice-sheet margins and at the beds of major outlet glaciers. A wave of erosion accompanies the expansion of local and regional ice to to a full continental ice sheet.

ice slopes, while complex flow changes will leave an irregular pattern. Areal scouring will be clearest on the lowlands and diminish upslope on upstanding massifs where summits may show no sign of glacial modification. This latter pattern reflects the effects of topography on the basal thermal regime and implies that the ice remained cold-based during local, regional, and continental stages of glaciation. Judging by the geomorphology of glaciated shields of the Northern Hemisphere, erosion will have removed some tens of meters of material, much of it initially as weathered regolith. This is sufficient to modify but not erase the preexisting river landscape, as argued for northern Europe (Lidmar-Bergström, 1982). However, given the longer duration of glaciation in Antarctica one would expect a greater depth to have been removed. The products of this erosion will be deposited offshore.

Antarctic Evidence

There is no direct evidence of areally scoured landscapes beneath the ice sheet. However, there are many observations from around the margins of Antarctica to indicate that such landscapes are likely to be widespread. Areas formerly covered by an earlier expanded ice sheet display extensive landscapes of areal scouring. This includes oases in East Antarctica, such as the Amery Oasis bordering the Lambert Glacier (Hambrey et al., 2007); the Ross Sea area where the scouring is most prominent near sea level and can be traced to elevations of 1000-2100 m along the Transantarctic Mountains front (Denton and Sugden, 2005); the inner parts of the offshore shelf in many parts of West Antarctica; and the offshore shelves surrounding islands off the Antarctic Peninsula and sub-Antarctic islands (Anderson, 1999). In all these situations meltwater channels testify to the activity of basal meltwater. Whereas all these observations are consistent with the view that warm-based ice occurred beneath thicker parts of the ice sheet when it was more extended and in maritime environments, what is surprising is that striated, scoured rock surfaces also define trimlines around mountains protruding above the ice sheet, even in the interior, such as the Ellsworth Mountains (Denton et al., 1992). The implication of shallow surface ice at the pressure melting point is that surface climatic conditions must have been within a few degrees of freezing point and thus several tens of degrees warmer than at present.

Landscapes of selective linear erosion are common in the mountainous rim of East Antarctica. In many areas glacial troughs are clearly delimited and excavated into a landscape-preserving fluvial valley form, often bearing diagnostic subaerial weathering forms such as regolith and tors. Such a description would apply to the landscapes bordering the Lambert Glacier (Hambrey et al., 2007), the Shackleton Mountains (Kerr and Hermichen, 1999), exposed escarpments in Dronning Maud Land (Näslund, 2001), the mountain blocks of the Transantarctic Mountains in Victoria Land (Sugden and Denton, 2004), and the plateaus of the Antarctic Peninsula area (Linton, 1963). As in the Northern Hemisphere such a description also applies on the scale of individual massifs. For example, upstanding nunataks in the Sarnoff Mountains of Marie Byrd Land are bounded by lower slopes with clear evidence of glacial scouring and yet their summits have retained an upper surface with tors, weathering pits, and block fields. In this case cosmogenic isotope analysis demonstrates that the mountains have been covered by ice of several glacial maxima and that weathering has continued sporadically in interglacials for ~1 million years (Sugden et al., 2005). The selectivity reflects the difference between the thicker, converging ice that scours as it flows round the massif and the thin diverging ice covering the summit that remains cold-based during each episode of overriding.

The implication of the above is that the hypothesis relating glacial modification of a preexisting landscape to the presence or absence of warm-based ice may be helpful in describing and understanding the landscape evolution of Antarctica. When the basal thermal regime remains the same beneath both local and full ice-sheet conditions, then the glacial transformation, or lack of it, will be clearest. What is exciting about the present time is that cosmogenic isotope analysis offers the opportunity to quantify such relationships.

CHRONOLOGY OF LANDSCAPE EVOLUTION

A number of dates help firm up the relative chronology of landscape evolution. Early studies of offshore sediments in Prydz Bay established the presence of fluvial sediments below the earliest glacial sediments, the latter dated to ~34 Ma (Cooper et al., 1991; O'Brien et al., 2001). The location, structure, and nature of the sediments suggest that the deposits have been derived from rivers flowing along the Lambert graben. Probably the deposits began to accumulate when rift separation began around 118 Ma (Jamieson et al., 2005).

Some large glacial troughs were cut by late Miocene times. In the Lambert Glacier area the offshore sedimentary evidence points to an ice sheet that discharged from a broad front on the offshore shelf until the late Miocene but experienced a switch to deposition within the overdeepened Lambert glacial trough subsequently. Ice-sheet modeling suggests that the deepening of the trough changed the dynamics of glacier flow and ablation to such an extent that calving velocity could match ice velocity and that the glacier was no longer able to advance through the deep water of the trough (Taylor et al., 2004). The implication is that the trough was excavated deeply by the late Miocene. Similar relationships occur in the glacially deepened mouths of the Dry Valleys in the McMurdo Sound area. Microfossils at the bottom of the Dry Valleys Drilling Project (DVDP-11) drillcore at the mouth of Taylor Valley are late Miocene in age (Webb and Wrenn, 1982). Marine shells deposited in a fjord in the glacial trough of Wright Valley are Pliocene in age and overlie a till of >13.6 Ma (Hall et al., 1993). These age relationships in glacial troughs related to ice flow from the interior of Antarctica demonstrate that they were cut by the late Miocene.

There is evidence from the McMurdo Sound area that the Antarctic ice sheet overrode marginal mountains and extended to the outer shelf at its maximum in the mid-Miocene. The case is argued out in a series of detailed papers in the Dry Valleys area, the key chronological fixes of which are:

• There is clear evidence that ice overrode all except perhaps the highest mountains in the Royal Society Range in the form of ice scouring on cols and subglacial meltwater channels that cross the mountains. The direction of flow is conformable with models of ice expansion to the outer edge of the offshore shelf (Sugden and Denton, 2004).

• $^{40}Ar/^{39}Ar$ dating of in situ volcanic ash deposits overrun by such ice and those ashes deposited on till sheets associated with such overriding ice constrains the event to between 14.8 Ma and 13.6 Ma (Marchant et al., 1993).

• Landscapes of areal scouring molded by the maximum ice sheet in front of the Royal Society are older than 12.4 ± 0.22 Ma, the age of the oldest undisturbed volcanic cone known to have erupted onto the land surface (Sugden et al., 1999).

• $^{40}Ar/^{39}Ar$ analyses of tephra show that the major meltwater feature represented by the Labyrinth in Wright Valley predates 12.4 Ma and that the last major outburst occurred some time between 14.4 Ma and 12.4 Ma (Lewis et al., 2006).

• ^{3}He ages of individual dolerite clasts in meltwater deposits from the overriding ice sheet reveal exposure ages between 8.63 ± 0.09 and 10.40 ± 0.04 Ma. Allowing for erosion, they are calculated to have been exposed for ~13 Ma (Margerison et al., 2005).

As yet there are few comparable terrestrial dates elsewhere in Antarctica, but it is worth drawing attention to work on the flanks of the Lambert Glacier in which there is biostratigraphical evidence of a pre-late Miocene phase of glacial erosion and deepening, followed by a 10-million-year period of exposure (Hambrey et al., 2007). It is tempting to equate the deepening to the same ice-sheet maximum.

In the McMurdo Sound area it is possible to establish that a phase of warm-based glaciation occurred before the mid-Miocene overriding ice. The critical evidence is that warm-based tills in the high mountains bounding the Dry Valleys, and indeed the Sirius Group deposits, have been modified by overriding ice. Typically there are erosional patches excavated into a preexisting till with material dragged out down-ice (Marchant et al., 1993) and ripple corrugations with a spacing of 25-50 m that are a coherent part of the overriding meltwater system (Denton and Sugden, 2005). One important fix on the switch from warm-based to

cold-based local glaciation has been reported from the Olympus Range in the McMurdo Sound area of the Transantarctic Mountains (Lewis et al., 2007). Here a classic warm-based till with meltout facies is overlain by weathered colluvium that is itself overlain by tills deposited by cold-based glaciers. The minimum date of transition is fixed by volcanic ashes interbedded between the two sets of tills and has an age of 13.94 Ma. Such a transition beneath small local glaciers is argued to represent an atmospheric cooling of 20-25°C. Moreover, the transition occurs before one or more major ice-sheet overriding events in the same area.

DISCUSSION

Here we attempt a synthesis of landscape evolution of Antarctica, based mainly on terrestrial evidence (Table 1). Inevitably the hypothesis is based on partial information and is biased toward the data-rich McMurdo Sound area of the Transantarctic Mountains. Nevertheless it seems helpful to try and generalize more broadly.

An early pulse of fluvial erosion was associated with the breakup of Gondwana and the creation of new lower base levels around the separating continental fragments. The timing of the pulse varied with the time of base level change around each segment. Erosion removed a wedge of material from around the margins of East Antarctica and the smaller continental fragments of West Antarctica. Escarpments

TABLE 1 The Chronology of Landscape Evolution in Antarctica Based Mainly on Terrestrial Evidence

Age	Evidence
>55-34 Ma	Passive continental margin erosion of coastal surfaces, escarpments, and river valleys, removing 4-7 km of rock at the coast and 1 km inland since rifting. Cool temperate forest and smectite-rich soils, at least at coast.
34 Ma	Initial glaciation of regional uplands with widespread warm-based ice, local radial troughs, and tills. Climate cooling.
34-14 Ma	Local, regional, and continental orbital ice-sheet fluctuations associated with progressive cooling, declining meltwater, and change to tundra vegetation. Local warm-based glaciers in mountains.
~14 Ma	Expansion of maximum Antarctic ice sheet to edge of continental shelf linked to sharp temperature decline of 20-25°C. Change from warm-based to cold-based local mountain glaciers. Selective erosion of continental-scale radial and offshore glacial troughs and meltwater routes.
13.6 Ma to Present	Ice sheet maintains hyperarid polar climate. Slight thickening of ice-sheet margins during Pliocene warming in East Antarctica. Outlet glaciers respond to sea-level change, especially in West Antarctica. Extremely low rates of subaerial weathering. Glacial erosion restricted to outlet glaciers and beneath thick ice.

its associated polar climate is demonstrated on land by the remarkably low erosion rates in the Transantarctic Mountains as revealed by $^{39}Ar/^{40}Ar$ dating of volcanic ashes and cones, by cosmogenic isotope analysis, and by the preservation of fragile deposits and buried ice (Brook et al., 1995; Ivy-Ochs et al., 1995; Summerfield, 1999; Marchant et al., 2002). Offshore the growth and decay of the maximum ice sheet is demonstrated by a widespread unconformity and dating evidence of a retreat of the ice in the Ross Sea area after 13.5 Ma (Anderson, 1999).

The implication of the above is that fluctuations of the Antarctic ice sheet in the Pleistocene are forced by changes in sea level. In East Antarctica the fluctuations are relatively minor. Outlet glaciers thicken and advance seaward in response to a lowering of sea level in the Northern Hemisphere, as demonstrated in the case of the outlets flowing through the Transantarctic Mountains (Denton et al., 1989). Slight thickening occurs in Mac Robertson Land (Mackintosh et al., 2007), but the ice does not extend far offshore (O'Brien et al., 2001; Leventer et al., 2006). In agreement with such a limited expansion, the coastal oases of the Bunger Hills and the Larsemann Hills appear to have remained ice-free during the last glacial cycle (Gore et al., 2001; Hodgson et al., 2001). In West Antarctica the Pleistocene behavior is markedly different. Here ice appears to have extended to the edge of the continental shelf and occupied deep troughs extending ~100 km from the present coast (Bentley and Anderson, 1998; Ó Cofaigh et al., 2005). In this case and following Mercer (1978), one can surmise that the grounded ice streams occupying the topography below sea level between the individual massifs are especially susceptible to sea-level changes.

It is interesting that the ice sheet achieved its present profile in Mac Robertson Land 6000 years ago (Mackintosh et al., 2007). This is the time when global sea level had largely completed its recovery following the final disappearance of the North American ice sheet. The coincidence supports the view that fluctuations in the Antarctic ice sheet in the Pleistocene are a response to sea-level changes driven primarily by the Northern Hemisphere ice sheets.

WIDER IMPLICATIONS

The thrust of this overview is that information about the evolution of passive continental margins and the processes and forms associated with mid-latitude Northern Hemisphere ice sheets is a useful guide for reconstructing the evolution of the landscape in Antarctica. At present there are insufficient constraints to do more than outline possibilities in a quali- tative way. Nonetheless even a preliminary view provides some insights into the debate about the relative importance of fluvial and glacial agents of erosion. Further, there seems ample scope for a focused modeling exercise increasingly founded on quantitative field data.

What is also encouraging about the emerging terrestrial

and erosion surfaces formed and were dissected to varying degrees by fluvial erosion. The degree of dissection increased with the complex geometry and small size of each fragment and was more pronounced near the coast. The interior of East Antarctica was characterized by large river basins, presum- ably more arid in the interior than at the coast. The climate was sufficiently warm to support beech forests around the coast. Soils contained the clay mineral smectite, derived from the chemical weathering associated with forests.

Around 34 Ma declining atmospheric carbon dioxide and the opening of significant seaways between Antarctica and the southern continents were factors in bringing the two conditions necessary for glaciers: cooling and increased precipitation from circumpolar storms. Glaciation began in a Patagonian-type climate, at least in the Transantarctic Mountains, and was centered on maritime mountains of East and West Antarctica and high continental mountains in East Antarctica. The record from the Cape Roberts cores of fluctu- ating ice-sheet extent, which is supported by marine oxygen isotope records, points to a dynamic ice sheet responding to orbital fluctuations in the same way as the Pleistocene ice sheets of the Northern Hemisphere. The preexisting regolith was progressively removed from the continent to create a subglacial landscape of areal scouring, probably in a complex series of flows as glacier extent and flow direc- tions oscillated between an interglacial and ice-maximum state. Modeling suggests a wave of erosion was associated with each expansion of ice from the mountain centers. The presence of warm-based local glaciers in the Transantarctic Mountains suggests relatively warm interglacial periods.

The terrestrial record agrees with the marine record in pinpointing a sharp temperature decline associated with the expansion of the Antarctic ice sheet over its continental shelf at ~14 Ma. Perhaps the expansion was triggered by a change in ocean circulation or declining atmospheric carbon dioxide. Perhaps it too could have been related to the internal dynam- ics of the ice sheet in that earlier glaciations had deposited shoals on the offshore shelf, reduced calving, and allowed the ice to advance to the outer edge, behavior well known in the case of fjord glaciers (Mercer, 1961).

The mid-Miocene maximum ice sheet eroded troughs on a continental-scale, cutting selectively through the mountain rim, deepening the interior parts of the offshore shelf (and subsea basins?) in West Antarctica. Land surfaces covered by thin diverging ice remained essentially unchanged. Perhaps the offshore deepening was such that in cycles of growth and decay the ice could no longer advance to the shelf edge, as demonstrated by the behavior of the Lambert Glacier. Alter- natively the change in ocean and atmospheric conditions after the mid-Miocene maximum to a hyperarid polar climate may have deprived the ice sheet of moisture. But after ~13.6 Ma the ice retreated to its present continental lair, at least in East Antarctica, and remained essentially intact. Coastal fjords were filled with shallow marine Miocene sediments. The retreat and subsequent stability of the full ice sheet and record of landscape evolution is that the main stages match the records obtained from deep-sea and inshore cores. At present, given the uncertainties associated with each dating technique, it is not possible to be sufficiently precise to establish cause and effect and thus understand better the links between the ocean, atmosphere, and ice sheet in influencing or responding to environmental change.

One important implication arising from this overview is the realization that fragile features in the landscape can be very old, whether they have been protected beneath ice or subject to a hyperarid climate with minimal erosion. It is possible for striations and moraines to survive for millions of years. Thus, it is possible that features such as trimlines with striations and associated moraines may date from glaciation prior to the mid-Miocene. Pleistocene changes in ice elevation may be indicated only by a sparse scatter of boulders. If so, then reconstructions of former Pleistocene ice thicknesses based solely on trimline and striation evidence can be misleading.

It is worth reflecting on the richness of the record of landscape evolution in Victoria Land. At least in part this must be due to ease of access and proximity to the permanent base of McMurdo Sound. If so, there is the prospect of equally rich archives in other parts of Antarctica. The challenge is to develop and explore these further to improve our understanding of landscape evolution and its contribution to the wider sciences.

ACKNOWLEDGMENTS

The authors are indebted to the U.K. Natural Environment Research Council for support. Author D.E.S. is grateful for support from the Division of Polar Programs of the U.S. National Science Foundation, the British Antarctic Survey, the Carnegie Trust for the Universities of Scotland and ACE (Antarctic Climate Evolution) Program. We are most grateful to Peter Barrett, David Elliot, Adrian Hall, and Andrew Mackintosh for their thoughtful and helpful comments on the manuscript.

REFERENCES

Anderson, J. B. 1999. *Antarctic Marine Geology.* Cambridge: Cambridge University Press.

Baroni, C., V. Noti, S. Ciccacci, G. Righini, and M. C. Salvatore. 2005. Fluvial origin of the valley system in northern Victoria Land (Antarctica) from quantitative geomorphic analysis. *Geological Society of America Bulletin* 117:212-228.

Barrett, P. J. 2007. Cenozoic climate and sea level history from glacimarine strata off the Victoria Land coast, Cape Roberts Project, Antarctica. In *Glacial Processes and Products*, eds. M. J. Hambrey, P. Christoffersen, N. F. Glasser, and B. Hubbard. *International Association of Sedimentologists Special Publication* 39:259-287.

Barrett, P. J., M. J. Hambrey, D. M. Harwood, A. R. Pyne, and P.-N. Webb. 1989. Synthesis. In *Antarctic Cenozoic History from CIROS-1 Drillhole, McMurdo Sound*, ed. P. J. Barrett. *DSIR Bulletin* 245:241-251. Wellington: DSIR Publishing.

Beaumont, C., H. Kooi, and S. Willett. 2000. Coupled tectonic-surface process models with applications to rifted margins and collisional orogens. In *Geomorphology and Global Tectonics*, ed. M. A. Summerfield, pp. 29-55. Chichester: Wiley.

Bentley, M. J., and J. B. Anderson. 1998. Glacial and marine evidence for the ice sheet configuration in the Weddell Sea-Antarctic Peninsula region during the Last Glacial Maximum. *Antarctic Science* 10:307-323.

Bishop, P., and G. Goldrick. 2000. Geomorphological evolution of the East Australian continental margin. In *Geomorphology and Global Tectonics*, ed. M. A. Summerfield, pp. 225-254. Chichester: Wiley.

Boulton, G. S. 1996. Theory of glacier erosion, transport and deposition as a consequence of subglacial sediment deformation. *Journal of Glaciology* 42:43-62.

Briner, J. P., G. H. Miller, P. Davis, and R. C. Finkel. 2006. Cosmogenic radionuclides from fjord landscapes support differential erosion by overriding ice sheets. *GSA Bulletin* 118:406-420.

Brook, E. J., E. T. Brown, M. D. Kurz, R. P. Ackert, G. M. Raisbeck, and F. Yiou. 1995. Constraints on age, erosion and uplift of Neogene glacial deposits in the Transantarctic Mountains determined from in situ cosmogenic ^{10}Be and ^{26}Al. *Geology* 23:1057-1152.

Cape Roberts Science Team. 2000. Summary of results. In *Studies from Cape Roberts Project: Initial Report on CRP-3, Ross Sea, Antarctica*, eds. P.J. Barrett, M. Sarti, and S. Wise. *Terra Antarctica* 7185-203. Siena: *Terra Antarctica Publication*.

Cooper, A. H., Stagg, and E. Geist. 1991. Seismic stratigraphy and structure of Prydz Bay, Antarctica: Implications from ODP Leg 119 drilling. In *Ocean Drilling Program Leg 119 Scientific Results*, eds. J. B. Barron and B. Larsen, pp. 5-25. College Station, TX: Ocean Drilling Program.

Cuffey, K. M., Conway, A. M. Gades et al. 2000. Entrainment at cold glacier beds. *Geology* 28:351-354.

De Santis, L., J. B. Anderson, G. Brancolini, and I. Zayatz. 1995. Seismic record of late Oligocene through Miocene glaciation on the central and eastern continental shelf of the Ross Sea. In *Geology and Seismic Stratigraphy of the Antarctic Margin*, eds. A. K. Cooper, P. F. Barker, and G. Brancolini. *Antarctic Research Series* 68:235-260. Washington, D.C.: American Geophysical Union.

DeConto, R. M., and D. Pollard. 2003. Rapid Cenozoic glaciation of Antarctica induced by declining atmospheric CO_2. *Nature* 421:245-249.

Denton, G. H., and D. E. Sugden. 2005. Meltwater features that suggest Miocene ice-sheet overriding of the Transantarctic Mountains in Victoria Land, Antarctica. *Geografiska Annaler* 87A:67-85.

Denton, G. H., J. G. Bockheim, S. C. Wilson, and M. Stuiver. 1989. Late Wisconsin and early Holocene glacial history, inner Ross Embayment, Antarctica. *Quaternary Research* 31:151-182.

Denton, G. H., M. L. Prentice, and L. H. Burckle. 1991. Cainozoic history of the Antarctic Ice Sheet. In *The Geology of Antarctica*, ed. R. J. Tingey, pp. 365-433. Oxford: Clarendon Press.

Denton, G. H., J. G. Bockheim, R. H. Rutford, and B. G. Andersen. 1992. Glacial history of the Ellsworth Mountains, West Antarctica. In *Geology and Palaeontology of the Ellsworth Mountains, West Antarctica*, eds. G. F. Webers, C. Craddock, and J. F. Splettstoesser. *Geological Society of America Memoir* 170:403-442.

Denton, G. H., D. E. Sugden, D. R. Marchant, B. L. Hall, and T. I. Wilch. 1993. East Antarctic Ice Sheet Sensitivity to Pliocene Climatic Change from a Dry Valleys Perspective. *Geografiska Annaler* 75A:155-204.

Drewry, D. J. 1982. Ice flow, bedrock and geothermal studies from radio-echo sounding inland of McMurdo Sound, Antarctica. In *Antarctic Geoscience*, ed. C. Craddock, pp. 977-983. Madison: University of Wisconsin Press.

Dunbar, G. B., T. R. Naish, P. J. Barrett, C. R. Fielding, and R. D. Powell. Forthcoming. Constraining the amplitude of late Oligocene bathymetric changes in western Ross Sea during orbitally-induced oscillations in the East Antarctic Ice Sheet. 1. Implications for glacimarine sequence stratigraphic models. *Palaeogeography, Palaeoclimatology, Palaeoecology*, doi:10.1016/j.palaeo.2007.08.018.

Ehrmann, W. U., M. Setti, and L. Marinoni. 2005. Clay minerals in Cenozoic sediments off Cape Roberts (McMurdo Sound, Antarctica) reveal the palaeoclimatic history, *Palaeogeography, Palaeoclimatology, Palaeoecology* 229:187-211.

Evans, I. S. 1969. The geomorphology and morphometry of glacial and nival areas. In *Water, Earth and Man*, ed. R. J. Chorley, pp. 369-380. London: Methuen.

Evatt, G. W., A. C. Fowler, C. D. Clark, and N. R. J. Hulton. 2006. Subglacial floods beneath ice sheets. *Philosophical Transactions of the Royal Society of London* 364:1769-1794.

Fitzgerald, P. G. 1992. The Transantarctic Mountains of southern Victoria Land: The application of fission track analysis to a rift shoulder uplift. *Tectonics* 11:634-662.

Gore, D. B., E. J. Rhodes, P. C. Augustinius, M. R. Leishman, E. A. Colhoun, and J. Rees-Jones. 2001. Bunger Hills, East Antarctica: Ice free at the Last Glacial Maximum. *Geology* 29:1103-1106.

Hall, B. L., G. H. Denton, D. R. Lux, and J. G. Bockheim. 1993. Late Tertiary Antarctic paleoclimate and ice-sheet dynamics inferred from surficial deposits in Wright Valley. *Geografiska Annaler* 75A:239-267.

Hambrey, M. J., N. F. Glasser, B. C. McKelvey, D. E. Sugden, and D. Fink. 2007. Cenozoic landscape evolution of an East Antarctic oasis (Radok Lake area, northern Prince Charles Mountains), and its implications for the glacial and climatic history of Antarctica. *Quaternary Science Reviews* 26:598-626.

Hodgson, D. A., P. E. Noon, W. Vyverman, C. L. Bryant, D. B. Gore, P. Appleby, M. Gilmour, E. Verleyen, A. Sabbe, V. J. Jones, J. C. Ellis-Evans, and P. B. Wood. 2001. Were the Larsemann Hills ice free through the Last Glacial Maximum? *Antarctic Science* 13:440-454.

Holbourn, A., W. Kuhn, M. Schulz, and H. Erlenkeuser. 2005. Impacts of orbital forcing and atmospheric carbon dioxide on Miocene ice-sheet expansion. *Nature* 438:483-487.

Holmlund, P., and J. O. Näslund. 1994. The glacially sculptured landscape in Dronning Maud Land, Antarctica, formed by wet-based mountain glaciation and not by the present ice sheet. *Boreas* 23:139-148.

Holtedahl, H. 1958. Some remarks on the geomorphology of continental shelves off Norway, Labrador and southeast Alaska. *The Journal of Geology* 66:461-471.

Holtedahl, H. 1967. Notes on the formation of fjords and fjord valleys. *Geografiska Annaler* 49A:188-203.

Horton, R. E. 1945. Erosional development of streams and their drainage basins: Hydrophysical approach to quantitative morphology. *Geological Society of America Bulletin* 56:275-370.

Huber, M., and D. Nof. 2006. The ocean circulation in the Southern Hemisphere and its climatic impacts in the Eocene. *Palaeogeography, Palaeoclimatology, Palaeoecology* 231:9-28.

Ives, J. D. 1966. Block fields, associated weathering forms on mountain tops and the nunatak hypothesis. *Geografiska Annaler* 48A:220-223.

Ivy-Ochs, S., C. Schlüchter, P. W. Kubik, B. Dittrich-Hannen, and J. Beer. 1995. Minimum [10]Be exposure ages of early Pliocene for the Table Mountain Plateau and the Sirius Group at Mount Fleming, Dry Valleys, Antarctica. *Geology* 23:1007-1010.

Jamieson, S. S. R., N. R. J. Hulton, D. E. Sugden, A. J. Payne, and J. Taylor. 2005. Cenozoic landscape evolution of the Lambert basin, East Antarctica: The relative role of rivers and ice sheets. *Global and Planetary Change* 45:35-49.

Jamieson, S. S. R., N. R. J. Hulton, and M. Hagdorn. Forthcoming. Modelling landscape evolution under ice sheets. *Geomorphology*, doi:10.1016/j.geomorph.2007.02.047.

Kennett, J. P. 1977. Cenozoic evolution of Antarctic glaciation, the circum-Antarctic Ocean and their impact on global oceanography. *Journal of Geophysical Research* 82:3843-3860.

Kerr, A., and W. D. Hermichen. 1999. Glacial modification of the Shackleton Range, Antarctica. *Terra Antartica* 6:353-360.

Lawver, L. A., and L. M. Gahagan. 2003. Evolution of Cenozoic seaways in the circum-Antarctic region. *Palaeogeography, Palaeoclimatology, Palaeoecology* 198(1-2):11-37.

Lawver, L. A., L. M. Gahagan, and M. F. Coffin. 1992. The development of seaways around Antarctica. *Antarctic Research Series* 56:7-30.

Leventer, A., E. Domack, R. Dunbar, J. Pike, C. Stickley, E. Maddison, S. Brachfeld, P. Manley, and C. McLennen. 2006. Marine sediment record from the East Antarctic margin reveals dynamics of ice sheet recession. *GSA Today* 16(12):4-10.

Lewis, A. R., D. R. Marchant, D. E. Kowalewski, S. L. Baldwin, and L. E. Webb. 2006. The age and origin of the Labyrinth, western Dry valleys, Antarctica: Evidence for extensive middle Miocene subglacial floods and freshwater discharge to the Southern Ocean. *Geology* 34:513-516.

Lewis, A. R., D. R. Marchant, A. C. Ashworth, S. R. Hemming, and M. L. Machlus. 2007. Major middle Miocene global climate change: Evidence from East Antarctica and the Transantarctic Mountains. *Geological Society of America Bulletin* 119(11):1449-1461, doi:10.1130/B26134.

Lidmar-Bergstrom, K. 1982. Pre-Quaternary geomorphological evolution in southern Fennoscandia. *Sveriges, geologiska undersökning C* 785:1-202.

Linton, D. L. 1949. Unglaciated areas in Scandinavia and Great Britain. *Irish Geography* 2:77-79.

Linton, D. L. 1963. The forms of glacial erosion. *Transactions of the Institute of British Geographers* 33:1-28.

Løken, O. H., and D. A. Hodgson. 1971. On the submarine geology along the east coast of Baffin Island. *Canadian Journal of Earth Sciences* 8:185-195.

Lythe, M., D. G. Vaughan, and BEDMAP Consortium. 2001. BEDMAP: A new thickness and subglacial topographic model of Antarctica. *Canadian Journal of Earth Sciences* 106:11335-11352.

Mackintosh, A., D. White, D. Fink, D. B. Gore, J. Pickard, and P. C. Fanning. 2007. Exposure ages from mountain dipsticks in Mac Robertson Land, East Antarctica, indicate little change in ice-sheet thickness since the Last Glacial Maximum. *Geology* 35:551-554.

Marchant, D. R., G. H. Denton, D. E. Sugden, and C. C. Swisher III. 1993. Miocene glacial stratigraphy and landscape evolution of the western Asgard Range, Antarctica. *Geografiska Annaler* 75A:303-330.

Marchant, D. R., A. R. Lewis, W. M. Phillips, E. Moore, R. Souchez, G. H. Denton, and D. E. Sugden. 2002. Formation of patterned ground and sublimation till over Miocene glacier ice, southern Victoria Land, Antarctica. *GSA Bulletin* 114:718-730.

Margerison, H. R., W. M. Phillips, F. M. Stuart, and D. E. Sugden. 2005. Cosmogenic [3]He concentrations in ancient flood deposits from the Coombs Hills, northern Dry Valleys, East Antarctica: Interpreting exposure ages and erosion rates. *Earth and Planetary Science Letters* 230:163-175.

Mercer, J. H. 1961. The response of fiord glaciers to changes in the firn limit. *Journal of Glaciology* 3:850-858.

Mercer, J. H. 1978. West Antarctic Ice Sheet and CO_2 greenhouse effect: A threat of disaster. *Nature* 271:321-325.

Naish, T. R. et al. 2001. Orbitally induced oscillations in the East Antarctic ice sheet at the Oligocene/Miocene boundary. *Nature* 413:719-723.

Näslund, J.-O. 2001. Landscape development in western and central Dronning Maud Land, East Antarctica. *Antarctic Science* 13:302-311.

Ó Cofaigh, C., R. D. Larter, J. A. Dowdeswell, C.-D. Hillenbrand, C. J. Pudsey, J. Evans, and P. Morris. 2005. Flow of the West Antarctic Ice Sheet on the continental margin of the Bellingshausen Sea at the Last Glacial Maximum. *Journal of Geophysical Research* 110(B11):103, doi:10.1029/2005JB003619.

O'Brien, P. E., A. K. Cooper, C. Richter et al. 2001. Leg 188 summary: Prydz Bay-Cooperation Sea, Antarctica. *Proceedings of the Ocean Drilling Program, Initial Reports* 188:1-65. College Station, TX: Ocean Drilling Program.

Payne, A. J. 1999. A thermomechanical model of ice flow in West Antarctica. *Climate Dynamics* 15:115-125.

Pekar, S. F., and R. M. DeConto. 2006. High-resolution ice-volume estimates for the early Miocene: Evidence for a dynamic ice sheet in Antarctica. *Palaeogeography, Palaeoclimatology, Palaeoecology* 231:101-109.

Perkins, D. 1984. Subglacial landscape in Antarctica. Unpublished Ph.D. thesis. University of Aberdeen.

Persano, C., F. M. Stuart, P. Bishop, and D. N. Barfod. 2002. Apatite (U-Th)/He age constraints on the development of the Great Escarpment on the southeastern Australian passive margin. *Earth and Planetary Science Letters* 200:79-90.

Prest, V. K. 1970. Quaternary geology of Canada. *Economic Geology Report* 1:676-764.

Priestley, R. E. 1909. Scientific results of the western journey. In *The Heart of the Antarctic*, vol. 2, ed. E. H. Shackleton, pp. 315-333. London: Heinemann.

Raine, J. I., and R. A. Askin. 2001. Terrestrial palynology of Cape Roberts drillhole CRP-3, Victoria Land Basin, Antarctica. *Terra Antartica* 8:389-400.

Reeh, N. 1991. Parameterization of melt rate and surface temperature on the Greenland ice sheet. *Polarforschung* 59(3):113-128.

Shevenell, A. E., J. P. Kennett, and D. W. Lea. 2004. Middle Miocene Southern Ocean cooling and Antarctic cryosphere expansion. *Science* 305:1766-1770.

Smith, A. M., T. Murray, K. W. Nicholls, K. Makinson, G. Aoalgeirsdottir, A. E. Behar, and D. G. Vaughan. 2007. Rapid erosion, drumlin formation, and changing hydrology beneath an Antarctic ice stream. *Geology* 35:127-130.

Sorlien, C. C., B. P. Luyendyk, D. S. Wilson, R. C. Decesari, L. R. Bartel, and J. B. Diebold. 2007. Oligocene development of the West Antarctic Ice Sheet recorded in eastern Ross Sea strata. *Geology* 35:467-470.

Stern, T. A., A. K. Baxter, and P. J. Barrett. 2005. Isostatic rebound due to glacial erosion within the Transantarctic Mountains. *Geology* 33:221-224.

Stickley, C. E., H. Brinkhuis, S. A. Schellenberg, A. Sluijs, U. Rohl, M. Fuller, M. Grauert, M. Huber, J. Warnaar, and G. L. Williams. 2004. Timing and nature of the deepening of the Tasmanian Gateway. *Paleoceanography* PA4027, doi:10.1029/2004PA001022.

Stokes, C. R., and C. D. Clark. 1999. Geomorphological criteria for identifying Pleistocene ice streams. *Annals of Glaciology* 28:67-74.

Strahler, A. N. 1958. Dimensional analysis applied to fluvially eroded landforms. *Bulletin of the Geological Society America* 69:279-300.

Stroeven, A. P., D. Fabel, C. Hättestrand, and J. Harbor. 2002. A relict landscape in the centre of Fennoscandian glaciation: Cosmogenic radionuclide evidence of tors preserved through multiple glacial cycles. *Geomorphology* 44:145-154.

Sugden, D. E. 1977. Reconstruction of the morphology, dynamics and thermal characteristics of the Laurentide ice sheet at its maximum. *Arctic and Alpine Research* 9:21-47.

Sugden, D. E. 1978. Glacial erosion by the Laurentide ice sheet. *Journal of Glaciology* 20:367-392.

Sugden, D. E., and G. H. Denton. 2004. Cenozoic landscape evolution of the Convoy Range to Mackay Glacier area, Transantarctic Mountains: Onshore to offshore synthesis. *GSA Bulletin* 116:840-857.

Sugden, D. E., M. A. Summerfield, G. H. Denton, T. I. Wilch, W. C. McIntosh, D. R. Marchant, and R. H. Rutford. 1999. Landscape development in the Royal Society Range, southern Victoria Land, Antarctica: Stability since the middle Miocene. *Geomorphology* 28:181-200.

Sugden, D. E., G. Balco, S. G. Cowdery, J. O. Stone, and L. C. Sass III. 2005. Selective glacial erosion and weathering zones in the coastal mountains of Marie Byrd Land, Antarctica. *Geomorphology* 67:317-334.

Summerfield, M. A. 2000. Geomorphology and global tectonics: Introduction. In *Geomorphology and Global Tectonics,* ed. M. A. Summerfield, pp. 3-12. Chichester: Wiley.

Summerfield, M. A., D. E. Sugden, G. H. Denton, D. R. Marchant, H. A. P. Cockburn, and F. M. Stuart. 1999. Cosmogenic isotope data support previous evidence of extremely low rates of denudation in the Dry Valleys region, southern Victoria Land. *Geological Society of London Special Publication* 162:255-267.

Taylor, G. 1922. *The Physiography of McMurdo Sound and Granite Harbour Region.* British Antarctic (Terra Nova) Expedition, pp. 1910-1913. London: Harrison.

Taylor, J., M. J. Siegert, A. J. Payne, M. J. Hambrey, P. E. O'Brien, A. K. Cooper, and G. Leitchenkov. 2004. Topographic controls on post-Oligocene changes in ice-sheet dynamics, Prydz Bay region, East Antarctica. *Geology* 32:197-200.

van de Flierdt, T., G. E. Gehrels, S. L. Goldstein, and S. R. Hemming. 2007. Pan-African Age of the Gamburtsev Mountains? In *Antarctica: A Keystone in a Changing World—Online Proceedings for the Tenth International Symposium on Antarctic Earth Sciences,* eds. Cooper, A. K., C. R. Raymond et al., USGS Open-File Report 2007-1047, Extended Abstract 176, http://pubs.usgs.gov/of/2007/1047/.

van der Wateren, F. M., T. J. Dunai, R. T. Van Balen, W. Klas, A. L. L. M. Verbers, S. Passchier, and U. Herpers. 1999. Contrasting Neogene denudation histories of different structural regions in the Transantarctic Mountains rift flank constrained by cosmogenic isotope measurements. *Global and Planetary Change* 23:145-172.

Vaughan, D. G., J. L. Bamber, M. Giovinetto, J. Russell, and A. P. R. Cooper. 1999. Reassessment of net surface mass balance in Antarctica. *Journal of Climate* 12:933-946.

Webb, P.-N. 1994. Paleo-drainage systems of East Antarctica and sediment supply to West Antarctic Rift System basins. *Terra Antartica* 1:457-461.

Webb, P.-N., and J. Wrenn. 1982. Upper Cenozoic biostratigraphy and micropaleontology of Taylor Valley, Antarctica. In *Antarctic Geoscience,* ed. C. Craddock, pp. 117-1122. Madison: University of Wisconsin Press.

Wellner, J. S., A. L. Lowe, S. S. Shipp, and J. B. Anderson. 2001. Distribution of glacial geomorphic features on the Antarctic continental shelf and correlation with substrate: Implications for ice behavior. *Journal of Glaciology* 47:397-411.

Zachos, J. C., J. Breza, and S. W. Wise. 1992. Early Oligocene ice sheet expansion on Antarctica, sedimentological and isotopic evidence from the Kergulen Plateau. *Geology* 20:569-573.

Cooper, A. K., P. J. Barrett, H. Stagg, B. Storey, E. Stump, W. Wise, and the 10th ISAES editorial team, eds. (2008). *Antarctica: A Keystone in a Changing World.* Proceedings of the 10th International Symposium on Antarctic Earth Sciences. Washington, DC: The National Academies Press.

A View of Antarctic Ice-Sheet Evolution from Sea-Level and Deep-Sea Isotope Changes During the Late Cretaceous-Cenozoic

K. G. Miller,[1] J. D. Wright,[1] M. E. Katz,[1,2] J. V. Browning,[1]
B. S. Cramer,[3] B. S. Wade,[4] S. F. Mizintseva[1]

ABSTRACT

The imperfect direct record of Antarctic glaciation has led to the delayed recognition of the initiation of a continent-sized ice sheet. Early studies interpreted initiation in the middle Miocene (ca 15 Ma). Most current studies place the first ice sheet in the earliest Oligocene (33.55 Ma), but there is physical evidence for glaciation in the Eocene. Though there are inherent limitations in sea-level and deep-sea isotope records, both place constraints on the size and extent of Late Cretaceous to Cenozoic Antarctic ice sheets. Sea-level records argue that small- to medium-size (typically $10\text{-}12 \times 10^6$ km^3) ephemeral ice sheets occurred during the greenhouse world of the Late Cretaceous to middle Eocene. Deep-sea $\delta^{18}O$ records show increases associated with many of these greenhouse sea-level falls, consistent with their attribution to ice-sheet growth. Global cooling began in the middle Eocene and culminated with the major earliest Oligocene (33.55 Ma) growth of a large (25×10^6 km^3) Antarctic ice sheet that caused a 55-70 m eustatic fall and a 1‰ $\delta^{18}O$ increase. This large ice sheet became a driver of climate change, not just a response to it, causing increased latitudinal thermal gradients and a spinning up of the oceans that, in turn, caused a dramatic reorganization of ocean circulation and chemistry.

INTRODUCTION

Glacial sediments on the Antarctic continent and its margins (Figure 1) (Barrett, 2007; Barrett et al., 1987; Birkenmajer et al., 2005; Cooper et al., forthcoming; Cooper and O'Brien, 2004; Ivany et al., 2006; Kennett and Barker, 1990; Leckie and Webb, 1986; LeMasurier and Rex, 1982; Strand et al., 2003; Troedson and Riding, 2002; Troedson and Smellie, 2002; Zachos et al., 1992) provide a direct record of ice sheets, but these records are temporally incomplete and often poorly dated and thus may not provide a complete and unequivocal history, especially of initiation of ice sheets. Deep-sea $\delta^{18}O$ records (Figures 1 and 2) provide well-dated evidence for changes in temperature and $\delta^{18}O_{seawater}$ due to ice-sheet growth, but separating these two effects is difficult (e.g., Miller et al., 1991). Global sea-level records provide evidence for large (tens of meters), rapid (<1 myr) changes in sea level (Figure 2) that can be explained only by changes in continental ice sheets, though the amplitudes of the changes have been poorly constrained until recently (Miller et al., 2005a). Each of these methods has its limitations, but by integrating results from all three we can begin to decipher the history of Antarctic ice sheets.

Over the past 30 years, study of glacial sediments and stable isotopes has progressively extended the initiation of ice sheets further back in time. For example, consider the history of Northern Hemisphere ice sheets (NHISs), better known as the "Ice Ages." Glacial deposits formed during advances of Laurentide ice led to the mistaken concept of only four Pleistocene glaciations (Flint, 1971), one that was eventually contradicted by deep-sea $\delta^{18}O$ records showing that there were:

[1]Department of Earth and Planetary Sciences, Rutgers University, Piscataway, NJ 08854, USA.

[2]Earth & Environmental Sciences, Rensselaer Polytechnic Institute, Troy, NY 12180, USA.

[3]Department of Geological Sciences, University of Oregon, Eugene, OR 97403, USA.

[4]Now at Department of Geology & Geophysics, Texas A&M University, College Station, TX 77843, USA.

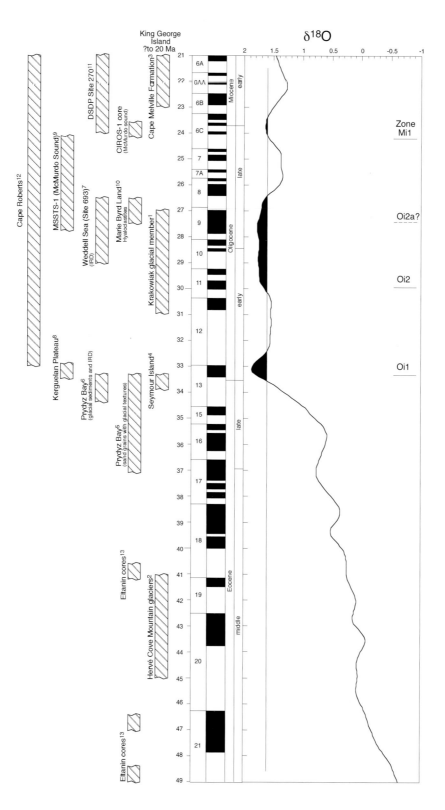

FIGURE 1 Oligocene benthic foraminiferal synthesis compared with the record of glaciomarine sediments (modified from Miller et al., 1991). Benthic foraminiferal stable isotope data were stacked and smoothed with a Gaussian convolution filter in order to remove periods less than 1.0 myr. Because filtering dampens the amplitude, an arbitrary line was placed through 1.6‰ and values higher than this were shaded. 1. Troedson and Smellie (2002); 2. Birkenmajer et al. (2005); 3. Troedson and Riding (2002); 4. Ivany et al. (2006); 5. Strand et al. (2003); 6. Cooper and O'Brien (2004); 7. Kennett and Barker (1990); 8. Zachos et al. (1992); 9. Barrett et al. (1987); 10. LeMasurier and Rex (1982); 11. Leckie and Webb (1986); 12. Barrett (2007).

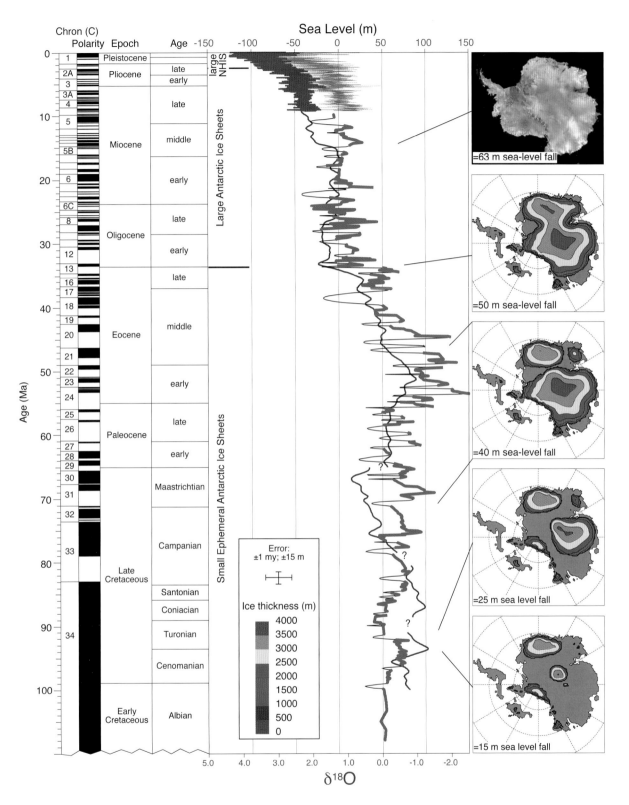

FIGURE 2 Global sea level (light blue) for the interval 7-100 Ma derived by new backstripped estimates; lowstands are indicated with thin black lines and are unconstrained by data (Miller et al., 2005a). Global sea level (purple) for the interval 0-7 Ma derived from δ¹⁸O after Miller et al. (2005a). Shown for comparison is a benthic foraminiferal δ¹⁸O synthesis from 0-100 Ma (red curve) with scale on bottom axis in ‰ (reported to *Cibicidoides* values [0.64‰ lower than equilibrium]). The portion of the δ¹⁸O curve from 0-65 Ma is derived using data from Miller et al. (1987); the Late Cretaceous synthesis is after Miller et al. (2004). Data from 7-100 Ma were interpolated to constant 0.1 myr interval and smoothed with a 21-point Gaussian convolution filter using Igor Pro™. Timescale of Berggren et al. (1995). Maps showing maximum sizes of ice sheets during peak glaciation for several intervals. Maps are after DeConto and Pollard (2003a,b) and show the amount of equivalent sea-level change proscribed for a given state (e.g., 25 m) and are calibrated to the δ¹⁸O synthesis (correlation lines) using sea-level data.

1. Eight large (~120 m sea-level lowerings), 100-kyr-scale ice ages over the past 800 kyr;

2. Sixty-two stages[5] representing 31 ice-sheet advances on ~20-, 41-, and ~100-kyr-scale during the Pleistocene; and

3. Over 100 named stages and 50 glacial advances since the late Pliocene "initiation" of NHIS (Emiliani, 1955; Hays et al., 1976; Shackleton, 1967).

A pulse of ice-rafted detritus (IRD) into the northern North Atlantic ca. 2.6 Ma (late Pliocene) is associated with a major $\delta^{18}O$ increase; this has been interpreted as the inception of NHISs (Shackleton et al., 1984). However, this inception reflects not initiation but an increase in the size of NHISs growth and decay (e.g., Larsen et al., 1994). Significant (at least Greenland-size) NHISs extend back at least to the middle Miocene (ca. 14 Ma; see summary in Wright and Miller, 1996) and recent data indicate that large NHISs may have existed since the middle Eocene (Eldrett et al., 2007; Moran et al., 2006).

The imperfect direct record of Antarctic glaciation has similarly led to the progressive extension of initiation of a continent-size ice sheet from 15 Ma (middle Miocene) back to 33.55 Ma (earliest Oligocene) (see summaries in Miller et al., 1991, 2005a,b; Zachos et al., 1996). In this contribution we suggest that continental ice sheets have been intermittently present on Antarctica through the Late Cretaceous, a time when Antarctica took up residence at the pole (http://www.ig.utexas.edu/research/projects/plates/).

Deep-sea isotope records have long been used to interpret Antarctic ice-sheet history. Based on deep-sea $\delta^{18}O$ records, early studies of Shackleton and Kennett (1975) and Savin et al. (1975) assumed that a continent-size ice sheet first appeared in Antarctica in the middle Miocene (ca. 15 Ma), though they noted that glaciation (in the form of mountain glaciers and sea ice) probably occurred back through the Oligocene. Also using isotope data, Matthews and Poore (1980) suggested that large ice sheets existed in Antarctica since at least the earliest Oligocene (33.5 Ma). The differences in interpretation partly illustrate problems in using $\delta^{18}O$ as an ice-volume proxy, because deep-sea $\delta^{18}O$ values also reflect deep-water temperature changes that generally mimic high-latitude surface temperatures. Miller and Fairbanks (1983, 1985) and Miller et al. (1987, 1991) provided the strongest isotopic evidence for the presence of ice sheets during the Oligocene; high $\delta^{18}O$ values measured in deep-sea cores (>1.8‰ in *Cibicidoides* spp. or >2.4‰ in *Uvigerina* spp.) require bottom-water temperatures colder than today if an ice-free world is assumed. Such low bottom-water temperatures are incompatible with an ice-free world; their isotopic synthesis (updated and presented

in Figures 1 and 2) suggest at least three major periods of Oligocene glaciation.

A campaign of drilling near Antarctica by the Ocean Drilling Program (ODP) in the late 1980s returned firm evidence that supported the $\delta^{18}O$ record for large ice sheets in the earliest Oligocene that included grounded tills and IRD at lower latitudes than today (see summaries by Miller et al., 1991; Zachos et al., 1992). We update the summary of the direct evidence for Eocene-Oligocene ice in the form of tills and glaciomarine sediments near the Antarctic (Figure 1), using more recent drilling by ODP (Cooper and O'Brien, 2004; Strand et al., 2003), the Cape Roberts drilling project (Barrett, 2007), plus studies that extend West Antarctic glaciation back through the early Oligocene (Seymour Island) (Ivany et al., 2006) and into the Eocene (King George Island; Birkenmajer et al., 2005; Troedson and Riding, 2002; Troedson and Smellie, 2002). The evidence for large, grounded ice sheets begins in the earliest Oligocene and continues through the Oligocene (Figure 1). Seismic stratigraphic studies summarized by Cooper et al. (forthcoming) also show intense glacial activity beginning in the Oligocene in both East and West Antarctica. There is excellent agreement among proxies that Antarctica was in fact an icehouse during the Oligocene and younger interval. Ice-volume changes have been firmly linked to global sea-level changes in the Oligocene and younger "icehouse world" of large, varying ice sheets (Miller et al., 1998); Pekar et al. (1996, 2002) recognized that the three to four major Oligocene glaciations of Miller et al. (1991) in fact reflected six myr-scale sea-level falls and attendant ice-growth events. The record of glaciomarine sediments documents that the ice sheets occurred in Antarctica (Figure 1), though an NHIS component cannot be precluded due to scarce Northern Hemisphere Oligocene records. Today 33.5 Ma is cited as the inception of the Antarctic ice sheet, though this supposition is now being challenged and pushed back into the Cretaceous (Miller et al., 1999, 2003, 2005a,b; Stoll and Schrag, 1996). Nevertheless, 33.55 Ma was probably the first time in the past 100 myr that the ice sheets reached the coast, allowing large icebergs to calve and reach distal locations such as the Kerguelen Plateau (Figure 1) (ODP Site 748) (Zachos et al., 1992).

There is evidence for glaciation in the older Antarctic record. Coring by ODP Legs 119 and 120 (Barron and Larsen, 1989; Breza and Wise, 1992) and seismic stratigraphic studies (Cooper et al., forthcoming) suggest the possibility of late Eocene (or even possibly middle Eocene) glaciers in Prydz Bay. Seismic stratigraphic studies (Cooper et al., forthcoming) also suggest the possibility of late Eocene glaciers in the Ross Sea. Other studies extend the record for West Antarctic glaciation back from 10 Ma to 45 Ma (Birkenmajer, 1991; Birkenmajer et al., 2005). Though Birkenmajer et al. (2005) interpreted the Eocene tills as evidence for mountain glaciers and not necessarily ice sheets, it points to the likelihood that the continental interior could have supported an ice sheet

---

[5]Although called "oxygen isotopic stages" by paleoceanographers for decades, the term "stage" is a stratigraphic term reserved for characterizing time-rock units (Hedberg, 1976). The proper term for "isotopic variations" are "zones in depth and chrons in time."

in the middle Eocene. Margolis and Kennett (1971), Wei (1992), and Wise et al. (1991) interpreted middle Eocene quartz grains in Eltanin cores from near Antarctica as reflecting IRD (Figure 1), though the evidence of this as IRD versus other transport mechanisms is not compelling.

Early studies recognized the importance of a cold, if not fully glaciated, Antarctica on deep-water circulation (Kennett, 1977; Kennett and Shackleton, 1976; Shackleton and Kennett, 1975). Kennett (1977) attributed the Eocene-Oligocene transition to the development of a nascent Antarctic Circumpolar Current that caused thermal isolation of Antarctica, development of sea ice (though not continental-scale glaciation), and an increase in Antarctic bottom water (AABW). The formation of AABW today is particularly sensitive to sea ice, and geologic evidence is clear that an erosional pulse of deep water in the Southern Ocean occurred near the Eocene-Oligocene transition (Kennett, 1977; Wright and Miller, 1996). It was also known that a major drop in the calcite compensation depth (CCD) occurred across the Eocene-Oligocene transition (van Andel, 1975) in concert with this change in deep-water circulation. However, our understanding of deep-water history and its relationship to the evolution of Antarctic ice volume has been unclear in part because of uncertainties in the proxies for deep water and continental ice sheets.

Here we review interpretations of Antarctic ice-volume changes using sea-level and oxygen isotopic records. Using recently published sea-level curves (Kominz et al., in review;

Miller et al., 2005a) and published isotope data (Figures 2 and 3), we argue for the likely presence of small- to medium-size ephemeral ice sheets in the greenhouse world of the Late Cretaceous to Eocene. Though there were ephemeral ice sheets in the greenhouse world, the Eocene-Oligocene transition represented the beginning of the icehouse with the largest cooling event of the last 100 myr, one that resulted in an ice sheet that reached the coast for the first time. We review published and recently submitted evidence for the nature and timing of paleoceanographic changes associated with the Eocene-Oligocene transition (Figure 4) and present new and published comparisons of global deep-water changes that resulted from this glaciation and attendant cooling (Figures 4 and 5).

THE CASE FOR ICE SHEETS IN THE GREENHOUSE WORLD (CRETACEOUS-EOCENE)

The sea-level curves of Exxon Production Research Company (EPR) (Haq et al., 1987; Vail et al., 1977) stimulated interest in large, rapid, global sea-level changes. Vail et al. (1977) reported numerous large (>100 m) Phanerozoic sea-level falls, including a 400 m drop in mid-Oligocene. On the Vail curve these falls were shown as virtually instantaneous. Subsequent study showed that this saw-toothed pattern with extremely rapid falls was an artifact of measurement of coastal onlap versus offlap (Thorne and Watts, 1984), but subsequent generations of the EPR curve covering the past

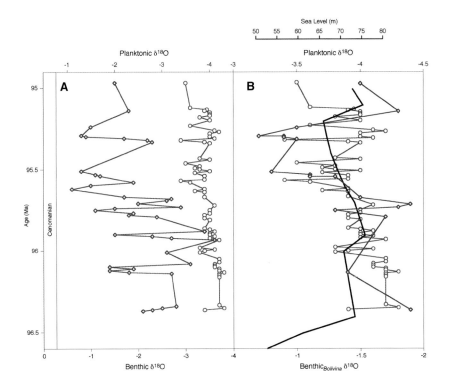

FIGURE 3 (A) Planktonic (red line) and benthic (blue line) $\delta^{18}O$ record of Moriya et al. (2007) plotted with a full 4.5‰ range. Benthic foraminifera $\delta^{18}O$ is based on the combined records of *Bolivina anambra*, *Gavelinella* spp., and *Neobulimina* spp. species. (B) Moriya et al.'s (2007) $\delta^{18}O$ record of planktonic (red line) and benthic foraminifera *Bolivina anambra* (blue line) plotted on enlarged scales with 1.5‰ ranges; note different scales for benthic (bottom scale) and planktonic (top scale) values. Only values of *Bolivina anambra* greater than –2‰ are included. Also shown is the sea-level record (black line) of Kominz et al. (in review) shifted in age by ~0.2 myr to maximize correlations.

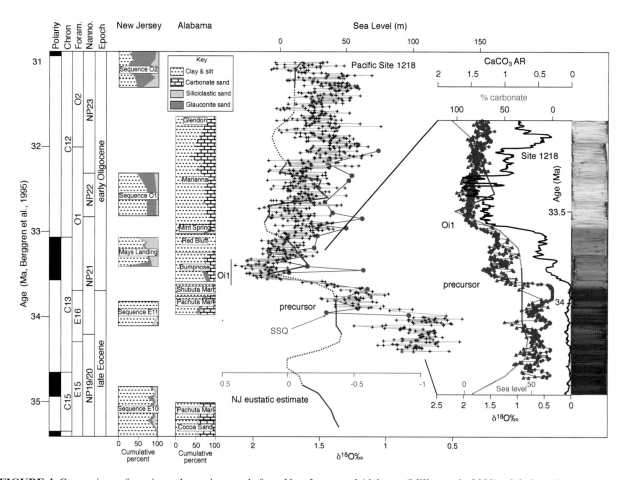

FIGURE 4 Comparison of continental margin records from New Jersey and Alabama (Miller et al., 2008), global sea-level estimates, and benthic foraminiferal $\delta^{18}O$ records from deep Pacific ODP Site 1218 (Coxall et al., 2005) and St. Stephens Quarry, Alabama (Miller et al., 2004). Shown at right is a blowup of ODP Site 1218 oxygen isotopes that includes carbonate percentage and mass accumulation rate (AR) data (Coxall et al., 2005) that shows two large drops associated with the precursor and Oi1 oxygen isotope increases that are reflected in the core photograph at right (light color = carbonate rich; dark = carbonate poor).

200 myr still showed 100 m falls in much less than 1 myr (Haq et al., 1987), including a mid-Oligocene fall of ~160 m. Though the EPR curves have been strongly criticized for their methodology and proprietary data (Christie-Blick et al., 1990; Miall, 1991), recently published sea-level estimates (Kominz et al., in review; Miller et al., 1998, 2005a) show that the timing of the EPR curves is largely correct. These recent estimates (discussed below) show that the EPR sea-level amplitudes are typically two to three times too high but still require tens of meters of change in much less than 1 myr.

Such rapid sea-level changes pose an enigma, because the only known mechanism for causing sea-level changes in excess of 10 m in less than 1 myr is glacioeustasy (Pitman and Golovchenko, 1983). No other known mechanisms (steric effects, storage in lakes, deep-water changes, groundwater, or sea ice) can explain these changes (see Figure 1 in Miller et al., 2005a). Temperature changes can explain rapid sea-level changes such as the changes happening today,

but the effect is small (i.e., a 10°C global warming would cause only a 10 m sea-level rise [Pitman and Golovchenko, 1983]). Changes in terrestrial storage in lakes and groundwater can only explain 5 m of sea-level change (Pitman and Golovchenko, 1983). Desiccating and refilling Mediterranean basins could explain very rapid (<1 kyr) changes, but this effect is small (~10 m) and it is impossible to explain the number of Late Cretaceous to early Eocene sea-level events with this mechanism (Pitman and Golovchenko, 1983). Thus only ice-volume changes can explain these large (tens of meters) sea-level changes, even in the supposedly ice-free world of the Late Cretaceous to Eocene.

Matthews and Poore (1980) first realized this enigma and Matthews (1984), based on sea-level records and his reinterpretation of the $\delta^{18}O$ record, postulated that intermittent ice sheets occurred in the mid-Cretaceous through the Paleogene. Based on a comparison of continental margin and $\delta^{18}O$ records, Miller et al. (1987, 1991) suggested that the growth and decay of continental ice sheets controlled

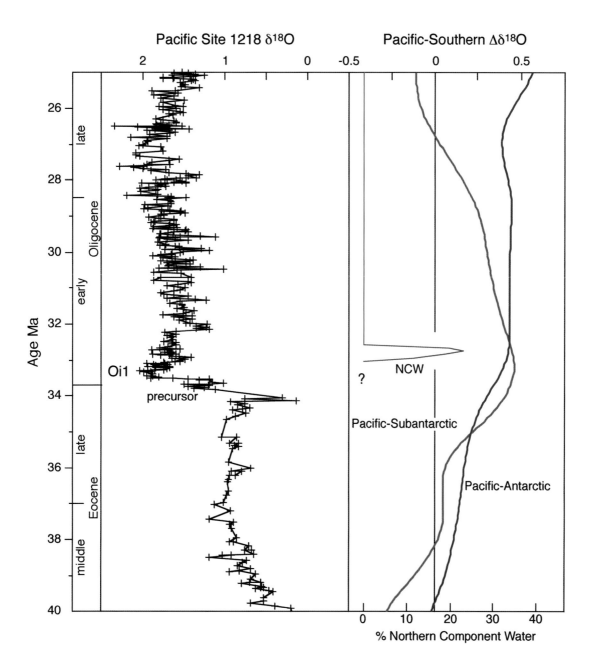

FIGURE 5 Comparison of benthic foraminiferal δ¹⁸O data from Pacific ODP Site 1218 (Coxall et al., 2005) with the percentage of northern component water (NCW) (Wright and Miller, 1996) and unpublished difference curve between a Pacific and southern ocean δ¹⁸O values. Values were interpolated to constant 0.1 myr intervals and smoothed with an 11-point Gaussian convolution filter using Igor Pro®. Data were divided into Maud Rise sites closest to Antarctica and subantarctic sites. Timescale after Berggren et al. (1995).

sea-level changes in the Oligocene. Subsequent studies have confirmed the existence of Oligocene ice sheets (see Figure 1, and summaries of Miller et al., 1991; Zachos et al., 1994) and led to the suggestion that small- to medium-size ice sheets existed in the middle to late Eocene (Browning et al., 1996). Yet other than Matthews (1984) the concept of ice sheets during the interval of peak global warmth, the greenhouse world of the Late Cretaceous to Eocene, was ignored.

Stoll and Schrag (2000) examined mid-Cretaceous isotopic variations and revived Matthews's idea by postulating that continental ice sheets grew and decayed in this greenhouse world, and recent sea-level studies have supported their supposition (Kominz et al., in review; Miller et al., 2005a; Van Sickel et al., 2004).

Studies from the New Jersey coastal plain provided a eustatic estimate for the past 100 myr, using inverse models termed "backstripping" (Kominz et al., 1998, in review; Miller et al., 2005a; Van Sickel et al., 2004). Backstripping progressively removes the effects of sediment compaction and loading from observed basin subsidence (e.g., Kominz et al., 1998) and thermal-flexural subsidence is modeled by fitting exponential curves to the remaining observed subsidence. The difference between the best fit exponential curve and subsidence is a result of either eustatic change or any subsidence unrelated to thermal subsidence (Kominz et al., 1998). We have applied this technique to coreholes from the New Jersey and Delaware coastal plains by dating sequences (unconformity bounded units) with a ±0.5 myr or better resolution and developing a water depth history by integrating biofacies and lithofacies analysis. The similarity of records (six sites for the Cenozoic and four for the Cretaceous) (Miller et al., 2004) indicates that the effects of thermal subsidence, loading, and water-depth variations have been successfully removed. Backstripping, seismicity, seismic stratigraphic data, and distribution patterns of sediments all indicate minimal tectonic effects on the Late Cretaceous to Tertiary New Jersey coastal plain (Miller et al., 2004, 2005a), though some minor (10-30 m) differences between New Jersey and Delaware must be attributed to local subsidence or uplift (Browning et al., 2006).

The timing of sequences in New Jersey and other margins is similar, suggesting a global cause. There are few other backstripped records to test the New Jersey eustatic estimate against, though a Russian platform backstripped eustatic estimate (Sahagian et al., 1996) shows remarkably similar changes to New Jersey in the interval of overlap from 100 Ma to 90 Ma (Miller et al., 2003, 2005a; Van Sickel et al., 2004). In addition, comparison of the New Jersey record with northwest Europe (Ali and Hailwood, 1995; Hancock, 1993), the U.S. Gulf Coast (Mancini and Tew, 1995), the EPR synthesis (Haq et al., 1987), and long-term sea-level predictions from Milankovitch forcing (Matthews and Frohlich, 2002) suggests that Late Cretaceous to Eocene sea-level falls were rapid, synchronous, and global (see summary figures in Miller et al., 2004). Therefore, it is unlikely that the New Jersey eustatic estimate can be attributed to tectonics. The only mechanism than can explain the observed rates of sea-level change (in excess of 25 m/myr) is the growth and decay of continental ice sheets that caused glacioeustatic changes. Error bars for the sea-level falls are discussed in detail by Kominz et al. (in review). The greatest uncertainty is that the onshore sites mostly miss the lowstands and thus are minimal estimates of the eustatic falls. The Late Cretaceous to Eocene falls were typically 25 m, with small events (15 m) near the detection level. The only exception was a large Campanian/ Maastrichtian boundary (ca. 71.5 Ma) sea-level fall of 40 m. It should be noted that the New Jersey eustatic estimate must be considered a testable model. A true global sea-level curve must be derived from numerous margins, not just one; thus further studies are needed to validate this record.

Based on the New Jersey sea-level records, Miller et al.

(2003, 2004, 2005a,b) postulated the existence of ice sheets in the greenhouse world of the Late Cretaceous to Eocene. Miller et al. (2005b) presented a new view of Earth's cryospheric evolution that reconciled warm, largely ice-free poles with cold periods that resulted in glacioeustatic lowerings. Their view was developed from work by DeConto and Pollard (2003a,b) who used a coupled global climate and ice-sheet model that accounts for bedrock loading, lapse rate effects of topography, surface mass balance, basal melting, and ice flow and computes both the sea level and $\delta^{18}O$ effects of ice growth. The DeConto and Pollard (2003a,b) model used declining atmospheric CO_2 values of ~3 to 2 times present day, similar to empirical estimates for declining CO_2 from the Eocene to Oligocene (Pagani et al., 2005). These model runs were used to estimate the size and geographic distribution of ice sheets (Figure 2) for Oligocene continental configurations, though the modeling results are applicable to the older record as long as Antarctica was a polar continent (i.e., for the entire interval considered here).

We estimate Antarctic ice volume using the New Jersey sea-level record (Figure 2). The typical 15-30 m greenhouse eustatic falls correspond to growth of ice volumes of $8-12 \times 10^6$ km³ using a 0.1‰/10 m calibration of sea level and ice volume (DeConto and Pollard, 2003a,b). The exception is the large Campanian/Maastrichtian boundary fall of 40 m that suggests growth of an ice sheet of 17×10^6 km³. As illustrated in Figure 2, these ice sheets did not reach the Antarctic coast. This together with the fact that this ice sheet probably existed for relatively short periods (see below) reconciles ice with coastal warmth in Antarctica. Though the temporal coverage is limited, there is ample evidence that Antarctic coastal climates were quite warm during much of the Late Cretaceous to Eocene (e.g., Askin, 1989; Francis and Poole, 2002); deep waters were relatively warm also (e.g., early Maastrichtian paleotemperatures of ~10°C at 1500 m paleodepth on the Maud Rise) (Barrera and Huber, 1990). Yet these coastal and offshore studies do not address regions of ice sheet nucleation predicted for the continental interiors (Figure 2) (DeConto and Pollard, 2003a,b).

These ice sheets were ephemeral and for the most part Antarctica lacked ice sheets during the greenhouse world except for these cool or cold periods, called "cool snaps" by Royer et al. (2004). It is not possible to say how long these cool or cold intervals lasted. Matthews and Frohlich (2002) computed an estimated Cretaceous sea-level and ice-volume curve based on Milankovitch forcing, and minima in this curve appear to correlate with sequence boundaries in New Jersey (Miller et al., 2005b). The dominant beats in the predicted curve are the ca. 2.4 myr and 405 kyr very long and long eccentricity cycles and the 1.2 myr tilt cycle. Yet these longer-term modulations predict that shorter-term periods (tilt and precession) must have been operative and it seems likely that the cool intervals only lasted during minima in insolation on the precession and tilt periods. Studies of myr-scale sea-level events in the Oligocene suggest that peak cold intervals lasted only 2-3 tilt (41 kyr) cycles (Coxall et al., 2005; Zachos et al., 1996). Thus we speculate that ice sheets existed only for brief intervals (~100 kyr for a typical 2.4 myr cycle) during the greenhouse.

We illustrated Antarctic cryospheric evolution using maps derived from the models of DeConto and Pollard (2003b). These maps show areas and thicknesses computed from the models, which also compute the equivalent water volume and sea level. We correlate each map and its attendant sea-level fall to equivalent sea-level falls estimated in New Jersey (e.g., the map for a 40 m fall is aligned with the 40 m 71.5 Ma fall) (Figure 2). For periods of small (~15 m) sea-level falls that typified most of the Late Cretaceous and early Eocene, models indicate that small isolated ice caps would have formed in the highest elevation of Dronning Maud Land, the Gamburtsev Plateau, and the Transantarctic Mountains (Figure 2) under high CO_2 conditions. For periods with moderate sea-level falls (25 m) that were typical of the larger events in greenhouse intervals (e.g., the mid-Turonian and mid-Cenomanian), a small ice-sheet threshold was reached owing to height and mass balance feedback; the three ice sheets (Dronning Maud Land, the Gamburtsev Plateau, and the Transantarctic Mountain) continued to grow but not coalesce, and were isolated from the coast. With a 40 m sea-level lowering, ice caps would have begun to coalesce (Figure 2); this configuration would have been achieved only during the largest of the greenhouse sea-level falls (e.g., the 71.5 Ma Campanian/Maastrichtian event). With sea-level falls of over 50 m, the three ice-sheet nodes would have united (Figure 2); this configuration would not have been achieved until CO_2 fell below a critical threshold (estimated at 2.8 times preanthropogenic levels) (DeConto and Pollard, 2003a,b) in the earliest Oligocene (33.55 Ma).

The elevation of Antarctica in the greenhouse world is an important unknown. The models of DeConto and Pollard (2003a) account for changes in elevation due to isostasy and mountain building and are appropriate for Oligocene and later Eocene configurations. However, the elevation of the continent may have been lower in the earliest Eocene and older. Uplift of the Transantarctic Mountains may have begun in the Cretaceous (Fitzgerald, 2002), though it is also possible that uplift did not begin until the Eocene (ten Brink et al., 1997). We concede that it may have been more difficult to nucleate ice sheets on a lower continent but maintain that the elevational history of Antarctica is poorly known.

The scenario for Late Cretaceous to Cenozoic ice-sheet history (Figure 2) is partly testable with $\delta^{18}O$ data because each sea-level event should be associated with a $\delta^{18}O$ increase due to ice growth. This prediction has been verified for the Oligocene to middle Miocene (Miller et al., 1996, 1998) and extended back to the middle to late Eocene, where sea-level changes are coupled with $\delta^{18}O$ increases (Browning et al., 1996), suggesting glacioeustatic control. Tripati et al. (2005) also used stable isotopic data to argue for middle and

late Eocene ice sheets. Cretaceous to early Eocene $\delta^{18}O$ data are sparse because of poor core recovery and diagenesis (i.e., isotopic studies of sections with >400 m burial are suspect) (1987), but some comparisons can be made. The large Campanian/Maastrichtian boundary sea-level fall is associated with a major $\delta^{18}O$ increase in both planktonic and benthic foraminifera (Miller et al., 1999, using data of Barrera and Savin, 1999; and Huber et al., 2002). Also, large (>0.75‰) mid-Cenomanian (ca. 96 Ma) and mid-Turonian (ca. 92-93 Ma) benthic foraminiferal $\delta^{18}O$ increases reported from western North Atlantic ODP Site 1050 (Huber et al., 2002) correlate with major sea-level falls in New Jersey. These large $\delta^{18}O$ increases at 92-93 Ma and 96 Ma cannot be entirely attributed to ice-volume changes because this would require ice sheets larger than those of modern times, and in fact much of the $\delta^{18}O$ signal must be attributed to deep-sea (hence inferred high-latitude; see below) temperature change. Miller et al. (2005b) estimated that about two-thirds of the $\delta^{18}O$ signal was attributable to temperature, suggesting about 0.25‰ was due to a change in $\delta^{18}O_{seawater}$, corresponding to a sea-level change of ~25 m and growth of about one-third of the Antarctic ice sheet.

A recent study has challenged correlations of sea-level falls and $\delta^{18}O$ increases for the Cenomanian. Moriya et al. (2007) tested evidence for Cenomanian glacioeustasy by generating high-resolution (26 kyr sampling) benthic and planktonic $\delta^{18}O$ records from ODP Site 1258 on the Demerara Rise (western tropical Atlantic Ocean). They concluded that there was no support for Cretaceous glaciation and called into question evidence for greenhouse ice sheets. We replotted their data (Figure 3) and make the following observations:

1. Benthic foraminiferal $\delta^{18}O$ values from the Demerara Rise (Moriya et al., 2007) show very large amplitudes (up to 3‰) that would require deep-water temperature changes of 13°C in a few 100 kyr (Figure 3). The extremely low benthic $\delta^{18}O$ values (less than –4‰, mean value of –1.8‰) correspond to maximum and average deep-water temperatures of 30°C and 19.5°C, respectively (assuming an ice-free $\delta^{18}O_{seawater}$ value of –1.2‰) (Moriya et al., 2007). Considering that the lowest benthic foraminiferal $\delta^{18}O$ values for the Cenomanian at western North Atlantic ODP Site 1050 (Figure 2) (Huber et al., 2002) are –1.4‰ to –1.6‰, corresponding to deep-water temperatures of 18-19°C, we suggest that Moriya et al.'s (2007) benthic foraminiferal records are overprinted by diagenesis, very large local $\delta^{18}O_{seawater}$ effects, or by analysis of multiple species with very large vital effects. Their $\delta^{18}O$ values on the genus *Neobulimina* particularly show considerable variations, with many values close to those of planktonic foraminifera. We culled their benthic foraminiferal $\delta^{18}O$, accepting only *Bolivina* data and rejecting all values lower than –2‰ as physically unrealistic; the resulting dataset shows a similar pattern to the planktonic $\delta^{18}O$ record (Figure 3b).

2. Though Moriya et al. (2007) claimed minimal change in planktonic foraminiferal $\delta^{18}O$ values, this is largely an artifact of the scale of 5‰ on which the data were plotted (e.g., Figure 3a). Plotting planktonic foraminiferal $\delta^{18}O$ values with a 1.5‰ scale (Figure 3b) show that values increase by ~0.5-0.6‰ in the mid-Cenomanian (~95.5 Ma) at precisely the same time as the sea-level fall delineated in New Jersey (Figure 3).

3. As noted by Miller et al. (2005a) and Moriya et al. (2007), $\delta^{18}O_{seawater}$ effects for these greenhouse ice sheets would have been small (~0.2-0.3‰), but this is detectable within measurement error. While we have observed and would reasonably expect cooler deep-water temperatures associated with these greenhouse ice-growth events (e.g., two-thirds of the Campanian/Maastrichtian fall must be due to cooling), there is no reason to require proportional coupling of cooling and ice volume as suggested by Moriya et al. (2007). In fact, we suspect that one reason that early Eocene ice-volume events have proven to be so elusive is that there was little change in deep-water temperature and the ice-volume signal is quite small (0.2-0.3‰).

4. Thus the Demerera Rise isotope data are consistent with (rather than disprove) the idea that there was a seawater $\delta^{18}O$ change on the order of ~0.2-0.3‰ associated with the mid-Cenomanian event.

We conclude that sea-level records indicate that small- to medium-size ($10\text{-}15 \times 10^6$ km³), ephemeral (lasting a few tilt cycles based on analogy with high-resolution Oligocene studies of Zachos et al., 1994) ice sheets occurred during the greenhouse world of the Late Cretaceous to middle Eocene. However, an Antarctic record of these glaciations exists only from the middle Eocene (Figure 1).

THE BIG CHILL INTO THE ICEHOUSE: THE EOCENE-OLIGOCENE TRANSITION

Following peak warmth in the early Eocene (ca. 50 Ma), benthic foraminiferal deep-water and by extension high-latitude surface waters cooled across the early to middle Eocene boundary (ca. 48-49 Ma) and in the late middle Eocene (ca. 44-41 Ma) (Figure 2). The sea-level record shows development of progressively larger ice sheets in the middle to late Eocene (Figure 2). The Eocene-Oligocene transition culminated with the largest $\delta^{18}O$ increase of the past 50 myr, the earliest Oligocene Oi1 $\delta^{18}O$ maximum (33.55 Ma).

Antarctica entered the icehouse in the earliest Oligocene (33.55 Ma). There is widespread evidence for large ice sheets (IRD and grounded diamictons [Figure 1]) and we suggest that Antarctica was nearly fully glaciated. A eustatic fall of 55-70 m occurred at 33.55 Ma in association with the $\delta^{18}O$ increase. This corresponds to an Antarctic ice sheet that was ~80-100 percent of the size of the present-day East Antarctic ice sheet (today a sea-level equivalent of ~63 m is stored in East Antarctica and 5-7 m in West Antarctica; see discus-

sion below). If fully compensated by isostasy, this 55-70 m represents ~82-105 m of change in water volume or apparent sea level (Pekar et al., 2002), which would correspond with 115-150 percent of the modern Antarctic ice sheet or 100-130 percent of the entire modern ice inventory.

These large ice estimates require storage of ice outside East Antarctica. Though earlier studies assumed that the West Antarctic ice sheet did not develop until the later Miocene (Kennett, 1977), recent studies (Ivany et al., 2006) suggest that glacial ice extended to sea level in this region by the earliest Oligocene; these were not just mountain glaciers. This is supported by seismic stratigraphic evidence summarized by Cooper et al. (forthcoming) for glacial influences in both East and West Antarctica beginning in the Oligocene.

These sea-level changes suggest at least moderate glaciation in the Northern Hemisphere in the earliest Oligocene. Today the Greenland ice sheet is composed of 6.5 m (Williams and Ferrigno, 1995) of sea-level equivalent. Tripati et al. (2005) have argued for very large (tens of meters) NHISs in the earliest Oligocene. They attribute the entire $\delta^{18}O$ increase observed in the deep Pacific to growth of ice sheets because Mg/Ca data associated with the Oi1 $\delta^{18}O$ increase. Mg/Ca data for Oi1 time indicate a 2°C warming that is an artifact of carbonate undersaturation in the deep sea (Lear and Rosenthal, 2006). However, though there is evidence for localized Northern Hemisphere glaciation in the Paleogene (Eldrett et al., 2007; Moran et al., 2006), the first appearance of IRD in the North Atlantic outside the Norwegian-Greenland Sea did not occur until 2.6 Ma (Shackleton et al., 1984). The Eocene IRD evidence presented by Eldrett et al. (2007) and Moran et al. (2006) is compelling but limited to the high-latitude Norwegian-Greenland Sea and Arctic and does not indicate widespread Northern Hemisphere glaciation. We thus doubt the presence of very large (tens of meters equivalent) ice sheets in the Northern Hemisphere prior to 2.6 Ma.

The suggestion that the Antarctic ice sheet was as large if not larger than today needs to be reconciled with paleobotanical data that suggest that ice-free refugia are needed for the Oligocene (e.g., Francis and Poole, 2002). Drilling in the Weddell Sea highlights this enigma; lowermost Oligocene grounded diamictons are overlain by strata containing Nothofagus leaves (Kennett and Barker, 1990). This is consistent with our scenario based on sea-level and stable isotopic studies (Coxall et al., 2005; Zachos et al., 1996) that earliest Oligocene ice sheets formed rapidly (<< 1 myr) in a series of Milankovitch tilt-driven (41 kyr) increases that lasted for a few hundred kyr, and then almost completely collapsed again, only to reform again at Oi1a time (ca. 33 Ma), but it leaves open the question of where these refugia were located during times of maximum glaciation. Under true icehouse conditions that began in the earliest Oligocene, the ice sheet reached the coastline (Figure 2), which limited its further growth. The sea-level estimates suggest a minimum drop of 55 m, which is illustrated by calving along much of the coastline except Wilkes Land (Figure 2); however, our upper limit eustatic lowering of 70-105 m would require ice volumes exceeding modern, and thus reaching the coast along much of the margin.

Using data of Miller et al. (2008), we present a detailed view of the climate changes that spanned the Eocene-Oligocene boundary (Figure 4). The Eocene-Oligocene transition is associated with a long-term (10^7 yr scale) CO_2 drawdown (Pagani et al., 2005) and related temperature change that triggered a precursor $\delta^{18}O$ increase at 33.8 Ma of ~0.5‰ identified at Pacific ODP Site 1218 (Coxall et al., 2005) and shelf site at St. Stephens Quarry, Alabama (Miller et al., 2008), followed by a 1.0‰ $\delta^{18}O$ increase at 33.55 Ma (= Oi1, earliest Oligocene). This is consistent with models that predict climate response of at least two nonlinear jumps associated with the transition (DeConto and Pollard, 2003b). The precursor increase is not associated with an observable change in sea level (i.e., a change greater than 15-20 m), suggesting that it was caused by a 2°C cooling, not ice-sheet growth (Miller et al., 2008). Global sea level dropped by ~55-82 m at 33.55 Ma, indicating that the deep-sea $\delta^{18}O$ increase was due to transfer of the water from the oceans to the Antarctic ice sheet, which was 80-130 percent of the modern size, as discussed above. The amount of cooling associated with Oi1 is still uncertain. Using the 55 m fall and a calibration of 0.1‰/10 m (DeConto and Pollard, 2003b) suggests that deep-water temperatures dropped 2°C; using the higher sea-level fall (82 m) suggests only 1°C of cooling.

The Antarctic ice sheet reached the coastline for the first time in the earliest Oligocene (Figure 2) and this large ice sheet became a driver of, not just a response to, climate change. The earliest Oligocene was characterized by increased latitudinal thermal gradients (Kennett, 1977). This increase caused: (1) enhanced wind intensities and a spinning up of the oceans, resulting in increased upwelling (Suko, 2006) and increased thermohaline circulation, with transient erosional pulses of North Atlantic deep water (Miller and Tucholke, 1983; Wright and Miller, 1996) and AABW (Kennett, 1977; Wright and Miller, 1996); (2) a major drop in the CCD as the deep oceans cooled, became better ventilated, and had reduced residence time and acidity (Figure 4) (Coxall et al., 2005; van Andel, 1975); and (3) a large increase in diatom diversity due to intensified latitudinal thermal gradients and upwelling (Falkowski et al., 2004; Finkel et al., 2005; Pak and Miller, 1992).

Seismic stratigraphy and mapping of hiatuses indicate a major strengthening of thermohaline circulation in the Oligocene. The Antarctic has generally been a source of deep water to the oceans (analogous to AABW) throughout much of the later Cretaceous and the Cenozoic (Pak and Miller, 1992; Mountain and Miller, 1992). The Eocene-Oligocene boundary saw intensification of AABW. Kennett (1977) and Wright and Miller (1996) showed that numerous hiatuses in the southern ocean are attributable to erosional pulses of

AABW, with a large one at the Eocene-Oligocene boundary. In the North Atlantic a major erosional pulse was associated with reflectors R4 (Miller and Tucholke, 1983) and A^u (Tucholke and Mountain, 1979), both of which date to near the Eocene-Oligocene boundary (Mountain and Miller, 1992). Northern Hemisphere high latitudes cooled concomitantly with polar cooling in the Southern Hemisphere, making the North Atlantic a viable source region for deep water (albeit briefly) in the earliest Oligocene.

Stable isotope reconstructions are generally consistent with the scenario derived from mapping of hiatuses and seismic disconformities. Carbon isotopic reconstructions of the Oligocene show low vertical and interbasinal differences (Diester-Haass and Zachos, 2003; Miller and Fairbanks, 1985; Salamy and Zachos, 1999; Wade and Pälike, 2004), limiting the use of this proxy for reconstructing deep-water changes. We attribute these low gradients to a general decrease in export production; although the spin-up of the oceans increased export productivity in eutrophic areas (Moore et al., 2004), available nutrients limited general export production, which was remarkably low in oligotrophic areas (Miller and Katz, 1987). Despite the low Oligocene $\delta^{13}C$ sensitivity (and the resulting flat line on Figure 5), $\delta^{13}C$ records indicate that there was a short early Oligocene interval (ca. 33-32 Ma) with high Atlantic-Pacific $\delta^{13}C$ gradients (Miller, 1992); Wright and Miller (1996) quantified the percentage of northern component water using interbasinal $\delta^{13}C$ gradients (Figure 5).

Oxygen isotopic records indicate that the Antarctic was the source of water with high $\delta^{18}O$ values in the late middle to late Eocene (Figure 5). ODP Site 689 on the Maud Rise (1800 m paleodepth) (Kennett and Stott, 1990; Thomas, 1990) records very high $\delta^{18}O$ values beginning with the late middle Eocene cooling and staying high through the Oligocene (Figure 5). This "cold spigot" indicates that AABW had a strong cold influence close to the continent, but this influence was reduced by mixing away from the proximal southern ocean (Miller, 1992). Subantarctic $\delta^{18}O$ records show that the cold water mass did not strongly influence regions further from the continent until the end of the Eocene (Figure 5). Very high subantarctic $\delta^{18}O$ values during the Eocene-Oligocene transition suggest a pulse of AABW that peaked with the Oi1 glaciation (Figure 5), consistent with the erosional record (Kennett, 1977; Wright and Miller, 1996). This increase in thermohaline circulation is coincident with the drop in the calcite compensation depth.

The drop in the CCD was caused by increased thermohaline circulation that caused a decrease in oceanic residence time and a decrease in deep-ocean acidity. Coxall et al. (2005) showed that the CCD drop occurred in two steps associated with the precursor and $\delta^{18}O$ increases (Figure 4); the drop in percentage of carbonate at the precursor event was as large or larger than at Oi1 time (Figure 5), but accumulation rate data indicate that the latter event was more significant. The cause of the drop in the CCD has been debatable. Any deepening of the CCD must be attributed to one of the following:

1.   An increase in deep-basin carbonate deposition at the expense of shallow carbonates (shelf-basin fractionation, as invoked by Coxall et al., 2005) for this drop. We argue that shelf-basin fractionation cannot be invoked to explain the CCD drop associated with the precursor $\delta^{18}O$ increase, because there is little or no sea-level fall during the precursor; in addition, they note that the amount of eustatic fall (55-82 m) associated with the 33.55 Ma Oi1 $\delta^{18}O$ increase is too small to account for the large deepening of the CCD.

2.   A global intensification of carbonate export to the deep sea caused by increased continental input to the oceans. Weathering rates increased in the earliest Oligocene (Robert and Kennett, 1997) and may have contributed to a global increase in carbonate production. However, global Oligocene sedimentation rates were low (Thunell and Corliss, 1986), and it is doubtful that input changes could explain the CCD drop. Rea and Lyle (2005) argued that the CCD drop could not be entirely due to shelf-basin shift and suggested that a sudden increase in weathering and erosion rates is unlikely to account for the change, thus implicating changes in deep-sea preservation.

3.   A large deep-water cooling. Cooling may have contributed to the CCD falls, but the minor amount of cooling at Oi1 time (1-2°C as discussed above) cannot fully explain this drop.

4.   A decrease of deep-ocean residence time. The links among deep-sea temperature, an increase in thermohaline circulation, and the drop in the CCD (Figure 4) are manifested as to the cause: a decrease in ocean acidity due to decreased residence time.

Following the earliest Oligocene event, Oligocene to middle Miocene ice sheets vacillated from near modern volumes to nearly fully deglaciated at times. This was a transitional period, with a wet-based Antarctic ice sheet (Marchant et al., 1993). Atmospheric CO_2 was at near-preanthropogenic levels (Pagani et al., 2005). A ~1.2-1.5‰ middle Miocene (ca. 14.8-12 Ma) $\delta^{18}O$ increase is associated with a major sea-level fall, which this event most likely represents the development of polar desert conditions with a permanent ice cap in Antarctica (Shackleton and Kennett, 1975) (Figure 2). There has been intense debate about whether this ice cap was indeed permanent, or whether it in fact disintegrated in the early Pliocene (Kennett and Hodell, 1996). We note that the amplitude of the myr-scale variability in both $\delta^{18}O$ and sea-level records appears to be lower after the middle Miocene event (Figure 2), suggesting relative stability.

QUO VADIM?

We present a review of Antarctic glacial history as interpreted from sea-level and oxygen isotopic records, updating and integrating previous reviews of sea-level change (Miller et al., 2005a), greenhouse ice sheets (Miller et al., 2005b), new data from the Eocene-Oligocene transition (Miller et al., 2008), and deep-sea circulation changes (Wright and Miller, 1996). The sea-level record was recently updated (Figure 2) by Kominz et al. (in review) who presented a thorough error analysis. We can make certain statements about sea-level changes:

1.   Very high-amplitude, myr-scale sea-level changes (up to 160 m) of EPR are not supported.

2.   Our sea-level record (Figure 2) is derived from studies of only one margin, except for the interval from 90 Ma to 100 Ma, which is corroborated by the Russian Platform (Sahagian et al., 1996). A true global sea-level curve must be derived from numerous margins and thus the curve presented must be considered a testable model. Further studies are needed to validate this record.

3.   Our best estimate of greenhouse sea-level amplitudes are 15-25 m on the myr scale, except for the ca. 71.5 Ma event, which was over 40 m. Icehouse amplitudes are as high as 55-70 m. We do not capture most lowstands in our sea-level record, and these estimates are minima.

4.   The New Jersey estimates are eustatic, having been corrected for the effects of loading. To estimate the actual volume of ice equivalent requires the making of assumptions about isostatic effects due to water loading. As noted by Pekar et al. (2002) a eustatic change of 55 m would require a change in water volume of 82 m, assuming full compensation. Similarly, in the younger record the change from 18 ka to present of 120 m measured in Barbados (Fairbanks, 1989) would have caused an 80 m eustatic change, assuming full compensation. On the myr scale considered here, compensation was likely partially achieved (Peltier, 1997) and thus the actual changes in water volume would have been higher than eustatic estimates.

5.   The link of sea-level falls and $\delta^{18}O$ increases has been documented for numerous icehouse increases. This relationship is expected because glacioeustasy is expected to drive sea-level change during this time. Further studies evaluating the link of $\delta^{18}O$ and sea level in the greenhouse are needed, though preliminary comparisons are intriguing. It must be remembered that we predict small (~0.2-0.3‰) $\delta^{18}O_{seawater}$ changes for greenhouse ice-volume changes.

Oxygen isotopic variations can be used to place constraints on Antarctic ice history. The firmest constraint is that ice sheets are required when benthic foraminiferal $\delta^{18}O$ values exceed 1.8‰ in *Cibicidoides* (Miller and Fairbanks, 1983; Miller et al., 1991). For the greenhouse world, $\delta^{18}O$ values suggest largely warm high latitudes, at least near the coast, where deep waters form. However, we need to be careful in assuming that warm coastal Antarctic temperatures indicate a continent devoid of ice sheets for the period 100-33.55 Ma. As models illustrate (DeConto and Pollard,

2003a,b), the Antarctic continent may have been remarkably diverse during the greenhouse with unglaciated coastal regions and small- to moderate-size ice sheets in the interior. Though much of the evidence may have been destroyed or buried by the modern ice sheet, the challenge now is to seek evidence for these greenhouse ice sheets. Intense study of the Antarctic continent has demonstrated the importance of ice sheets on sedimentation over the past 33 myr and provided hints of cold conditions even in the Eocene (Figure 1). Comparison with the Arctic is illustrative. Until the most sensitive area of the Arctic was drilled by IODP (Moran et al., 2006), it was assumed that the Arctic was not glaciated until the later Cenozoic and bipolar glaciation was a relatively recent event. In contrast, sampling on the Lomonsov Ridge has extended the glacial record of the Arctic back to ~46 Ma (Moran et al., 2006), and we predict that future studies of Antarctica will result in extension of small- to medium-size ice sheets back into the Late Cretaceous.

ACKNOWLEDGMENTS

We thank J. P. Kennett and S. F. Pekar for reviews and A. Cooper for suggestions. The ideas presented here were developed with help from previous reviews of sea-level change (Kominz et al., in review; Miller et al., 2005a), greenhouse ice sheets (Miller et al., 2005a), the Eocene-Oligocene transition (Miller et al., 2008), and deep-sea circulation changes (Wright and Miller, 1996); we thank co-authors of those papers, who are not listed here, for helping to develop our views of Antarctic glacial history. Supported by National Science Foundation grants EAR06-06693 (Miller) and OCE06-23256 (Katz, Miller, Wade, and Wright).

REFERENCES

Ali, J., and E. A. Hailwood. 1995. Magnetostratigraphy of upper Paleocene through lower middle Eocene strata of northwest Europe. In *Geochronology, Time Scales and Global Stratigraphic Correlation,* eds. W. A. Berggren, D. V. Kent, M.-P. Aubry, and J. Hardenbol, pp. 275-279. Special Publication. Tulsa, OK: Society for Sedimentary Geology.

Askin, R. A. 1989. Endemism and heterochroneity in the Late Cretaceous (Campanian) to Paleocene palynofloras of Seymour Island, Antarctica: Implications for origins, dispersal and palaeoclimates of southern floras. In *Origins and Evolution of the Antarctic Biota,* ed. J. A. Crame, pp. 107-119. Special Publication. Geological Society of London.

Barrera, E., and B. T. Huber. 1990. Evolution of Antarctic waters during the Maestrichtian: foraminifer oxygen and carbon isotope ratios, ODP Leg 113. In *Proceedings of the Ocean Drilling Program, Scientific Results,* vol. 113, eds. P. F. Barker and J. P. Kennett, pp. 813-823. College Station, TX: Ocean Drilling Program.

Barrera, E., and S. M. Savin. 1999. Evolution of late Campanian-Maastrichtian marine climates and oceans. *Geological Society of America Special Papers* 332:245-282.

Barrett, P. J. 2007. Cenozoic climate and sea level history from glacimarine strata off the Victoria Land coast, Cape Roberts Project, Antarctica. In *Glacial Processes and Products,* eds. M. J. Hambrey, P. Christoffersen, N. F. Glasser, and B. Hubbart. International Association of Sedimentologists Special Publication 39:259-287.

Barrett, P. J., D. P. Elston, D. M. Harwood, B. C. McKelvey, and P.-N. Webb. 1987. Mid-Cenozoic record of glaciation and sea level change on the margin of the Victoria Land Basin, Antarctica. *Geology* 15:634-637.

Barron, J. A., and B. Larsen. 1989. *Proceedings of the Ocean Drilling Program, Initial Reports,* vol. 119. College Station, TX: Ocean Drilling Program.

Berggren, W. A., D. V. Kent, C. C. Swisher, and M.-P. Aubry. 1995. A revised Cenozoic geochronology and chronostratigraphy. In *Geochronology, Time Scales and Global Stratigraphic Correlations: A Unified Temporal Framework for an Historical Geology,* eds. W. A. Berggren, D. V. Kent, M.-P. Aubry, and J. Hardenbol, pp. 129-212. Special Publication. Tulsa, OK: Society for Sedimentary Geology.

Birkenmajer, K. 1991. Tertiary glaciation in the South Shetland Islands, West Antarctica: Evaluation of data. In *Geological Evolution of Antarctica,* eds. M. R. A. Thomson, J. A. Crame, and J. W. Thomson, pp. 629-632. Cambridge: Cambridge University Press.

Birkenmajer, K., A. Gazdzicki, K. P. Krajewski, A. Przybycin, A. Solecki, A. Tatur, and H. I. Oon. 2005. First Cenozoic glaciers in West Antarctica. *Polish Polar Research* 26:3-12.

Breza, J. R., and S. W. Wise, Jr. 1992. Lower Oligocene ice-rafted debris on the Kerguelen Plateau: Evidence for East Antarctic continental glaciation. In *Proceedings of the Ocean Drilling Program, Scientific Results,* vol. 120, eds. S. W. Wise, Jr. and R. Schlich, pp. 161-178. College Station, TX: Ocean Drilling Program.

Browning, J. V., K. G. Miller, and D. K. Pak. 1996. Global implications of lower to middle Eocene sequences on the New Jersey coastal plain: The icehouse cometh. *Geology* 24:639-642.

Browning, J. V., K. G. Miller, P. P. McLaughlin, M. A. Kominz, P. J. Sugarman, D. Monteverde, M. D. Feigenson, and J. C. Hernàndez. 2006. Quantification of the effects of eustasy, subsidence, and sediment supply on Miocene sequences, Mid-Atlantic margin of the United States. *Geological Society of America Bulletin* 118:567-588.

Christie-Blick, N., G. S. Mountain, and K. G. Miller. 1990. Seismic stratigraphic record of sea-level change. In *Sea-Level Change,* pp. 116-140. Washington, D.C.: National Academy Press.

Cooper, A. K., and P. E. O'Brien. 2004. Leg 188 synthesis: Transitions in the glacial history of the Prydz Bay region, East Antarctica, from ODP drilling. In *Proceedings of the Ocean Drilling Program, Scientific Results,* vol. 188, eds. A. K. Cooper, P. E. O'Brien, and C. Richter, pp. 1-42. College Station, TX: Ocean Drilling Program.

Cooper, A. K., G. Brancolini, E. Escutia, Y. Kristoffersen, R. Larter, G. Leitchenkov, P. O'Brien, and W. Jokat. Forthcoming. Cenozoic climate history from seismic-reflection and drilling studies on the Antarctic continental margin. In *Antarctic Climate Evolution.* Amsterdam: Elsevier.

Coxall, H. K., P. A. Wilson, H. Pälike, C. H. Lear, and J. Backman. 2005. Rapid stepwise onset of Antarctic glaciation and deeper calcite compensation in the Pacific Ocean. *Nature* 433:53-57.

DeConto, R. M., and D. Pollard. 2003a. A coupled climate-ice sheet modeling approach to the early Cenozoic history of the Antarctic ice sheet. *Paleogeography, Paleoclimatology, Paleoecology* 198:39-52.

DeConto, R. M., and D. Pollard. 2003b. Rapid Cenozoic glaciation of Antarctica induced by declining atmospheric CO_2. *Nature* 421:245-249.

Diester-Haass, L., and J. Zachos. 2003. The Eocene-Oligocene transition in the Equatorial Atlantic (ODP Site 925); paleoproductivity increase and positive $\delta^{13}C$ excursion. In *Greenhouse to Icehouse: The Marine Eocene-Oligocene Transition,* eds. D. R. Prothero, L. C. Ivany, and E. A. Nesbitt. pp. 397-416, New York: Columbia University Press.

Eldrett, J. S., I. C. Harding, P. A. Wilson, E. Butler, and A. P. Roberts. 2007. Continental ice in Greenland during the Eocene and Oligocene. *Science* 446:176-179.

Emiliani, C. 1955. Pleistocene temperatures. *Journal of Geology* 63:538-578.

Fairbanks, R. G. 1989. A 17,000-year glacio-eustatic sea level record: Influence of glacial melting rates on the Younger Dryas event and deep-ocean circulation. *Nature* 342:637-642.

Falkowski, P. G., M. E. Katz, A. Knoll, A. Quigg, J. A. Raven, O. Schofield, and M. Taylor. 2004. The evolutionary history of eukaryotic phytoplankton. *Science* 305:354-360.

Finkel, Z. V., M. E. Katz, J. D. Wright, O. M. E. Schofield, and P. G. Falkowski. 2005. Climatically-driven evolutionary change in the size of diatoms over the Cenozoic. *Proceedings of the National Academy of Sciences U.S.A.* 102:8927-8932.

Fitzgerald, P. G. 2002. Tectonics and landscape evolution of the Antarctic plate since Gondwana breakup, with an emphasis on the West Antarctic rift system and the Transantarctic Mountains. In *Antarctica at the Close of a Millennium.* Proceedings of the 8th International Symposium on Antarctic Earth Science, eds. J. A. Gamble, D. N. B. Skinner, and S. Henrys. *Bulletin of the Royal Society of New Zealand* 35:453-469.

Flint, R. F. 1971. *Glacial and Quaternary Geology.* New York: John Wiley.

Francis, J. E., and I. Poole. 2002. Cretaceous and early Tertiary climates of Antarctica: Evidence from fossil wood. *Paleogeography, Paleoclimatology, Paleoecology* 182:47-64.

Hancock, J. M. 1993. Transatlantic correlations in the Campanian-Maastrichtian stages by eustatic changes of sea-level. In *High Resolution Stratigraphy,* eds. E. A. Hailwood and R. B. Kidd, pp. 241-256. Geological Society Special Publication.

Haq, B. U., J. Hardenbol, and P. R. Vail. 1987. Chronology of fluctuating sea levels since the Triassic (250 million years ago to present). *Science* 235:1156-1167.

Hays, J. D., J. Imbrie, and N. J. Shackleton. 1976. Variations in the earth's orbit: Pacemaker of the ice ages. *Science* 194:1121-1132.

Hedberg, H. D. 1976. *International Stratigraphic Guide.* New York: Wiley-Interscience.

Huber, B. T., R. D. Norris, and K. G. MacLeod. 2002. Deep sea paleotemperature record of extreme warmth during the Cretaceous. *Geology* 30:123-126.

Ivany, C. L., S. Van Simaeys, E. W. Domack, and S. D. Samson. 2006. Evidence for an earliest Oligocene ice sheet on the Antarctic Peninsula. *Geology* 34:377-380.

Kennett, J. P. 1977. Cenozoic evolution of Antarctic glaciation, the Circum-Antarctic Ocean, and their impact on global paleoceanography. *Journal of Geophysical Research* 82:3843-3860.

Kennett, J. P., and P. F. Barker. 1990. Latest Cretaceous to Cenozoic climate and oceanographic developments in the Weddell Sea, Antarctica: An ocean-drilling perspective. In *Proceedings of the Ocean Drilling Program, Scientific Results,* vol. 113, eds. P. F. Barker and J. P. Kennett, pp. 937-960. College Station, TX: Ocean Drilling Program.

Kennett, J. P., and D. A. Hodell. 1996. Stability or instability of Antarctic ice sheets during warm climates of the Pliocene? *GSA Today* 5:1-22.

Kennett, J. P., and N. J. Shackleton. 1976. Oxygen isotopic evidence for the development of the psychrosphere 38 Myr ago. *Nature* 260:513-515.

Kennett, J., and L. Stott. 1990. Proteus and Proto-Oceanus: Ancestral Paleogene oceans as revealed from Antarctic stable isotopic results, ODP Leg 113. In *Proceedings ODP Scientific Results,* vol. 113, eds. P. Barker and J. Kennett, pp. 865-880. College Station, TX: Ocean Drilling Program.

Kominz, M. A., K. G. Miller, and J. V. Browning. 1998. Long-term and short-term global Cenozoic sea-level estimates. *Geology* 26:311-314.

Kominz, M. A., J. V. Browning, K. G. Miller, P. J. Sugarman, S. Mizintseva, A. Harris, and C. R. Scotese. In review. Late Cretaceous to Miocene sea-level estimates from the New Jersey and Delaware coastal plain coreholes: An error analysis. *Basin Research.*

Larsen, H. C., A. D. Saunders, P. D. Clift, J. Beget, W. Wei, S. Spezzaferri, and ODP Leg 152 Scientific Party. 1994. Seven million years of glaciation in Greenland. *Science* 264:952-955.

Lear, C. H., and Y. Rosenthal. 2006. Benthic foraminiferal Li/Ca: Insights into Cenozoic seawater carbonate saturation state. *Geology* 34:985-988.

Leckie, M., and P.-N. Webb. 1986. Late Paleogene and early Neogene foraminifers of Deep Sea Drilling Project Site 270, Ross Sea, Initial Reports DSDP 90:1093-1142. Washington, D.C.: U.S. Government Printing Office.

LeMasurier, W. E., and D. C. Rex. 1982. Volcanic record of Cenozoic glacial history in Marie Byrd Land and western Ellsworth Land: Revised chronology and evaluation of tectonic factors. In *Antarctic Geoscience,* ed. C. Craddock, pp. 725-732. Madison: University of Wisconsin Press.

Mancini, E. A., and B. H. Tew. 1995. Geochronology, biostratigraphy and sequence stratigraphy of a marginal marine to marine shelf stratigraphic succession: Upper Paleocene and lower Eocene, Wilcox Group, eastern Gulf Coastal Plain, U.S.A. In *Geochronology, Time Scales and Global Stratigraphic Correlations: A Unified Temporal Framework for an Historical Geology,* eds. W. A. Berggren, D. V. Kent, M.-P. Aubry, and J. Hardenbol, pp. 281-293. Special Publication 54. Tulsa, OK: Society for Sedimentary Geology.

Marchant, D. R., G. H. Denton, and C. C. Swisher. 1993. Miocene-Pliocene-Pleistocene glacial history of Arena Valley, Quartermain Mountains, Antarctica. *Geografiska Annaler* 75A:269-302.

Margolis, S. V., and J. P. Kennett. 1971. Cenozoic paleoglacial history of Antarctica recorded in subantarctic deep-sea cores. *American Journal of Science* 271:1-36.

Matthews, R. K. 1984. Oxygen-isotope record of ice-volume history: 100 million years of glacio-eustatic sea-level fluctuation. *Memoirs of the American Association of Petroleum Geologists* 36:97-107.

Matthews, R. K., and C. Frohlich. 2002. Maximum flooding surfaces and sequence boundaries: Comparisons between observations and orbital forcing in the Cretaceous and Jurassic (65-190 Ma). *GeoArabia* 7:503-538.

Matthews, R. K., and R. Z. Poore. 1980. Tertiary $\delta^{18}O$ record and glacio-eustatic sea-level fluctuations. *Geology* 8:501-504.

Miall, A. D. 1991. Stratigraphic sequences and their chronostratigraphic correlation. *Journal of Sedimentary Petrology* 61:497-505.

Miller, K. G. 1992. Middle Eocene to Oligocene stable isotopes, climate, and deep-water history: The Terminal Eocene Event? In *Eocene-Oligocene Climatic and Biotic Evolution,* eds. D. Prothero and W. A. Berggren, pp. 160-177. Princeton: Princeton University Press.

Miller, K. G., and R. G. Fairbanks. 1983. Evidence for Oligocene-middle Miocene abyssal circulation changes in the western North Atlantic. *Nature* 306:250-252.

Miller, K. G., and R. G. Fairbanks. 1985. Oligocene to Miocene carbon isotope cycles and abyssal circulation changes. In *The Carbon Cycle and Atmospheric CO2: Natural Variations Archean to Present,* eds. E. T. Sundquist and W. S. Broecker, pp. 469-486. Washington, D.C.: American Geophysical Union.

Miller, K. G., and M. E. Katz. 1987. Oligocene-Miocene benthic foraminiferal and abyssal circulation changes in the North Atlantic. *Micropaleontology* 33:97-149.

Miller, K. G., and B. E. Tucholke. 1983. Development of Cenozoic abyssal circulation south of the Greenland-Scotland Ridge. In *Structure and Development of the Greenland-Scotland Ridge,* eds. M. H. P. Bott, S. Saxov, M. Talwani, and J. Thiede, pp. 549-589. New York: Plenum Press.

Miller, K. G., R. G. Fairbanks, and G. S. Mountain. 1987. Tertiary oxygen isotope synthesis, sea level history, and continental margin erosion. *Paleoceanography* 2:1-19.

Miller, K. G., J. D. Wright, and R. G. Fairbanks. 1991. Unlocking the Ice House: Oligocene-Miocene oxygen isotopes, eustasy, and margin erosion. *Journal of Geophysical Research* 96:6829-6848.

Miller, K. G., G. S. Mountain, the Leg 150 Shipboard Party, and Members of the New Jersey Coastal Plain Drilling Project. 1996. Drilling and dating New Jersey Oligocene-Miocene sequences: Ice volume, global sea level, and Exxon records. *Science* 271:1092-1094.

Miller, K. G., G. S. Mountain, J. V. Browning, M. A. Kominz, P. J. Sugarman, N. Christie-Blick, M. E. Katz, and J. D. Wright. 1998. Cenozoic global sea-level, sequences, and the New Jersey transect: Results from coastal plain and slope drilling. *Reviews of Geophysics* 36:569-601.

Miller, K. G., E. Barrera, R. K. Olsson, P. J. Sugarman, and S. M. Savin. 1999. Does ice drive early Maastrichtian eustasy? *Geology* 27:783-786.

Miller, K. G., P. J. Sugarman, J. V. Browning, M. A. Kominz, J. C. Hernàndez, R. K. Olsson, J. D. Wright, M. D. Feigenson, and W. Van Sickel. 2003. Late Cretaceous chronology of large, rapid sea-level changes: Glacioeustasy during the greenhouse world. *Geology* 31:585-588.

Miller, K. G., P. J. Sugarman, J. V. Browning, M. A. Kominz, R. K. Olsson, M. D. Feigenson, and J. C. Hernàndez. 2004. Upper Cretaceous sequences and sea-level history, New Jersey coastal plain. *Geological Society of America Bulletin* 116:368-393.

Miller, K. G., M. A. Kominz, J. V. Browning, J. D. Wright, G. S. Mountain, M. E. Katz, P. J. Sugarman, B. S. Cramer, N. Christie-Blick, and S. F. Pekar. 2005a. The Phanerozoic record of global sea-level change. *Science* 310:1293-1298.

Miller, K. G., J. D. Wright, and J. V. Browning. 2005b. Visions of ice sheets in a greenhouse world. *Marine Geology* 217:215-231.

Miller, K. G., J. V. Browning, M.-P. Aubry, B. S. Wade, M. E. Katz, A. A. Kulpecz, and J. D. Wright. 2008. Eocene-Oligocene global climate and sea-level changes: St. Stephens Quarry, Alabama. *Geological Society of America Bulletin* 120:34-53.

Moore, T. C., Jr., J. Backman, I. Raffi, C. Nigrini, A. Sanfilippo, H. Pälike, and M. Lyle. 2004. Paleogene tropical Pacific: Clues to circulation, productivity and plate motion. *Paleoceanography* 19, doi:10.1029/2003PA000998.

Moran, K., J. Backman, H. Brinkhuis, S. C. Clemens, T. Cronin, G. R. Dickens, F. Eynaud, J. Gattacceca, M. Jakobsson, R. W. Jordan, M. Kaminski, J. King, N. Koc, A. Krylov, N. Martinez, J. Matthiessen, D. McInroy, T. M. Moore, J. Onodera, M. O'Regan, H. Palike, B. Rea, D. Rio, T. Sakamoto, D. C. Smith, R. Stein, K. St John, I. Suto, N. Suzuki, K. Takahashi, M. Watanabe, M. Yamamoto, J. Farrell, M. Frank, P. Kubik, W. Jokat, and Y. Kristoffersen. 2006. The Cenozoic palaeoenvironment of the Arctic Ocean. *Nature* 441:601-605.

Moriya, K., P. A. Wilson, O. Friedrich, J. Erbacher, and H. Kawahata. 2007. Testing for ice sheets during the mid-Cretaceous greenhouse using glassy foraminiferal calcite from the mid-Cenomanian tropics on Demerara Rise. *Geology* 35:615-618.

Mountain, G. S., and K. G. Miller. 1992. Seismic and geologic evidence for early Paleogene deep-water circulation in the western North Atlantic. *Paleoceanography* 7:423-439.

Pagani, M., J. Zachos, K. H. Freeman, S. Bohaty, and B. Tipple. 2005. Marked change in atmospheric carbon dioxide concentrations during the Oligocene. *Science* 309:600-603.

Pak, D. K., and K. G. Miller. 1992. Paleocene to Eocene benthic foraminiferal isotopes and assemblages: Implications for deepwater circulation. *Paleoceanography* 7:405-422.

Pekar, S., and K. G. Miller. 1996. New Jersey Oligocene "Icehouse" sequences (ODP Leg 150X) correlated with global δ^{18}O and Exxon eustatic records. *Geology* 24:567-570.

Pekar, S. F., N. Christie-Blick, M. A. Kominz, and K. G. Miller. 2002. Calibration between eustatic estimates from backstripping and oxygen isotopic records for the Oligocene. *Geology* 30:903-906.

Peltier, W. R. 1997. Postglacial variations in the level of the sea: Implications for climate dynamics and solid-earth geophysics. *Reviews of Geophysics* 36:603-689.

Pitman, W. C., III, and X. Golovchenko. 1983. The effect of sea-level change on the shelf edge and slope of passive margins. Special Publication 33, pp. 41-58. Society of Economic Paleontologists and Mineralogists.

Rea, D. K., and M. W. Lyle. 2005. Paleogene calcite compensation depth in the eastern 340 subtropical Pacific; answers and questions. *Paleoceanography* 20:1-9.

Robert, C., and J. P. Kennett. 1997. Antarctic continental weathering changes during Eocene-Oligocene cryosphere expansion: Clay mineral and oxygen isotope evidence *Geology* 25:587-590.

Royer, D., R. A. Berner, I. P. Montanez, N. J. Tabor, and D. J. Beerling. 2004. CO2 as a primary driver of Phanerozoic climate. *GSA Today* 14:4-10.

Sahagian, D., O. Pinous, A. Olferiev, V. Zakaharov, and A. Beisel. 1996. Eustatic curve for the Middle Jurassic-Cretaceous based on Russian platform and Siberian stratigraphy: Zonal resolution. *American Association of Petroleum Geologists Bulletin* 80:1433-1458.

Salamy, K. A., and J. C. Zachos. 1999. Late Eocene-Early Oligocene climate change on southern ocean fertility: Inferences from sediment accumulation and stable isotope data. *Paleogeography, Paleoclimatology, Paleoecology* 145:79-93.

Savin, S. M., R. G. Douglas, and F. G. Stehli. 1975. Tertiary marine paleotemperatures. *Geological Society of America Bulletin* 86:1499-1510.

Shackleton, N. J. 1967. Oxygen isotope analyses and Pleistocene temperatures re-assessed. *Nature* 215:15-17.

Shackleton, N. J., and J. P. Kennett. 1975. Paleotemperature history of the Cenozoic and the initiation of Antarctic glaciation, oxygen and carbon isotope analyses in DSDP sites 277, 279, and 281. In *Init. Reports DSDP*, eds. J. P. Kennett, R. E. Houtz et al., pp. 743-755. Washington, D.C.: U.S. Government Printing Office.

Shackleton, N. J., J. Backman, H. Zimmerman, D. V. Kent, M. A. Hall, D. G. Roberts, D. Schnitker, J. G. Baldauf, A. Desprairies, R. Homrighausen, P. Huddlestun, J. B. Keene, A. J. Kaltenback, K. A. O. Krumsiek, A. C. Morton, J. W. Murray, and J. Westberg-Smith. 1984. Oxygen isotope calibration of the onset of ice-rafting and history of glaciation in the North Atlantic region. *Nature* 307:620-623.

Stoll, H. M., and D. P. Schrag. 1996. Evidence for glacial control of rapid sea level changes in the Early Cretaceous. *Science* 272:1771-1774.

Stoll, H. M., and D. P. Schrag. 2000. High resolution stable isotope records from the Upper Cretaceous rocks of Italy and Spain: Glacial episodes in a greenhouse planet? *Geological Society of America Bulletin* 112:308-319.

Strand, K., S. Passchier, and J. Nasi. 2003. Implications of quartz grain microtextures for onset Eocene/Oligocene glaciation in Prydz Bay, ODP Site 1166, Antarctica. *Paleogeography, Paleoclimatology, Paleoecology* 198:101-111.

Suko, I. 2006. The explosive diversification of the diatom genus *Chaetoceros* across the Eocene/Oligocene and Oligocene/Miocene boundaries in the Norwegian Sea. *Marine Micropaleontology* 58:259-269.

ten Brink, U., R. Hackney, S. Bannister, T. Stern, and Y. Makovsky. 1997. Uplift of the Transantarctic Mountains and the bedrock beneath the East Antarctic ice sheet. *Journal of Geophysical Research* 102: 27603-27621.

Thomas, E., ed. 1990. *Late Cretaceous through Neogene Deep-Sea Benthic Foraminifers (Maud Rise, Weddell Sea, Antarctica)*. College Station, TX: Ocean Drilling Program.

Thorne, J. A., and A. B. Watts. 1984. Seismic reflectors and unconformities at passive continental margins. *Nature* 311:365-368.

Thunell, R. C., and B. H. Corliss. 1986. Late Eocene-early Oligocene carbonate sedimentation in the deep sea. In *Terminal Eocene Events*, eds. C. Pomerol and I. Premoli-Silva, pp. 363-380. Amsterdam: Elsevier.

Tripati, A., J. Backman, H. Elderfield, and P. Ferretti. 2005. Eocene bipolar glaciation associated with global carbon cycle changes. *Nature* 436:341-346.

Troedson, A. L., and J. B. Riding. 2002. Upper Oligocene to lowermost Miocene strata of King George Island, South Shetland Islands, Antarctica: Stratigraphy, facies analysis and implications for the glacial history of the Antarctic Peninsula. *Journal of Sedimentary Research* 72:510-523.

Troedson, A. L., and J. L. Smellie. 2002. The Polonez Cove Formation of King George Island, Antarctica: Stratigraphy, facies and implications for mid-Cenozoic cryosphere development. *Sedimentology* 49:277-301.

Tucholke, B. E., and G. S. Mountain. 1979. Seismic stratigraphy, lithostratigraphy and paleosedimentation patterns in the North American basin. In *Deep Drilling Results in the Atlantic Ocean: Continental Margins and Paleoenvironment*, eds. M. W. Talwani, W. Hay, and W. B. F. Ryan, pp. 58-86. Washington, D.C.: American Geophysical Union.

Vail, P. R., R. M. Mitchum, R. G. Todd, J. M. Widmier, S. Thompson III, J. B. Sangree, J. N. Bubb, and W. G. Hatlelid. 1977. Seismic stratigraphy and global changes of sea level. In *Seismic Stratigraphy—Applications to Hydrocarbon Exploration,* ed. C. E. Payton. *Memoirs of the American Association of Petroleum Geologists* 26:49-205. Tulsa, OK: AAPG.

van Andel, T. H. 1975. Mesozoic/Cenozoic calcite compensation depth and global distribution of calcareous sediments. *Earth and Planetary Science Letters* 26:187-194.

Van Sickel, W. A., M. A. Kominz, K. G. Miller, and J. V. Browning. 2004. Late Cretaceous and Cenozoic sea-level estimates: Backstripping analysis of borehole data, onshore New Jersey. *Basin Research* 16:451-465.

Wade, B. S., and H. Pälike. 2004. Oligocene climate dynamics. *Paleoceanography* 19:PA4019. doi:10.1029/2004PA001042.

Wei, W. 1992. Calcareous nannofossil stratigraphy and reassessment of the Eocene glacial record in subantarctic piston cores of the southeast Pacific. In *Proceedings of the ODP Scientific Results*, eds. S. W. Wise, Jr. and R. Schlich, pp. 1093-1104. College Station, TX: Ocean Drilling Program.

Williams, R. S., and J. G. Ferrigno. 1995. Satellite image atlas of glaciers of the world—Greenland. *U.S. Geological Survey Professional Paper* 1386-C: 141 pp, http://pubs.usgs.gov/pp/p1386c/p1386c.pdf.

Wise, S. W., Jr., J. R. Breza, D. M. Harwood, and W. Wei. 1991. Paleogene glacial history of Antarctica. In *Controversies in Modern Geology: Evolution of Geological Theories in Sedimentology, Earth History and Tectonics,* eds. D. W. Miller, J. A. McKenzie, and H. Weissert, pp. 133-171. London: Academic Press.

Wright, J. D., and K. G. Miller. 1996. Control of North Atlantic deep water circulation by the Greenland-Scotland Ridge. *Paleoceanography* 11:157-170.

Zachos, J. C., J. Breza, and S. W. Wise. 1992. Earliest Oligocene ice-sheet expansion on East Antarctica: Stable isotope and sedimentological data from Kerguelen Plateau. *Geology* 20:569-573.

Zachos, J. C., L. D. Stott, and K. C. Lohmann. 1994. Evolution of early Cenozoic marine temperatures. *Paleoceanography* 9:353-387.

Zachos, J. C., T. M. Quinn, and K. A. Salamy. 1996. High resolution (104 yr) deep-sea foraminiferal stable isotope records of the earliest Oligocene climate transition. *Paleoceanography* 9:353-387.

Cooper, A. K., P. J. Barrett, H. Stagg, B. Storey, E. Stump, W. Wise, and the 10th ISAES editorial team, eds. (2008). *Antarctica: A Keystone in a Changing World.* Proceedings of the 10th International Symposium on Antarctic Earth Sciences. Washington, DC: The National Academies Press.

Late Cenozoic Climate History of the Ross Embayment from the AND-1B Drill Hole: Culmination of Three Decades of Antarctic Margin Drilling

T. R. Naish,[1,2] R. D. Powell,[3] P. J. Barrett,[1] R. H. Levy,[4] S. Henrys,[1] G. S. Wilson,[5] L. A. Krissek,[6] F. Niessen,[7] M. Pompilio,[8] J. Ross,[9] R. Scherer,[3] F. Talarico,[10] A. Pyne,[1] and the ANDRILL-MIS Science team[11]

ABSTRACT

Because of the paucity of exposed rock, the direct physical record of Antarctic Cenozoic glacial history has become known only recently and then largely from offshore shelf basins through seismic surveys and drilling. The number of holes on the continental shelf has been small and largely confined to three areas (McMurdo Sound, Prydz Bay, and Antarctic Peninsula), but even in McMurdo Sound, where Oligocene and early Miocene strata are well cored, the late Cenozoic is poorly known and dated. The latest Antarctic geological drilling program, ANDRILL, successfully cored a 1285-m-long record of climate history spanning the last 13 m.y. from subsea-floor sediment beneath the McMurdo Ice Shelf (MIS), using drilling systems specially developed for operating through ice shelves. The cores provide the most complete Antarctic record to date of ice-sheet and climate fluctuations for this period of Earth's history. The >60 cycles of advance and retreat of the grounded ice margin preserved in the AND-1B record the evolution of the Antarctic ice sheet since a profound global cooling step in deep-sea oxygen isotope records ~14 m.y.a. A feature of particular interest is a ~90-m-thick interval of diatomite deposited during the warm Pliocene and representing an extended period (~200,000 years) of locally open water, high phytoplankton productivity, and retreat of the glaciers on land.

[1]Antarctic Research Centre, Victoria University of Wellington, Wellington, New Zealand (t.naish@gns.cri.nz, peter.barrett@vuw.ac.nz, alex. pyne@vuw.ac.nz).

[2]Geological and Nuclear Sciences, Lower Hutt, New Zealand (t.naish@ gns.cri.nz, s.henrys@gns.cri.nz).

[3]Department of Geology and Environmental Geosciences, Northern Illinois University, DeKalb, IL, USA (ross@geol.niu.edu).

[4]ANDRILL Science Management Office, University of Nebraska-Lincoln, 126 Bessey Hall, Lincoln, NE 68588-0341, USA (rlevy2@unl. edu).

[5]Department of Geology, University of Otago, PO Box 56, Dunedin, New Zealand (gary.wilson@otago.ac.nz).

[6]Department of Geosciences, The Ohio State University, Columbus, OH, USA (krissek@mps.ohio-state.edu).

[7]Department of Marine Geophysics, Alfred Wegener Institute, Postfach 12 01 61, Columbusstrasse, D-27515, Bremerhaven, Germany (fniessen@ awi-bremerhaven.de).

[8]Istituto Nazionale di Geofisica e Vulcanologia, Via della Faggiola, 32, [9]I-56126 Pisa, Italy (pompilio@pi.ingv.it).

New Mexico Geochronology Research Laboratory, Socorro, NM 87801, USA (jirhiker@nmt.edu).

[10]Università di Siena, Dipartimento di Scienze delle Terra, Via Laterina 8, I-53100 Siena, Italy (talarico@unisi.it).

[11]See http://www.andrill.org/support/references/appendixc.html.

HISTORICAL OVERVIEW

The remarkable late Cenozoic record of glacial history in the Ross Embayment recovered in late 2006 by the ANDRILL-MIS Project is the culmination of work begun over three decades ago to document and understand the more recent glacial history of Antarctica by drilling close to the margin. Ironically, although the last 3 million years of Earth's climate is often said to be the best studied interval of the Cenozoic, the contribution of Antarctic ice volume changes is the most poorly understood. Until this most recent hole was drilled, the middle Cenozoic record was better known from several holes in both the McMurdo region and Prydz Bay (Table 1). This is in general because the older strata were exposed closer to the coast through basin uplift, where younger strata had been removed by Neogene erosion, but also because in offshore basins glacial debris from the last glacial advance

TABLE 1 Antarctic Coastal and Continental Shelf Rock-Drilling Sites, 1973 to 2006

Project	Year	Site	Lat	Long	Elev (+) or Wat.Dep (-)	Depth cored	% recov.	Oldest core	Reference
Ross Sea									
DSDP 28	1973	270	77°26'S	178°30'W	-634 m	423 m	62%	gneiss - E Paleozoic	Hays, Frakes et al., 1975
		271	77°26'S	178°30'W	-562 m	233 m	7%	diamict clasts - E Pliocene	
		272	77°26'S	178°30'W	-629 m	439 m	37%	diamict - E Miocene	
		273	77°26'S	178°30'W	-491 m	333 m	25%	diamict - E Miocene	
McMurdo Sound area - onshore									
DVDP	1973	1	77°50'S	166°40'E	67 m	201 m	98%	basalt - L Quat	Kyle, 1981
	1973	2	77°51'S	166°40'E	47 m	179 m	96%	basalt - L Quat	
	1973	3	77°51'S	166°40'E	48 m	381 m	90%	basalt - L Quat	
	1974	10	77°35'S	163°31'E	3 m	182 m	83%	diamict - L Miocene	Powell, 1981
	1974	11	77°35'S	163°25'E	80.2 m	328 m	94%	diamict - L Miocene	
	1974	12	77°38'S	162°51'E	75.1 m	185 m	98%	migmatite - E Paleozoic	
McMurdo Sound area - offshore									
DVDP 15	1975	15	77°26'S	164°23'E	-122 m	62 m	52%	black sand - E Pleist	Barrett and Treves, 1981
MSSTS	1979	1	77°34'S	163°23'E	-195 m	230 m	62%	mudstone - L Oligocene	Barrett, 1986
CIROS	1986	1	77°05'S	164°30'E	-197 m	702 m	98%	boulder congl - L Eocene	Barrett, 1989
CIROS	1984	2	77°41'S	163°32'E	-211 m	168 m	67%	gneiss - E Paleozoic	Barrett and Hambrey, 1992
CRP	1997	1	77°00'S	163°45'E	-154 m	148 m	86%	diamict - E Miocene	CRST, 1998
	1998	2	77°00'S	163°43'E	-178 m	624 m	95%	mudstone - Oligocene	CRST, 1999
	1999	3	77°00'S	163°43'E	-295 m	939 m	97%	sandstone - Devonian	CRST, 2000
ANDRILL	2006	1	77°55'S	167°01'E	-840 m	1285 m	98%	basalt - E Miocene	Naish et al., 2006
Prydz Bay									
ODP 119	1988	739	67°17'S	75°05'E	-412 m	487 m	34%	diamict - L Eo-E Oligocene	Barron, Larsen et al., 1988
	1988	740	68°41'S	76°43'E	-808 m	226 m	32%	red beds - ?Triassic	
	1988	741	68°23'S	76°23'E	-551 m	128 m	26%	sandst, siltst -?E Cretaceous	
	1988	742	67°33S	75°24'E	-416 m	316 m	53%	mudst, diamict - ?Eo-Olig	
	1988	743	66°55'S	74°42'E	-989 m	97 m	22%	diamict - Pleistocene	
ODP 188	2000	1166	67°42'S	74°47'E	-475 m	381 m	19%	claystone - L Cretaceous	O'Brien, Cooper, Richter et al., 2001
Antarctic Peninsula									
ODP 178	1998	1097	66°24'S	70°45'W	-563 m	437 m	14%	diamict - E Pliocene	Barker, Camerlenghi, Acton et al., 1999
	1998	1098	64°52'S	64°12'W	-1010 m	47 m	99%	mud - Holocene	
	1998	1099	64°57'S	64°19'W	-1400 m	108 m	102%	mud - Holocene	
	1998	1100	66°53'S	65°42'W	-459 m	111 m	5%	diamict - Pleistocene	
	1998	1102	66°48'S	65°51'W	-431 m	15 m	6%	diamict - Pleistocene?	
	1998	1103	64°00'S	65°28'W	-494 m	363 m	12%	diamict - L Miocene	
SHALDRIL	2005	1	62°17'S	58°45'W	-488 m	108 m	87%	mud - L Pleistocene	http://shaldril.rice.edu/
SHALDRIL	2006	3	63°51'S	54°39'W	-340 m	20 m	32%	mudst - L Eo/E Oligocene	Anderson et al., 2007
	2006	5	63°15'S	52°22'W	-506 m	23 m	40%	muddy sand - mid Miocene	
	2006	6	63°20'S	52°22'W	-532 m	21 m	n/a	muddy sand - E Pliocene	
	2006	12	63°16'S	52°50'W	-442 m	4 m	64%	mudst - Oligocene	

prevented penetration and recovery of the older record from gravity coring (Anderson, 1999). Plainly deep-drilling technology was required from an appropriate location. Here we summarize the history and rationale behind this successful outcome.

By the early 1970s the International Geophysical Year (1956-1958) had already resulted in a vast increase in knowledge of rock types, ages, and history of the continent itself (Bushnell and Craddock, 1970), but the history of its ice sheet was still a mystery. This changed in early 1973 with the voyages of the *Glomar Challenger* to the South Indian Ocean and Ross Sea (Leg 28) (Hayes et al., 1975) and the Tasman Sea and Southwest Pacific Ocean (Leg 29) (Kennett et al., 1974). The first of these cruises showed that the Antarctic ice sheet, which had previously been seen as a Quaternary feature, had been in existence since at least the Oligocene. The second yielded the first oxygen isotope measurements on deep-sea calcareous microfossils, providing the first evidence of dramatic ocean cooling and global ice volume increase at the Eocene-Oligocene boundary (Shackleton and Kennett, 1974).

The 1970s was also a period in which detailed paleontological and chronological studies of cores from continuously deposited deep-sea sediments displaced the long-held view of four major Quaternary glaciations (Flint, 1971), showing much more frequent cycles of climate, ice volume, and sea level in the late Quaternary (Hays et al., 1976) in response to variations in Earth's orbital parameters calculated by Milankovitch. How far back in time these should be evident was not clear, but there seemed no reason why they should not have influenced the Antarctic ice sheet from the time of its inception.

While deep ocean sediments were useful for their continuous record of past ocean chemistry, providing an ice volume-temperature signal, they could not provide information on the extent of ice or regional climate in the high latitudes. This could only come out of sediment cores from the Antarctic margin, where the direct influence of ice advance and retreat (and perhaps also sea-level fall and rise) could be obtained.

In the early 1970s two drilling platforms were available. The *Glomar Challenger* operated by the Deep Sea Drilling Project and a land-based system put together for the Dry Valley Drilling Project, an initiative of the United States, Japan, and New Zealand to explore the late Cenozoic history of the McMurdo Dry Valleys (Smith, 1981). Over the next three decades both of these platforms (and their successors) ran in parallel (Table 1, Figure 1) with varying degrees of success.

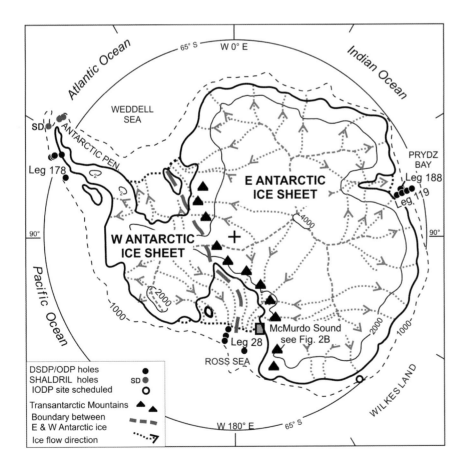

FIGURE 1 Locations of geological drill sites on land and on the Antarctic continental shelf. Details and references are given in Table 1. McMurdo Sound drill sites are shown in Figure 2A.

The ship-based system had the advantage that it could be deployed at a number of places around the Antarctic margin, but was limited by ice conditions and poor recovery of glacimarine sediments from near-shore shelf basins. Nevertheless, the cores taken have provided useful constraints on the Cenozoic history of the Antarctic ice margin and climate from the Ross Sea, Prydz Bay, and Antarctic Peninsula sectors (Table 1). The land-based system, once adapted for sea ice, had the advantage that it could yield long and continuous core with near-complete recovery. However, it was rather cumbersome and required an ice platform that was firmly tied to land. Although the system has been improved through the MSSTS-1, CIROS, and Cape Roberts Projects (Table 1), continuing to exploit the fast-ice rim around McMurdo Sound as a drilling platform, attractive drilling targets beneath fast sea ice elsewhere on the Antarctic margin have yet to be identified.

The results from both ship-based and sea-ice-based drilling have provided a framework for the Cenozoic history of the Antarctic ice sheet (Kennett and Warnke, 1992, 1993; Barrett, 1999), but more detail has become known for Oligocene and early Miocene times, especially from the 1500 m of strata cored off Cape Roberts (Naish et al., 2001; Barrett, 2007), than in subsequent times. The first results of the shift from sea-ice- to shelf-ice-based drilling reported below add a great deal to the late Cenozoic story.

THE ANDRILL MCMURDO ICE-SHELF PROJECT

The aim of the MIS Project was to obtain a continuous sediment core through approximately 1200 m of Neogene (~0-10 Ma) glacimarine and volcanic sediment that had accumulated in the Windless Bight region of the MIS (Figure 2A). The present-day MIS forms the northwest part of Ross Ice Shelf where it has been pinned by Ross Island for the last ~10 ka (McKay et al., 2007), and is nourished by ice sourced locally and from the East Antarctic ice sheet (EAIS) outlet glaciers in the southern Transantarctic Mountains (TAM). The drill site was situated above a flexural moat basin formed in response to Quaternary volcanic loading of the crust by Ross Island, superimposed on regional subsidence associated with Neogene extension of the Terror Rift (Horgan et al., 2005; Naish et al., 2006) (Figure 2B).

Between October 29 and December 26, 2006, a single 1284.87-m-deep drill core (AND-1B) was recovered from the bathymetric and depocentral axis of the moat in 943 m of water from an ice-shelf platform. The drilling technology employed a sea-riser system in a similar fashion to the Cape Roberts Project (CRP), but utilized a combination of soft sediment coring in upper soft sediments and continuous wire line diamond-bit coring. Innovative new technology, in the form of a hot-water drill and over-reamer, was used to make an access hole through 85 m of ice and to keep the riser free during drilling operations.

The MIS project has two key scientific objectives:

1.   Provide new knowledge on the late Neogene behavior and variability of the Ross Ice Shelf and Ross Ice Sheet and the West Antarctic ice sheet, and their influence on global climate, sea-level, and ocean circulation.

2.   Provide new knowledge on the Neogene tectonic evolution of the West Antarctic Rift System, Transantarctic Mountains, and associated volcanism.

A key outcome of the project will be to provide age control for, and determine the environmental significance of, seismic reflectors that have been mapped regionally within the Victoria Land Basin (Fielding et al., 2007; Henrys et al., 2007) in order to assess the regional impact of global climatic and local tectonic events. A second key outcome of the project will be to use paleoclimatic proxies and boundary conditions to help constrain numerical climate and dynamical ice-sheet models. This paper presents an overview of the MIS Project, and the details are reported in the Initial Results volume (Naish et al., 2007).

TECTONIC AND STRATIGRAPHIC SETTING

Ross Island lies at the southern end of the Victoria Land Basin (VLB), a ~350-km-long, half-graben hinged on its western side at the TAM front (Figure 2). Major rifting in the VLB has occurred since the latest Eocene, perhaps having been initiated in the Cretaceous (Cooper and Davey, 1985; Brancolini et al., 1995), and has accommodated up to 10 km of sediment. A new rift history, based on the CRP drill cores linked to a new regional seismic stratigraphic framework (Fielding et al., 2007; Henrys et al., 2007), indicates that crustal stretching during the Oligocene syn-rift phase produced rapid subsidence, followed by thermally controlled slower subsidence in the early Miocene. Renewed rifting within the center of the VLB beginning in the late Miocene has continued through to present day. This forms the Terror Rift (Cooper et al., 1987) and is associated with alkalic igneous intrusions and extrusive volcanism (e.g., Beaufort Island and Ross Island). Quaternary loading of the crust by the Ross Island volcanoes has added significantly to subsidence near Ross Island, and the development of an enclosing moat (Stern et al., 1991). The Terror Rift has accommodated up to 3 km of Neogene sediment beneath Windless Bight. Here the load-induced subsidence caused by Ross Island has contributed significantly to the generation of accommodation space, especially during the last 2 m.y. (Horgan et al., 2005).

Neogene strata have now been extensively mapped in southern McMurdo Sound from the Drygalski Ice Tongue south to Ross Island. These Neogene strata show a thickening and eastward-dipping succession extending under Ross Island in the vicinity of the MIS Project drill site. Analysis of these strata, which have now been sampled by MIS project drilling, will contribute significantly to the young tectonic history of the West Antarctic Rift System.

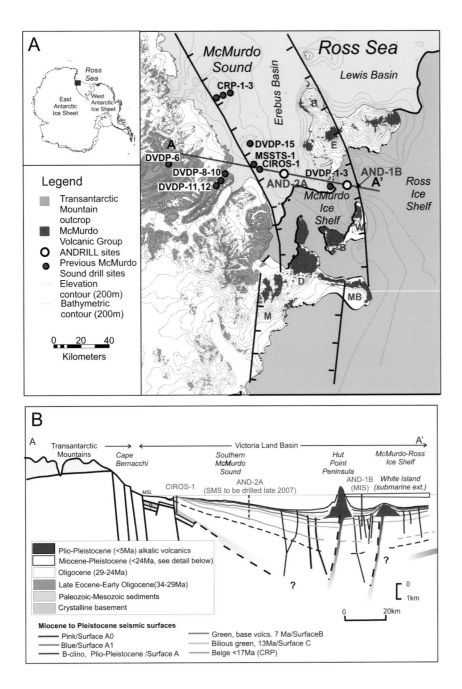

FIGURE 2 (A) Location of key geographical, geological, and tectonic features in southern McMurdo Sound. Volcanic centers of the Erebus Volcanic Province include Mt. Erebus (E), Mt. Terror (T), Mt. Bird (B), White Island (W), Black Island (B), Mt. Discovery (D), Mt. Morning (M), and Minna Bluff (MB). Boundary faults of the southern extension of Terror Rift are also shown. Location of ANDRILL Program drill sites and previous programs (DVDP, CIROS, MSST) are shown. (B) Schematic structural-stratigraphic cross-section across the VLB (located in [A] as "A-A") shows the stratigraphic context of the MIS and SMS drill sites with respect to previous drilling in Southern McMurdo Sound. The cross-section is compiled from interpreted seismic reflection data, previous drill core data from MSSTS-1 and CIROS-1 (Barrett, 1986, 1989), and models for the evolution of the VLB.

CHRONOSTRATIGRAPHY OF "AND-1B"

A preliminary age model for the upper 700 m of drill core constructed from diatom biostratigraphy (Scherer et al., 2007) and radiometric ages on volcanic material (Ross et al., 2007) allows a unique correlation of ~70 percent of the magnetic polarity stratigraphy with the Geomagnetic Polarity Time Scale (Wilson et al., 2007). The age model provides several well-constrained intervals displaying relatively rapid (<1m/k.y.) and continuous accumulation of sediment punctuated by several 0.5 m.y. to 1.0 m.y. stratal hiatuses

representing more than half of the last 7 Ma. Thus the AND-1B record provides several highly resolved "windows" into the development of the Antarctic ice sheets during the late Cenozoic. Strata below ~620 mbsf are late Miocene in age (5-13 m.y.). At the time of writing, the chronostratigraphic data available for this interval include three radiometric ages on volcanic clasts from near 1280 mbsf constraining the age for the base of the AND-1B drill core to <13.5 m.y. Work continues to improve the age control on the lower part of the cored interval.

RELATIONSHIP TO REGIONAL SEISMIC STRATIGRAPHY

Prior to drilling the MIS target, five distinctive reflectors marking regional stratal discontinuities had been mapped through a grid of seismic data in the vicinity of the drill site (HPP and MIS lines) (Naish et al., 2006) (Figure 2), and linked to marine seismic reflection data and reflector nomenclature in McMurdo Sound (Fielding et al., 2007). We have used whole-core velocity measurements and VSP first arrival travel-time picks (see Hansaraj et al., 2007; Naish et al., 2007) to derive a time-depth conversion curve to convert the seismic reflection section to depth. We have summarized in Figure 3 and below our correlation of the regional seismic stratigraphy with the AND-1B drill core.

1. **Rg (Surface C, bilious green reflector):** This regionally extensive discontinuity is correlated with top of a ~60-m-thick interval late Miocene volcanic sandstone (LSU 7) and the base of a 150-m-thick, high-velocity (3000 ms^{-1}) interval of diamictite cycles (LSU 6.4). ^{40}Ar/^{39}Ar dates on ashes beneath Rg indicate that this discontinuity is <13.8 Ma.

2. **Rh (Surface B, dark green):** This regionally extensive discontinuity is correlated with the base of a ~180-m-thick interval of late Miocene-early Pliocene, pyrite-cemented, high-velocity volcanic sandstone and mudstone (LSU 5). Regionally the green reflector correlates with the base of volcanic bodies in the VLB north of Ross Island. It is also correlated with the base of White Island volcano (Figure 2) dated at ~7.6 Ma (Alan Cooper, University of Otago, unpublished data).

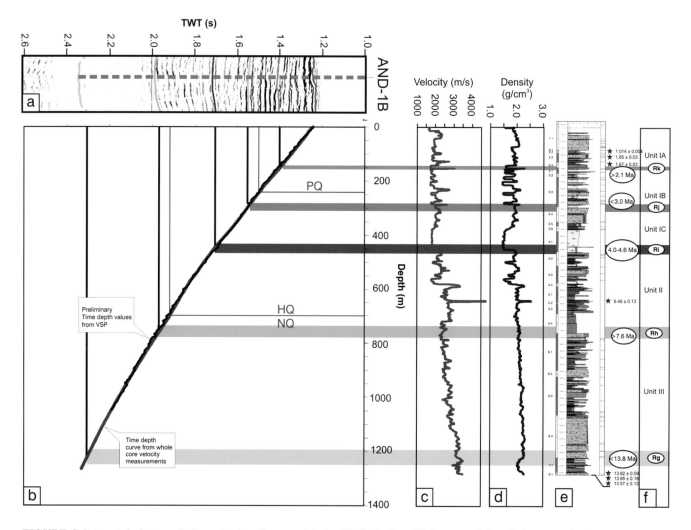

FIGURE 3 Integrated plot correlating seismic reflectors with the lithologic log. We have used the whole-core velocity (c) to derive a time-depth conversion curve (b) together with VSP arrival times to map the seismic reflection profile from MIS-1 (a) to drill hole depth and correlate with core lithologies and lithostratigraphic units (e). Seismic stratigraphic units identified are from the MIS Science Logistics and Implementation Plan (SLIP) (Naish et al., 2006) and have been mapped regionally (Fielding et al., 2007).

3. Ri (Surface A2, b-clino, red reflector): This regionally extensive reflector marks the base of a ~100-m-thick seismically opaque interval that separates high-amplitude reflections of the underlying unit. It corresponds with the base of prograding clinoforms north of Ross Island, and locally marks the base of flexure associated with Ross Island volcanic loading (Horgan et al., 2005). In AND-1B, Surface A2 correlates with the boundary between the ~90-m-thick, low-density, low-velocity (1700 ms⁻¹), early Pliocene diatomite interval (LSU 4.1), and the higher-velocity (<2500 ms⁻¹) diamictites of LSU 4.2 beneath. Diatom assemblages indicate the age of this surface lies between 5.0 Ma and 4.0 Ma. Regionally this reflector has been traced into western VLB, where it is correlated biostratigraphically in MSSTS-1 to core at about 20 mbsf that yields a Pliocene age of 4.6-4.0 Ma based on diatom microfossils (Naish et al., 2006).

4. Rj (Surface A1, turquoise reflector): This regionally extensive reflector marks the base of a ~150-m-thick unit of strongly alternating high- and low-amplitude reflections. These dramatic cycles in density and velocity reflect regular alternations between diatomite and diamictite in late Pliocene (LSU 3). The turquoise reflector separates strata above that are younger than ~3.0 Ma and below greater than ~3.5 Ma.

STRATIGRAPHIC ARCHITECTURE

The 1285-m-long AND-1B drill core provides the first high-resolution, late Neogene record from the Antarctic margin (Figure 4) as well as the first long geological record from under a major ice shelf. Details of the lithostratigraphic subdivision, facies analysis, and sequence stratigraphy are presented in an Initial Results volume (Krissek et al., 2007).

Glacial-Interglacial Cyclostratigraphy

At the time of writing, 60 unconformity-bounded glacimarine sedimentary cycles, of probable Milankovitch duration have been identified, representing repeated advances and retreats of an ice sheet across the drill site during the late Neogene. Bounding unconformities, glacial surfaces of erosion (GSEs), are typically sharp and planar and mark dislocations between enclosing facies. Facies immediately beneath display a range of intraformational deformation features, including physical mixing of lithologies, clastic intrusions, faulting, and soft-sediment deformation. Facies above these surfaces are typically diamictites and conglomerates, and are interpreted as subglacial tillites or near grounding-line

FIGURE 4 Lithostratigraphy, chronostratigraphy, and sequence stratigraphy of the AND-1B drillcore. Cyclic variations in lithologies reflect periodic fluctuations of the ice margin in western Ross Embayment during the last 13 Ma. Sources: * = After Ross et al. (2007); ^ = After Wilson et al. (2007).

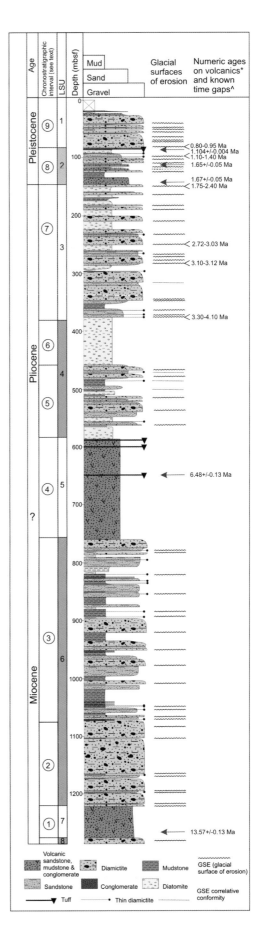

glacimarine deposits. In many cycles the facies succession reflects retreat of the grounding line through ice shelf into open-ocean environments at the interglacial minimum, followed by ice readvance characterized by progressively more glacially influenced facies in the upper parts, culminating in a GSE at the glacial maximum.

Major Chronostratigraphic Intervals and Their Cyclic Character

Here we summarize the sedimentary succession cored in AND-1B by subdividing it into nine chronostratigraphic intervals largely on the basis of characteristic facies cycles, with their glacial and climatic implications (Figure 4). The latter are addressed in more detail by Powell et al. (2007). Three of the intervals lack cyclicity: interval 1 (a volcanic sandstone), interval 4 (largely volcanic ash), and interval 6 (the ~90-m-thick diatomite).

1. Late Miocene volcanic sandstone (1275.24-1220.15 mbsf), LSU 7.
2. Late Miocene diamictite-dominated sedimentary cycles (1220.15-1069.2 mbsf), LSU 6.4.
3. Late Miocene, diamictite/mudstone and sandstone sedimentary cycles (1069.2-759.32 mbsf), LSU 6.1-6.3.
4. Late Miocene-early Pliocene? lapilli tuff, lava flow, and volcanic sandstone and mudstone (759.32-586.45 mbsf), LSU 5.
5. Early Pliocene diamictite/diatomite sedimentary cycles (586.45-459.24 mbsf), LSU 4.2-4.4.
6. Early Pliocene diatomite (459.24-382.98mbsf), LSU 4.1.
7. Late Pliocene diamictite and diatomite sedimentary cycles (459.25-146.79 mbsf), LSU 3.
8. Late Pliocene-early Pleistocene diamictite/volcanic mudstone and sandstone cycles (146.79-82.72 mbsf), LSU 2.
9. Middle-late Pleistocene, diamictite-dominated sedimentary cycles (82.72-0mbsf), LSU 1.

Late-Miocene Diamictite-Dominated Cycles of LSU 6.4 (1220.15-1069.2 mbsf)

This interval is dominated by thick massive diamictite with thin stratified mudstone interbeds, representing alternations between basal glacial and glacimarine deposition. The diamictites, which form over 90 percent of the interval, contain medium- to high-grade metamorphic basement clasts known from the Byrd Glacier region (Talarico et al., 2007). At this stage we have been conservative in our recognition of grounding-line features in the core that might be used to identify boundaries of the sedimentary cycles. More detailed study is likely to find features such as sets of alternating sharp-based massive and stratified diamictites, representing subglacial and pro-grounding-line environments. The dominance of subglacial deposition and the occurrence of gran-

itoid and metamorphic rocks from the Byrd Glacier region imply the long-term existence of a grounded Ross Ice Sheet with short ice-shelf phases, and ice streaming from the southern Transantarctic Mountains. Once a magnetostratigraphy is developed for this section of the core together with new ^{40}Ar/^{39}Ar ages, further sedimentological and petrographic work should provide an important history of Antarctic ice sheet behavior during the "big" late Miocene, Mi-glaciations of Miller et al. (1991).

Late Miocene, Diamictite/Mudstone-Sandstone Sedimentary Cycles of LSU 6.1-6.3 (1069.2-759.32 mbsf)

This section is characterized by cycles of subglacial and grounding-line diamictites that pass upward into a glacimarine retreat succession of redeposited conglomerate, sandstone and mudstone. These units are overlain by a more distal hemipelagic terrigenous mudstone with out-size clasts and lonestones. The retreat facies succession is then followed by a proglacial advance facies assemblage in which clast abundance increases together with the occurrence of submarine outwash facies, which is truncated by a glacial surface of erosion. In contrast with the underlying diamictite-dominated interval, over 60 percent of this unit comprises strata representing periods of both ice-shelf cover and open ocean, though still with significant ice rafting, over the drill site. These observations imply a warmer climate for this whole period, and the association increased subglacial meltwater.

Late Miocene-early Pliocene Volcanic Sandstone and Mudstone of LSU 5 (759.32-586.45mbsf)

This interval is dominated by subaqueously redeposited volcanic sediments, many with near-primary volcanic characteristics. The volcaniclastic sediments are organized into sediment gravity flow deposits (mostly proximal turbidites), indicating a nearby active volcanic center delivering primary volcanic material into a deep basin (several hundred meters water depth). A series of fining upward, altered and degraded pumice lapilli tuffs occur at 623, 603, and 590 mbsf, and a pure fine volcanic glass sand occurs at 577 mbsf. All of these have been targeted for argon geochronology. A plagioclase Hawaiite lava with chilled margins and "baked sediments" under its lower contact interrupts the volcanic sediment gravity flows at 646.49-649.30 mbsf, and has been dated by ^{40}Ar/^{39}Ar at 6.38 Ma. The source of this lava and the redeposited volcanic units is thought to be nearby because of the freshness of the glass and angular nature of clasts in the breccias. The unit is virtually devoid of out-size clasts or other features indicative of iceberg rafting or grounding-line proximity, suggesting that much of the 120-m-thick interval was deposited rapidly in open water. At this stage we consider this volcanic interval to represent a single eruptive sequence with very high rates of accumulation.

Early Pliocene Diamictite and Diatomite Sedimentary Cycles of LSU 4.2-4.4 (586.45-459.24 mbsf)

These cycles can be summarized as comprising, in ascending stratigraphic order: (1) a sharp-based massive diamictite with variable amounts of volcanic glass (subglacial to grounding zone) passing up into (2) stratified diamictite, sandstone, and mudstone with dispersed clasts (grounding-line to distal glacimarine) followed by (3) stratified diatomite (open ocean), which in turn commonly passes up into glacimarine mudstone and sandstone with dispersed clasts (perhaps an indication of an approaching grounding line). The interval that includes the lower part of the diamictite and the underlying glacimarine lithologies is often physically intermixed and displays deformation associated with glacial overriding and shearing but, interestingly, appears to represent little significant erosion by the advance. These cycles show a dramatic change from hemipelagic open-water sedimentation to open-water biogenic sediments (diatomite) at interglacial ice minima.

Early Pliocene Diatomite of LSU 4.1 (459.24-382.98 mbsf)

This interval is almost solely diatomite, indicating an extended period of high-productivity open water over the site. It is of early Pliocene age (~4.2 Ma) and is likely to span a period of about 4.2-4.0 m.y., involving a number of glacial-interglacial cycles. The onset of an inferred warmer period is captured in the 10 m transition beginning at 460 mbsf from a clast-rich to clast-poor muddy diamictite through terrigenous mud to diatomite, with a decline in mud component, outsize clasts, and frequency of cm- to dm-thick gravity flow deposits. Further research is planned in order to quantify the contemporaneous climate and will involve examination of the terrestrial microfossils and estimation of sea-surface temperatures from geochemical and biological proxies. This diatomite appears to coincide with the widely recognized early Pliocene global warming and accompanying higher sea levels (Dowsett et al., 1999; Ravelo et al., 2004), for which there is evidence elsewhere on the Antarctic margin (e.g., Whitehead et al., 2005). However, well-dated deposits and geomorphological evidence from the adjacent McMurdo Dry Valleys indicate a polar climate persisted there throughout this period (Marchant et al., 1996).

Middle to Late Pliocene Diamictite and Diatomite Sedimentary Cycles of LSU 3 (382.98-146.79 mbsf)

Within this interval the core is strongly cyclic in nature, and is characterized by 12 glacial-interglacial cycles that is typically composed of a sharp-based lower interval of diamictite in the upper few meters that passes upward into a ~5- to 10-m-thick unit of biosiliceous ooze or biosiliceous-bearing mudstone (diatomite). The lower part of diamictites and the upper few meters of diatomites are sheared and deformed, presumably from grounded ice. The transitions from diamictite (glacial and glacimarine) to diatomite (open ocean) are dramatic, in many cases occurring in less than a meter of core. Powell et al. (2007) discuss these abrupt facies transitions in terms of rapid retreat of an ice sheet or collapse of an ice shelf. Our preliminary age model suggests that the cycles in this interval may have formed in response to orbital forcing of the ice sheet. This interval will be the focus of further work as we evaluate its significance in terms of the global climatic deterioration coincident with the development of large ice sheets on the northern hemisphere continents (e.g., Shackleton et al., 1984; Maslin et al., 1999).

Late Pliocene-Early Pleistocene Diamictite/Volcanic Mudstone-Sandstone Cycles of LSU2 (146.79-82.72 mbsf)

This interval displays lithologic alternations between ice-proximal facies and open-water volcaniclastic facies. A GSE at 150.90 mbsf is correlated with the Rk-pink reflector and spans as much as 0.7 m.y. between 2.4 Ma and 1.7 Ma. Above this the interval from 146.79 mbsf to 134.48 mbsf represents another glacial-interglacial cycle; that above the diamictite is almost entirely made up of redeposited volcanic (basaltic) sandstone. The sandstone is organized into a series of normally graded turbidites. The interval from 134.48 mbsf to 120 mbsf is characterized by an open-water assemblage of interstratified mudstone and volcanic sandstone that lies stratigraphically above weakly stratified diamictite alternating with sparsely fossiliferous claystone and mudstone, more typical of grounding line and iceberg zone systems. A muddy-sandy volcanic breccia with near-primary volcanic material at 117 mbsf has yielded a preliminary $^{40}Ar/^{39}Ar$ age of ~1.6 Ma on basaltic volcanic glass.

A dramatic change upward into massive diamictite from 109.7 mbsf to 97.2 mbsf indicates the return of the ice sheet to the proximity of the drill site. Our preliminary age model suggests that up to ~0.4 m.y., between 1.5 Ma and 1.1 Ma is missing between GSE at 109.7 mbsf and at least 2 GSEs in the overlying interval of amalgamated diamictite above. From 97.2 mbsf to 92.5 mbsf diamictite and volcanic sandstone passes downward into interstratified bioturbated volcanic sandstone and mudstone. The interval from 92.5 mbsf to 86.6 mbsf contains the youngest biosiliceous-bearing sediments in the AND-1B core. The presence of numerous volcaniclastic units and biosiliceous sediments in this interval indicate an extended period of open-water conditions with no sea ice beyond the calving line. An immediately overlying, fining upward lapilli tuff between 85.86 mbsf and 85.27 mbsf has yielded a high-precision $^{40}Ar/^{39}Ar$ age on sanidine phenocrysts of 1.01 Ma within a short normal polarity interval interpreted as the Jaramillo Subchron. Thus the underlying biosiliceous interglacial sediments are tentatively correlated with warming during the "super-interglacial" associated with Marine Isotope Stage 31. This interval will be the focus of a future concentrated effort to better characterize the impact of this warm period on ice-sheet behavior.

Middle-Late Pleistocene, Diamictite-Dominated Sedimentary Cycles of LSU 1 (82-0 mbsf)

Between 83 mbsf and 27 mbsf there are at least 8 cycles of diamictite and thinner bioturbated and stratified intervals of mudstone, volcanic sandstone, and muddy conglomerate. Like the late Miocene diamictite-dominated cycles, the high proportion of subglacial and grounding-line proximal deposits, together with the occurrence of granitoid and metamorphic rocks known from the Byrd Glacier region (Talarico et al., 2007), implies existence of a grounded Ross Ice Sheet in the middle and late Pleistocene. The implied presence of a grounded ice sheet for much of this time is intriguing as it also corresponds with a period of Earth history dominated by 80-120 ka fluctuations in large Northern Hemisphere ice sheets. Further work will focus on (1) how this ice sheet responded to the late Pleistocene interglacials, which in Antarctic ice-core records indicated polar temperatures warmer than today (EPICA Community, 2004; Jouzel et al., 2007), and (2) the role of orbital forcing on Antarctic ice sheets (e.g., Huybers and Wunsch, 2005; Raymo et al., 2006) in regulating Pleistocene climate.

IMPLICATIONS FOR ANTARCTIC GLACIAL AND CLIMATE HISTORY

The AND-1B core has the potential to contribute significant new knowledge about the dynamics of the West Antarctic ice sheet (WAIS) and Ross Ice Shelf and Ice Sheet system, as well as contributing to understanding the behavior of the EAIS outlet glaciers during the late Cenozoic. New chronological data will allow certain intervals of the core to be correlated with other proxy climate records (e.g., ice and marine isotope records) and hence the sensitivity of the ice sheet to a range of past global climate changes to be evaluated. The significant results thus far are that the Ross Ice Sheet, fed by WAIS and EAIS outlet glaciers, has undergone significant cyclic variations in extent and timing during the late Neogene. A relatively colder and more stable ice sheet dominated the Ross Embayment in the early late Miocene between 13 Ma and 10 Ma, becoming more dynamic in the latest Miocene (9-7 Ma) with subglacial water discharge, still with periodic grounded ice in the Ross Embayment. These conditions were followed by a period around 4 Ma when the Ross Embayment was relatively ice-free, with highly productive, warmer oceanic conditions, followed by a return to cycles of advance and retreat of grounded ice. From middle Pleistocene to Recent the ice sheet is characterized by a change back to more stable, colder conditions. Our preliminary analysis of the more than 25 Pliocene sedimentary cycles indicates significant glacial-interglacial variability, with regular oscillations between subglacial/ice proximal and open-ocean ice distal environments, including extended periods of interglacial warmth when the ice was not calving into the ocean. Our environmental reconstructions to date imply changes in ice-sheet volume that must have contributed significantly to eustasy (e.g., 10-20 m).

Cold early Miocene and Pleistocene till-dominated intervals with clasts originating in the TAM to the south imply that grounded ice from the big outlet glaciers to the south was reaching McMurdo Sound during these times. Today Byrd Glacier-sourced ice in the shelf flows east of Ross Island. Glaciological reconstructions (e.g., Denton and Hughes, 2000) require significant ice volume from WAIS to direct the flow lines of the southern outlet glaciers into the McMurdo region during glacial periods, and also to maintain an ice shelf during ensuing interglacial retreats. Thus, we view the sedimentary cycles representing primarily the expansion and contraction of WAIS in concert with fluctuations in the flow of TAM outlet glaciers. The TAM outlet glaciers alone do not provide enough ice volume to maintain an ice sheet or ice shelf in the Ross Embayment.

Work is planned in the near future to further understand and constrain ice-sheet dynamics, especially during periods of hypothesized global warmth that will provide useful analogues for constraining the behavior of Antarctic ice sheets in the context of future climate change. To achieve this we will integrate our core data—such as clast provenance, biological and geochemical proxies—with the new generation of ice-sheet and climate models.

SUMMARY

Repetitive vertical successions of facies imply at least 60 fluctuations of probable Milankovitch duration between subglacial, ice-proximal, and ice-distal open marine environments. These have been grouped into three main facies associations that correspond to glacial-interglacial variability during climatically distinct periods of the late Neogene:

1. Cold polar climate with MIS site dominated by grounded ice but some retreat to ice-shelf conditions (late Miocene, ~13-10 Ma and Pleistocene, ~1-0 Ma).
2. Warmer climate with MIS site dominated by ice-shelf and open-water conditions (hemipelagites), with occasional periods of grounded ice (early-late Miocene, ~9-6 Ma).
3. Warmer climate with extended periods of open-ocean conditions (pelagic diatomites) with periods of sub-ice shelf and grounded ice deposition (Pliocene, ~5-2 Ma).

The ~90-m-thick early Pliocene (~4.2 Ma) interval of diatomite shows no apparent glacial cyclicity and represents an extended period of ice-free conditions indicative of a reduced WAIS. Late Pliocene (~2.6-2.2 Ma) glacial inter-glacial cycles characterized by abrupt alternations between subglacial/ice-proximal facies and open marine diatomites imply significant WAIS volume fluctuations around the time of the early Northern Hemisphere glaciations. A ~4-m-thick interval of diatomaceous mudstone in the middle Pleistocene also represents similar open-ocean interglacial conditions. The last million years is dominated by deposition from grounded ice with periods of sub-ice-shelf sedimentation like the present day.

ACKNOWLEDGMENTS

The ANDRILL project is a multinational collaboration of the Antarctic Programs of Germany, Italy, New Zealand, and the United States. Antarctica New Zealand is the project operator and has developed the drilling system in collaboration with Alex Pyne at Victoria University of Wellington and Webster Drilling and Enterprises Ltd. Antarctica New Zealand supported the drilling team at Scott Base and Raytheon Polar Services supported the science team at McMurdo Station and the Crary Science and Engineering Laboratory. Scientific support was provided by the ANDRILL Science Management Office, University of Nebraska-Lincoln. Scientific studies are jointly supported by the U.S. National Science Foundation, NZ Foundation for Research, the Italian Antarctic Research Program, the German Science Foundation, and the Alfred-Wegener-Institute.

REFERENCES

Anderson, J. B. 1999. *Antarctic Marine Geology.* Cambridge: Cambridge University Press.

Anderson, J. B., J. Wellner, S. Wise, S. Bohaty, P. Manley, T. Smith, F. Weaver, and D. Kulhanek. 2007. Seismic and chronostratigraphic results from SHALDRIL II, Northwestern Weddell Sea. In *Antarctica: A Keystone in a Changing World—Online Proceedings for the Tenth International Symposium on Antarctic Earth Sciences,* eds. Cooper, A. K., C. R. Raymond et al., USGS Open-File Report 2007-1047. Short Research Paper 094, doi:10.3133/of2007-1047.srp094.

Barker, P. F., A. Camerlenghi, G. D. Acton et al. 1999. Proc. Ocean Drilling Program, Initial Report 178, http://www-odp.tamu.edu/publications/178_IR/178TOC.HTM.

Barrett, P. J., ed. 1986. Antarctic Cenozoic history from the MSSTS-1 drill hole, McMurdo Sound, Antarctica. *NZ DSIR Bulletin,* 237 pp.

Barrett, P. J. ed. 1989. Antarctic Cenozoic history from the CIROS-1 drillhole, McMurdo Sound, Antarctica. *NZ DSIR Bulletin,* 245 pp.

Barrett, P. J. 1999. Antarctic climate history over the last 100 million years. *Terra Antartica* 3:53-72.

Barrett, P. J. 2007. Cenozoic climate and sea level history from glacimarine strata off the Victoria Land coast, Cape Roberts Project, Antarctica. In *Glacial Processes and Products,* eds. M. J. Hambrey, P. Christoffersen, N. F. Glasser, and B. Hubbart. *International Association of Sedimentologists Special Publication* 39:259-287.

Barrett, P. J., and M. J. Hambrey. 1992. Plio-Pleistocene sedimentation in Ferrar Fiord, Antarctica. *Sedimentology* 39:109-123.

Barrett, P. J., and S. B. Treves. 1981. Sedimentology and petrology of core from DVDP 15, western McMurdo Sound. In *Dry Valley Drilling Project,* ed. L. D. McGinnis. *Antarctic Research Series* 81:281-314. Washington, D.C.: American Geophysical Union.

Barron, J., B. Larsen et al. 1989. Proc. Ocean Drilling Program, Initial Report 119. College Station, TX: Ocean Drilling Program.

Brancolini, G., et al. 1995. Descriptive text for the seismic stratigraphic atlas of the Ross Sea, Antarctica. In *Geology and Seismic Stratigraphy of the Antarctic Margin,* eds. A. K. Cooper, P. F. Barker, and G. Brancolini, *Antarctic Research Series* 68:271-286. Washington, D.C.: American Geophysical Union.

Bushnell, V. C., and C. Craddock, eds. 1970. *Antarctic Map Folio Series.* New York: American Geographical Society, Map 64-29.

Cape Roberts Science Team (CRST). 1998. Initial Report on CRP-3. *Terra Antartica* 5:1-187.

Cape Roberts Science Team (CRST). 1999. Studies from the Cape Roberts Project, Ross Sea, Antarctica. Initial Report on CRP-2/2A. *Terra Antartica* 6(with supplement):1-173.

Cape Roberts Science Team (CRST). 2000. Studies from the Cape Roberts Project, Ross Sea, Antarctica. Initial Report on CRP-3. *Terra Antartica* 7(with supplement):1-209.

Cooper, A. K., and F. J. Davey. 1985. Episodic rifting of the Phanerozoic rocks of the Victoria Land basin, western Ross Sea, Antarctica. *Science* 229:1085-1087.

Cooper, A. K., F. J. Davey, and J. C. Behrendt. 1987. Seismic stratigraphy and structure of the Victoria Land Basin, Western Ross Sea, Antarctica. In *The Antarctic Continental Margin: Geology and Geophysics of the Western Ross Sea,* eds. A. K. Cooper and F. J. Davey. Earth Science Series 5B:27-77. Houston, TX: Circum-Pacific Council Energy Mineral Resources.

Denton, G. H., and T. J. Hughes. 2000. Reconstruction of the Ross ice drainage system, Antarctica, at the last glacial maximum. *Geografiska Annaler* 82:143-166.

Dowsett, H. J., J. A. Barron, R. Z. Poore, R. S. Thompson, T. M. Cronin, S. E. Ishman, and D. A. Willard. 1999. *Middle Pliocene Paleoenvironmental Reconstruction: PRISM2.* U.S. Geological Survey Open File Report 99-535, http://pubs.usgs.gov/openfile/of99-535/.

EPICA Community Members. 2004. Eight glacial cycles from an Antarctic ice core, *Nature* 429:623-628.

Fielding, C. R., J. Whittaker, S. A. Henrys, T. J. Wilson, and T. R. Naish. 2007. Seismic facies and stratigraphy of the Cenozoic succession in McMurdo Sound, Antarctica: Implications for tectonic, climatic and glacial history. *Palaeogeography, Palaeoclimatology, Palaeoecology.* In *Antarctica: A Keystone in a Changing World—Online Proceedings for the Tenth International Symposium on Antarctic Earth Sciences,* eds. Cooper, A. K., C. R. Raymond et al., USGS Open-File Report 2007-1047. Short Research Paper 090, doi:10.3133/of2007-1047.srp090.

Flint, R. F. 1971. *Glacial and Quaternary Geology.* New York: Wiley.

Hansaraj, D., S. A. Henrys, T. R. Naish, and ANDRILL MIS-Science Team. 2007. McMurdo Ice Shelf seismic reflection data and correlation to the AND-1B drill hole. In *Antarctica: A Keystone in a Changing World —Online Proceedings for the Tenth International Symposium on Antarctic Earth Sciences,* eds. Cooper, A. K., C. R. Raymond et al., USGS Open-File Report 2007-1047, Extended Abstract 101, http://pubs.usgs.gov/of/2007/1047/.

Hayes, D. E., L. A. Frakes et al. 1975. *Initial Reports of the Deep Sea Drilling Project,* vol. 28. Washington, D.C.: U.S. Government Printing Office.

Hays, J. D., J. Imbrie, and N. J. Shackleton. 1976. Variations in the earth's orbit: Pacemaker of the ages. *Science* 194:1121-1132.

Henrys, S. A., T. J. Wilson, J. M. Whittaker, C. R. Fielding, J. M. Hall, and T. Naish. 2007. Tectonic history of mid-Miocene to present southern Victoria Land Basin, inferred from seismic stratigraphy in McMurdo Sound, Antarctica. In *Antarctica: A Keystone in a Changing World—Online Proceedings for the Tenth International Symposium on Antarctic Earth Sciences,* eds. Cooper, A. K., C. R. Raymond et al., USGS Open-File Report 2007-10477, Short Research Paper 049, doi:10.3133/of2007-1047.srp049.

Horgan, H., T. Naish, S. Bannister, N. Balfour, and G. Wilson. 2005. Seismic stratigraphy of the Ross Island flexural moat under the McMurdo-Ross Ice Shelf, Antarctica, and a prognosis for stratigraphic drilling. *Global Planetary Change* 45:83-97.

Huybers, P., and C. Wunsch. 2005. Obliquity pacing of the late Pleistocene glacial terminations. *Nature* 434:491-494.

Jouzel, J., et al. 2007. Orbital and Millennial Climate variability over the past 800,000 years. *Science* 317:793-796.

Kennett, J. P., and D. A. Warnke, eds. 1992. *The Antarctic Paleoenvironment: A Perspective on Global Change,* Part 1. Antarctic Research Series, vol. 56. Washington, D.C.: American Geophysical Union.

Kennett, J. P., and D. A. Warnke, eds. 1993. *The Antarctic Paleoenvironment: A Perspective on Global Change,* Part 2. Antarctic Research Series, vol. 60. Washington, D.C.: American Geophysical Union.

Kennett, J. P., R. E. Houtz et al. 1974. *Initial Reports of the Deep Sea Drilling Project,* vol. 29. Washington, D.C.: U.S. Government Printing Office.

Krissek, L. A., G. Browne, L. Carter, E. Cowan, G. Dunbar, R. McKay, T. Naish, R. Powell, J. Reed, T. Wilch, and the ANDRILL-MIS Science Team. 2007. Sedimentology and stratigraphy of the ANDRILL McMurdo Ice Shelf (AND-1B) core. In *Antarctica: A Keystone in a Changing World—Online Proceedings for the Tenth International Symposium on Antarctic Earth Sciences,* eds. Cooper, A. K., C. R. Raymond et al., USGS Open-File Report 2007-1047, Extended Abstract 148, http://pubs.usgs.gov/of/2007/1047/.

Kyle, P. R. 1981. Geological history of Hut Point Peninsula as inferred from DVDP 1, 2 and 3 drill cores and surface mapping. In *Dry Valley Drilling Project,* ed. L. D. McGinnis. *Antarctic Research Series* 81:427-445. Washington, D.C.: American Geophysical Union.

Marchant, D. R., G. H. Denton, C. C. Swisher, III, and N. Potter, Jr. 1996. Late Cenozoic Antarctic paleoclimate reconstructed from volcanic ashes in the Dry Valleys region of southern Victoria Land. *Geological Society of America Bulletin* 108:181-194.

Maslin, M. A., Z. Li, M.-F. Loutre, and A. Berger. 1999. The contribution of orbital forcing to the progressive intensification of Northern Hemisphere glaciation. *Quaternary Science Reviews* 17:411-426.

McKay, R., Dunbar, G., Naish, T. R., Barrett, P., Carter, L., Harper, M. 2007. Retreat of the Ross Ice Shelf since the Last Glacial Maximum derived from sediment cores in deep basins surrounding Ross Island. *Paleoclimatology, Paleogeography, Paleoecology.* A Sediment Model and Retreat History for the Ross Ice (Sheet) Shelf in the Western Ross Sea Since the Last Glacial Maximum. In *Antarctica: A Keystone in a Changing World—Online Proceedings for the Tenth International Symposium on Antarctic Earth Sciences,* eds. Cooper, A. K., C. R. Raymond et al., USGS Open-File Report 2007-1047, Extended Abstract 159, http://pubs.usgs.gov/of/2007/1047/.

Miller, K. G., J. D. Wright, and R. G. Fairbanks. 1991. Unlocking the ice house: Oligocene-Miocene oxygen isotopes, eustasy, and margin erosion. *Journal of Geophysical Research* 96:6829-6848.

Naish, T. R., et al. 2001. Orbitally induced oscillations in the East Antarctic ice sheet at the Oligocene/Miocene boundary. *Nature* 413:719-723.

Naish, T. R., R. H. Levy, R. D. Powell, and the ANDRILL MIS Science and Operations Teams. 2006. *ANDRILL McMurdo Ice Shelf Scientific Logistical Implementation Plan.* ANDRILL Contribution No. 7. Lincoln: University of Nebraska-Lincoln.

Naish, T. R., R. D. Powell, R. H. Levy, and the ANDRILL-MIS Science Team. 2007. Initial science results from AND-B, ANDRILL McMurdo Ice Shelf Project, Antarctica. *Terra Antartica* 14(2).

O'Brien, P. E., A. K. Cooper, C. Richter et al. 2001. Proc. Ocean Drilling Program, Initial Report 188, http://www-odp.tamu.edu/publications/188_IR/188ir.htm.

Powell, R. D. 1981. Sedimentation conditions in Taylor Valley inferred from textural analyses of DVDP cores. In *Dry Valley Drilling Project,* ed. L. D. McGinnis. *Antarctic Research Series* 81:331-350. Washington, D.C.: American Geophysical Union.

Powell, R. D., T. R. Naish, L. A. Krissek, G. H. Browne, L. Carter, E. A. Cowan, G. B. Dunbar, R. M. McKay, T. I. Wilch, and the ANDRILL-MIS Science team. 2007. Antarctic ice sheet dynamics from evidence in the ANDRILL-McMurdo Ice Shelf Project drillcore (AND-1B). In *Antarctica: A Keystone in a Changing World—Online Proceedings for the Tenth International Symposium on Antarctic Earth Sciences,* eds. Cooper, A. K., C. R. Raymond et al., USGS Open-File Report 2007-1047, Extended Abstract 201, http://pubs.usgs.gov/of/2007/1047/.

Ravelo, A. C., D. H. Andreasen, L. Mitchell, A. O. Lyle, and M. W. Wara. 2004. Regional climate shifts caused by gradual global cooling in the Pliocene epoch. *Nature* 429:263-267.

Raymo, M. E., L. E. Lisecki, and K. H. Nisancioglu. 2006. Plio-Pleistocene ice volume, Antarctic climate, and the global $\delta^{18}O$ record. *Science* 313:492-495.

Ross, J., W. C. McIntosh, and N. W. Dunbar. 2007. Preliminary $^{40}Ar/^{39}Ar$ results from the AND-1B core. In *Antarctica: A Keystone in a Changing World—Online Proceedings for the Tenth International Symposium on Antarctic Earth Sciences,* eds. Cooper, A. K., C. R. Raymond et al., USGS Open-File Report 2007-1047, Extended Abstract 093, http://pubs.usgs.gov/of/2007/1047/.

Scherer, R., D. Winter, C. Sjunneskog, and P. Maffioli. 2007. The diatom record of the ANDRILL-McMurdo Ice Shelf project drillcore. In *Antarctica: A Keystone in a Changing World—Online Proceedings for the Tenth International Symposium on Antarctic Earth Sciences,* eds. Cooper, A. K., C. R. Raymond et al., USGS Open-File Report 2007-1047, Extended Abstract 171, http://pubs.usgs.gov/of/2007/1047/.

Shackleton, N. J., and J. P. Kennett. 1974. Paleotemperature history of the Cenozoic and the initiation of Antarctic glaciation: Oxygen and Carbon isotope analyses in DSDP Sites 277, 279 and 281. In *Initial Reports of the Deep Sea Drilling Project,* vol. 29, eds. J. P. Kennett, R. E. Houtz et al., pp. 743-756. Washington, D.C.: U.S. Government Printing Office.

Shackleton, N. J., et al. 1984. Oxygen isotope calibration of the onset of ice-rafting and history of glaciation in the North Atlantic region. *Nature* 307:620-623.

Smith, P. M. 1981. The role of the Dry Valley Drilling Project in Antarctic and international science policy. In *Dry Valley Drilling Project,* ed. L. D. McGinnis. *Antarctic Research Series* 81:1-5. Washington, D.C.: American Geophysical Union.

Stern, T. A., F. J. Davey, and G. Delisle. 1991. Lithospheric flexure induced by the load of the Ross Archipelago, southern Victoria Land, Antarctica. In *Geological Evolution of Antarctica* eds. M. R. A. Thomson, A. Crame, and J. W. Thomson, pp. 323-328. Cambridge: Cambridge University Press.

Talarico, F., et al. 2007. Clast provenance and variability in MIS (AND-1B) core and their implications for the paleoclimatic evolution recorded in the Windless Bight, southern McMurdo Sound area (Antarctica). In *Antarctica: A Keystone in a Changing World—Online Proceedings for the Tenth International Symposium on Antarctic Earth Sciences,* eds. Cooper, A. K., C. R. Raymond et al., USGS Open-File Report 2007-1047, Extended Abstract 118, http://pubs.usgs.gov/of/2007/1047/.

Whitehead, J. M., S. Wotherspoon, and S. M. Bohaty. 2005. Minimal Antarctic sea ice during the Pliocene. *Geology* 33:137-140.

Wilson, G. S., et al. 2007. Preliminary chronostratigraphy for the upper 700 m (late Miocene-Pleistocene) of the AND-1B drillcore recovered from beneath the McMurdo Ice Shelf, Antarctica. In *Antarctica: A Keystone in a Changing World—Online Proceedings for the Tenth International Symposium on Antarctic Earth Sciences,* eds. Cooper, A. K., C. R. Raymond et al., USGS Open-File Report 2007-1047, Extended Abstract 092, http://pubs.usgs.gov/of/2007/1047/.

Cooper, A. K., P. J. Barrett, H. Stagg, B. Storey, E. Stump, W. Wise, and the 10th ISAES editorial team, eds. (2008). *Antarctica: A Keystone in a Changing World.* Proceedings of the 10th International Symposium on Antarctic Earth Sciences. Washington, DC: The National Academies Press.

A Pan-Precambrian Link Between Deglaciation and Environmental Oxidation

T. D. Raub and J. L. Kirschvink[1]

ABSTRACT

Despite a continuous increase in solar luminosity to the present, Earth's glacial record appears to become more frequent, though less severe, over geological time. At least two of the three major Precambrian glacial intervals were exceptionally intense, with solid evidence for widespread sea ice on or near the equator, well within a "Snowball Earth" zone produced by ice-albedo runaway in energy-balance models. The end of the first unambiguously low-latitude glaciation, the early Paleoproterozoic Makganyene event, is associated intimately with the first solid evidence for global oxygenation, including the world's largest sedimentary manganese deposit. Subsequent low-latitude deglaciations during the Cryogenian interval of the Neoproterozoic Era are also associated with progressive oxidation, and these young Precambrian ice ages coincide with the time when basal animal phyla were diversifying. However, specifically testing hypotheses of cause and effect between Earth's Neoproterozoic biosphere and glaciation is complicated because large and rapid True Polar Wander events appear to punctuate Neoproterozoic time and may have episodically dominated earlier and later intervals as well, rendering geographic reconstruction and age correlation challenging except for an exceptionally well-defined global paleomagnetic database.

INTRODUCTION

Despite a 30 percent increase in solar luminosity during the past 4.6 billion years, we have solid geological evidence that liquid water was usually present on the surface. If the sun were to suddenly shift to even a 5-10 percent lower luminosity, our oceans would rapidly freeze over. We infer that this climatic regulation is due in large part to a combination of greenhouse gasses—principally H_2O, CO_2, and CH_4—which have varied over time. For one of these, CO_2, there is a clear inorganic feedback mechanism helping regulate climate (Walker et al., 1981), as CO_2 removal by silicate weathering increases with temperature, a process that can act on a 10^6- to 10^7-year timescale.

Geologists observe that a major shift in redox state of Earth's atmosphere happened sometime between 2.45 and 2.22 Ga ago, as signaled by the loss of a mass-independent fractionation signal in sulfur isotopes, the disappearance of common detrital pyrite and uraninite from stream deposits, and the appearance of true continental redbeds, documented by a reworked paleosol that cements together coherent hematitic chips magnetized in random directions (Evans et al., 2001). The sedimentary sulfate minerals barite and gypsum also become more prevalent in evaporative environments post ~2.3 Ga, as seen in the Barr River Formation of the Huronian Supergroup of Ontario (see Figure 1).

The reappearance of sedimentary sulfates after the Gowganda and Makganyene Glaciations at about 2.2 Ga follows a nearly 800 myr absence in the rock record (Huston and Logan, 2004), arguing that enough oxygen was then present in the atmosphere to oxidize pyrite to sulfate in quantities that sulfate-reducing organisms could not completely destroy.

Numerous hints in the rock record suggest a general relationship between changes in atmospheric redox state and severe glaciation. Most dramatically, the sedimentary package deposited immediately after the Paleoproterozoic low-latitude Makganyene glaciation in South Africa contains a banded iron formation-hosted manganese deposit that is the richest economic unit of this mineral known on Earth; Mn

---

[1]Division of Geological and Planetary Sciences, California Institute of Technology, Pasadena, CA 91125, USA.

FIGURE 1 Gypsum casts, mud cracks, and ripples from the Barr River Formation north of Elliot Lake, Ontario, Canada.

can only be precipitated from seawater by molecular oxygen (Kirschvink et al., 2000; Kopp et al., 2005). Similarly, Neoproterozoic glacial events are associated with apparent bursts of oxygenation and may have stimulated evolutionary innovations like the Ediacara fauna and the rise of Metazoa. We argue here that Precambrian glaciations are generally followed by fluctuations in apparent redox parameters, consistent with a postulate by Liang et al. (2006) that significant quantities of peroxide-generated oxidants are formed and released through glacial processes.

LOW-LATITUDE GLACIATION AS A SNOWBALL EARTH

Despite assertions to the contrary (Lovelock, 2006), climatic regulatory mechanisms have not always maintained large open areas of water on Earth's surface. Substantial evidence exists that large-scale continental ice sheets extended well into the tropics, yielding sea ice at the equator (Embleton and Williams, 1986; Evans et al., 1997; Sohl et al., 1999; Sumner et al., 1987). The deposition of banded iron oxide formations (BIFs) associated with glacial sediments implies both sealing off of air-sea exchange and curtailing the input of sulfate to the oceans, which otherwise would be reduced biologically to sulfide, raining out Fe as pyrite. The Snowball Earth hypothesis (Kirschvink, 1992) accounts for the

peculiarities of low-latitude tillites, BIFs, abrupt and broadly synchronous glacial onset and termination, and many other features of these events (Evans, 2000; Hoffman, 2007; Hoffman and Schrag, 2002; Hoffman et al., 1998). No alternative hypothesis even attempts to explain as many diverse features of the Precambrian glacial record.

Initially, the most fundamental result driving the Snowball Earth hypothesis was a soft-sediment fold test on a varvite-like member of the ~635 Ma Marinoan-age Elatina formation in South Australia, which implied incursion of sea ice into subtropical latitudes (Figure 2) (Sumner et al., 1987). A few years later, Evans et al. (1997) demonstrated similarly robust results from the ~2.22 Ga Makganyene glaciation in South Africa, indicating that at least two intervals of geological time, separated by more than a billion years, experienced low-latitude glaciation. Comparison of less robust paleomagnetic data for all Precambrian glaciations with well-documented paleolatitudes for Phanerozoic glacial deposits yields an interesting schism. With the possible exception of the Archean Pongola event, there is a total absence of evidence for polar or subpolar glaciation throughout the Precambrian, while marine glacial sedimentation never breaches the tropics through the Phanerozoic (Evans, 2003). While the counterintuitive Precambrian polar glacial gap must be largely an artifact of the paleogeographic and

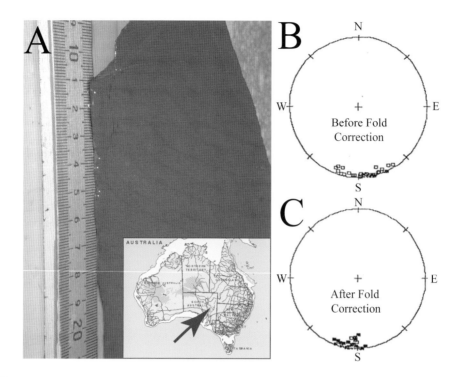

FIGURE 2 Soft-sediment paleomagnetic fold test on the rhythmite member of the Elatina Formation, Pichi-Richi Pass, Australia. The initial paleomagnetic study on this member by Embleton and Williams (1986) displayed a nearly equatorial remanent magnetization held in hematite, but lacked a geological field test to verify that the characteristic magnetization was acquired at or near the time of deposition. As part of the Precambrian Paleobiology Research Group (PPRG) at the University of California, Los Angeles, in 1986, Bruce Runnegar provided J. L. K. with an oriented block sample of this unit (Figure 2A), which displayed an apparent soft-sediment deformation feature. Careful subsampling and demagnetization of this block by then undergraduate student Dawn Sumner (now at the University of California, Davis) revealed a horizontally aligned, elliptical distribution of directions consistent with the earlier result (Figure 2B). However, correction for the bedding deformation significantly tightened the distribution, making it Fisherian and passing the McElhinny (1964) fold test at P <0.05. This result, along with an equally interesting result from a layer deformed by a glacial drop stone in the Rapitan Banded Iron Formation of Canada, was published as an American Geophysical Union abstract (Sumner et al., 1987); this led directly to the Snowball Earth Hypothesis (Kirschvink, 1992), and had the desired effect of stimulating further studies confirming the primary, low-latitude nature of the Elatina glacial event (Sohl et al., 1999; Williams et al., 1995).

rock preservation records (Evans, 2006), the data consensus points to an anomalously severe glacial mode in Proterozoic time relative to the Phanerozoic Era.

Evans (2003) suggests that this shift in Earth's glacial mode reflects the evolution of macroscopic continental life, especially of lichen and fungi through the Ediacaran-Cambrian transition (see also Peterson et al., 2005). Such organisms might modulate the silicate-weathering feedback to disfavor climate extremes, although the specifics of whether endolithic organisms promote or hinder physical and chemical weathering is surprisingly still ambiguous (see Beerling and Berner, 2005).

This fundamental Precambrian-Phanerozoic shift in Earth's glacial mode also appears to manifest itself in the relation of glacial events to a plate-tectonic supercontinent cycle. Figure 3 relates a simplified compilation of Earth's glacial record to a schematic representation of Earth's super-continents through time. Whereas the Paleoproterozoic and Neoproterozoic low-paleolatitude glacial events occupied

intervals dominated by dispersal of cratonic fragments from previous supercontinents (Kenorland and Rodinia, respectively), all Phanerozoic glacial events appear related to episodes of continental amalgamation. (Possible Ordovician glaciation could mark the formation of Gondwanaland; Carboniferous-Permian glaciation marks the assembly of Pangea; and the Miocene-present glacial epoch arguably presages the formation of a future supercontinent termed "SuperAsia" after the likely centroid of amalgamation.)

The characteristic length-scale of each supercontinent was centered at the "equator" and spread, as a yellow box, over the lifespan of that supercontinent. Blue waxing triangles indicate intervals of dominant supercontinent amalgamation, and red waning triangles indicate intervals of dominant supercontinent fragmentation and dispersal. A purple zone between the Paleoproterozoic supercontinent, Nuna, and the Mesoproterozoic-Neoproterozoic supercontinent, Rodinia, indicates basic uncertainty as to whether Nuna broke apart and reassembled into Rodinia, or whether

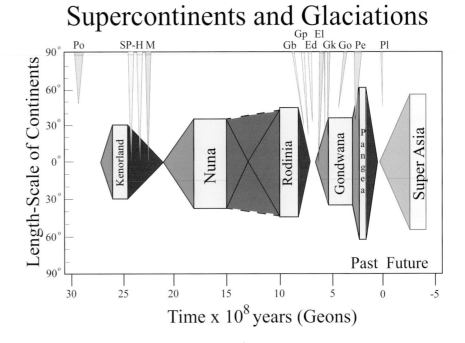

FIGURE 3 Character of glaciations and plate tectonics versus Earth history, measured in geons (100-million-year blocks of geological time). Surface areas for each of the demonstrated or likely supercontinents in Earth history were estimated, and a characteristic length scale for each supercontinent was defined as the square root of its surface area, converted from kilometers to degrees of arc. The vertical axis represents a characteristic meridian on Earth, running from 90 degrees north latitude to 90 degrees south.

a single supercontinent simply grew monotonically over that interval. The future supercontinent SuperAsia is predicted to begin its formal lifespan ~250 million years from now, when the oceanic lithosphere at the edge of the Atlantic ocean will have reached a foundering density and produced subduction zones for enough time to reunite South America with southern Africa and North America with northern Africa and Eurasia. Presumably Australia will have long since crumpled a neo-Himalayan orogenic belt still higher between its northern margin and southeast Asia-eastern India.

Maximum equatorward extents of ice ages were estimated from the paleomagnetic database (dark icicle fill) or using artistic license (light icicle fill) where paleomagnetic data do not yet exist (e.g., for the Neoproterozoic Ghubrah event). An icicle was dropped from the North Pole to that maximum equatorward latitude, with thickness approximating a plausible duration for each glacial event. Precambrian glaciations are abbreviated as follows: Po = Pongola; H-SP = Huronian and Snowy Pass Supergroups (at least two glaciations not correlative to South African Makganyene glaciation); M = Makganyene; Gp = Gariep; Gb = Ghubrah; Ed = Edwardsburg; El = Elatina-Ghaup; Gk = Gaskiers-Egan; Go = Gondwana; Pe = Permian; Pl = Pleistocene.

Whereas Precambrian glaciations appear restricted to intervals of supercontinent fragmentation and dispersal, Phanerozoic glaciations appear more generally associated with supercontinent amalgamation and intervals of orogenesis. All glaciations are plausibly connected with minor or major episodes of environmental oxidation or atmospheric oxygenation.

For most of Phanerozoic time, an integrated geochemi-

cal box model called GEOCARBSULF (Berner, 2006) predicts monotonic increases in atmospheric oxygen concentration spanning the late Ordovician, Carboniferous-Permian, and Miocene-present intervals of geologic time (for a recent discussion of the Paleozoic data, see Huey and Ward, 2005), with precipitous declines at the end-Permian. We suggest that a recent model for ice-based peroxide formation (Liang et al., 2006) contributes significantly to this Phanerozoic glaciation-oxygenation association, and extends even more significantly through the more severe Precambrian glacial episodes as well.

PONGOLA: EARTH'S OLDEST KNOWN GLACIATION

The middle Archean Pongola Supergroup exposed in Swaziland and parts of South Africa contains massive diamictite of the Klipwal and Mpatheni Members of the Delfkom Formation of the Mozaan Group (Young et al., 1998), which is constrained to be younger than underlying volcanics of the Nsuze group dated at 2985 ± 1 (Hegner et al., 1994) and older than a 2837 ± 5 Ma quartz porphyry sill (Gutzmer et al., 1999). The diamictites contain a diverse clast composition with striated and faceted pebbles, and occasional dropstones that attest to a glacial origin.

Although all sedimentary redox indicators throughout the Pongola Supergroup argue for widespread anoxia, studies of sulfur isotopes that indicate that mass-independent fractionation (MIF) decreases during and/or after the glacial intervals have been interpreted to support the presence of atmospheric oxygen (Bekker et al., 2005; Ohmoto et al., 2006). Although this is the conventional interpretation, *senso stricto* this is not required. The presence of significant MIF argues for O_2 levels below that needed to form an ozone-UV shield, whereas the absence of MIF could indicate either a volcanic sulfur source or increased ocean and atmosphere mixing. In fact, before the studies were done, Kopp et al. (2005) predicted that a drop in sulfur MIF would be present in the Pongola sediments simply from increased ocean and atmosphere mixing expected for a time of glaciation compared with an ice-free world. Nonetheless, relative oxidation of the oceans that could also draw down atmospheric SO_2 levels, even if unassociated with molecular oxygenation, remains a viable explanation for the geochemical blips associated with Pongola glaciation.

Nhelko (2004) studied the paleomagnetism of the Pongola diamictite and found an unusually strong and stable magnetization held in detrital magnetite, presumably derived from pulverizing basaltic-composition clasts present in the diamictite. He estimated the paleolatitude of deposition at ~48°, with positive fold and conglomerate tests on the characteristic, two-polarity magnetization. As other Snowball Earth lithostratigraphic markers such as cap carbonates and carbonate clasts are generally absent, there is as yet no suggestion that Earth's oldest glaciation might have been a low-latitude, global event.

PALEOPROTEROZOIC GLACIAL INTERVALS

At least three (and potentially many more) discrete intervals of glacial activity punctuate the geological record between about 2.45 and 2.22 Ga (e.g., Hambrey and Harland, 1981). Of these, the best-known and best-preserved belong to the Huronian Supergroup of Canada and the Transvaal Supergroup of southern Africa.

In Canada the classic Huronian succession includes the Ramsey Lake, Bruce, and Gowganda diamictites, separated from one another by thick successions of interbedded marine and fluvial sediments. A single carbonate unit (the Espanola formation) overlies the middle, Bruce Formation glacial horizon, with a gradual (not abrupt) transition from the diamictite to carbonate in the Elliot Lake region (abrupt transitions are seen elsewhere but could represent postglacial transgressions or unconformities). Basal volcanics have been U-Pb dated at ~2.45 Ga, and the entire glacial succession is cut by dikes and sills of the Nipissing swarm, providing an upper age constraint of ~2.22 Ga. Sedimentary indicators of a generally reducing surface environment are common in and around the Ramsey Lake and Bruce diamictites, but the first appearance of continental redbeds appears just after the Gowganda event. This is either strong evidence for surface redox conditions reaching the ferrous-ferric transition, or else the evolution of terrestrial iron-oxidizing organisms. As with the Archean Pongola event, MIF range of sulfur isotopes is diminished briefly after each glacial unit, hinting at but not proving transient oxidation events. Unfortunately, the paleolatitude of Huronian sedi-

mentation is not known, as all paleomagnetic components identified so far have failed field stability tests (Hilburn et al., 2005).

Glaciogenic units in southern Africa's Transvaal Supergroup include the Duitschland, Timeball Hill, and Makganyene formations. Hannah et al. (2004) obtained a Re-Os pyrite isochron from the Timeball Hill Formation yielding an age of 2.32 Ga for the unit, while Cornell et al. (1996) obtained a Pb-Pb isochron indicating ~2.22 Ga for the age of Ongeluk Formation volcanics that interfinger with the top of the Makganyene diamictite. The youngest detrital zircons from thin sedimentary interbeds between flows of the Ongeluk volcanics are ~2.23 Ga (Dorland, 2004), corroborating the Pb-Pb isochron for the volcanics themselves. Paleomagnetic data from the Ongeluk volcanics indicate that the Makganyene is a low-latitude Snowball Earth event (Evans et al., 1997; Kirschvink et al., 2000). Sedimentary redox indicators in the Duitschland and most of the Timeball Hill imply reducing conditions, but the uppermost units of the Timeball Hill formation contain a hematitic oolitic unit, which if primary, hints again that the redox potential of the atmosphere and ocean system reached the ferrous/ferric transition (which is energetically only halfway between the hydrogen and oxygen redox potentials).

In Canada a paleosol at Ville St. Marie, Quebec (Rainbird et al., 1990) contains granule and pebble clasts with reddened rims at approximately the stratigraphic level of the Lorrain Formation. In South Africa the final pulse of the glaciogenic succession records ice-rafted dropstones in the basal units of the massive banded iron and sedimentary manganese in the Hotazel Formation. Together with the superjacent, randomly magnetized hematitic breccia paleosol (Evans et al., 2001) (see "Background"), there is unequivocal evidence for significant oceanic oxidation as well as atmospheric oxygenation in the immediate aftermath of low-paleolatitude Snowball Earth glaciation.

STURTIAN AND MARINOAN

After at least a ~1-billion-year absence through late Paleoproterozoic time, all of the Mesoproterozoic Era, and the first half of the Neoproterozoic Era, BIFs reappear at <720 Ma, intimately associated with early glacial deposits of the "Cryogenian" interval (Klein and Beukes, 1993). At least three discrete glaciations punctuate the latter half of Neoproterozoic time (Evans, 2000), and current correlation schemes appear to permit five or more distinct events. The older among these tend to be associated with hematite-enriched BIFs interrupting otherwise suboxic-to-anoxic, organic-rich sediments, again suggesting penetration of oxidants to anomalous water depths accompanying deglaciation (Klein and Beukes, 1993).

The younger two of the Neoproterozoic deglaciations occupy the newly defined Ediacaran Period (Knoll et al., 2006), at its base (~635 Ma, Condon et al., 2005; Hoffmann et al., 2004) and approximately its middle (~580 Ma, Bowring et al., 2003). At ~635 Ma the basal Ediacaran "Marinoan" low-latitude event is only rarely associated with banded iron and sedimentary manganese formation (in Brazil's Urucum province), but it is frequently associated with reddened carbonate and shale dominating immediately postglacial sea-level transgression (e.g., Halverson et al., 2004).

Patterns of sulfur isotopic fractionation in carbonate-associated sulfate change across the Marinoan glaciation, such that seawater sulfate concentration was minimal during and after early Cryogenian glaciations, but significant following Marinoan glaciation (Hurtgen et al., 2005). Consistent with this trend, the postglacial transgressive sequences containing reddened carbonate and shale immediately after Marinoan deglaciation eventually culminate in black shale horizons with microbialaminate textures and isotopic signatures consistent with sulfate-reducing bacterial mat communities (e.g., Calver and Walter, 2000; Calver et al., 2004; see also Hoffman et al., 2007).

MID-EDIACARAN EGAN/GASKIERS GLACIATION

While the basal Ediacaran deglaciation marks the end of an unambiguously low latitude, likely Snowball Earth event, the middle interval of Ediacaran successions in northwest Australia and in Newfoundland is punctuated by a glacial event of uncertain severity. Correlation between the Egan glaciation in Australia's Ediacaran carbonate belt (Corkeron, 2007) and the Gaskiers glaciation in Newfoundland's Avalon terrane (Bowring et al., 2003) is not established, however both glacial events are younger than the Marinoan glaciation, and both are associated with anomalous carbonate facies in otherwise siliciclastic-dominated successions (Corkeron, 2007; Myrow and Kaufman, 1999).

As with the basal-Ediacaran Marinoan deglaciation, the mid-Ediacaran Gaskiers deglaciation is associated with postglacial reddening, culminating in pyrite-rich black shale at a presumed maximum flooding level. Silicate-hosted iron increases from pre-glacial to postglacial time, suggesting a step-function increase in atmospheric oxygen (Canfield et al., 2006).

Because the megascopic Ediacara fauna appear in the thick turbidite deposits following the Gaskiers deglaciation, back-of-the-envelope calculations suggest that the aftermath of the last Precambrian glaciation marked the first moment in Earth history when atmospheric oxygen levels exceeded ~15 percent of the present atmospheric level (Canfield et al., 2006). However, the Ediacara fauna have not yet been found in Newfoundland in the same, continuous stratigraphic section as the Gaskiers deglaciation, so the precise cause and effect of postglacial oxygenation and the evolution of complex life remains ambiguous.

CORRELATION CAVEATS FOR EDIACARAN-CAMBRIAN EVENTS

In a comprehensive study of inorganic and organic carbon, and sulfide as well as carbonate-associated-sulfate sulfur isotopes nearly spanning the Ediacaran Period, Fike et al. (2006) infer at least 25 million years of increasing bacterial sulfate reduction in the oceans following the Marinoan "Snowball" deglaciation. A sudden event known in Oman as the Shuram anomaly then quickly oxidized a previously isolated dissolved organic carbon reservoir, and the remainder of the Ediacaran Period experienced increasing levels of sulfur dissimilation reactions, permitted by enhanced oxygen concentrations (Fike et al., 2006).

Although the Shuram anomaly might correlate to the Gaskiers glacial event, in line with the general deglaciation-oxygenation association sketched in this paper, its age is strictly underconstrained, with widely varying estimates (e.g., see Condon et al., 2005, and Le Guerroue et al., 2006). Because the Ediacaran Period is ubiquitously punctuated with paleomagnetic anomalies suggesting multiple, rapid true polar wander events (Evans, 1998; Evans, 2003; Raub et al., 2007) which might also oxidize vast quantities of organic carbon (Kirschvink and Raub, 2003; Raub et al., 2007), glaciations are not the only available and attractive correlation targets for major isotopic excursions. In fact, decreased generation time and increased frequency of mutation fixation accompanying niche isolation and global warming in the aftermath of rapid true polar wander bursts has been proposed as an explicit mechanism linking true polar wander to the evolution of Ediacara and Metazoa (Kirschvink and Raub, 2003). In that respect, even the direct link between the final Precambrian, "Gaskiers" deglaciation and the evolution of animal phyla must be regarded as still hypothesized more than proven.

THE PEROXIDE PUMP: A MECHANISM FOR DEGLACIAL OXYGENATION

Many glaciologists have noted a semiregular oscillation in the quantity of hydrogen peroxide contained in Antarctic and Greenland ice cores, with concentrations increasing dramatically during the interval of enhanced ozone hole due to anthropogenic emissions (Frey et al., 2005, 2006; Hutterli et al., 2001, 2004). Similar peroxide peaks are inferred for the polar regions of Mars and the ice sheet encasing Jupiter's moon, Europa (Carlson et al., 1999).

Liang et al. (2006) generalize the phenomenon of peroxide snow produced by photolysis of water vapor above a cold ice sheet and applied 1-D mass-continuity models of peroxide production to hypothetical glacial scenarios, including Snowball Earths.

With modern volcanic outgassing and dry adiabatic lapse rates, and at modern atmospheric pressure and UV incidence, a ~10-million-year-long Snowball glacial event easily might rain out and capture in ice ~0.1 to 1.0 bar of molecular oxygen-equivalent hydrogen peroxide. The sensitivity of this astonishing result trends toward higher peroxide production for a depressed hydrologic cycle and lower global mean temperature, both plausible in a Snowball Earth scenario. UV-depletion of stratospheric ozone and enhanced molecular hydrogen escape to space (both correlated, among other factors, to decreased geomagnetic field intensity) would also increase peroxide mixing rates at Earth's surface.

We suggest that the model and mechanism of Liang et al. (2006) can explain a pan-Precambrian association in the geologic record of deglaciation with trace or significant environmental oxidation and, during the aftermath of at least the two most unambiguous Snowball Earth events, atmospheric oxygenation. We note that the Phanerozoic record of relative atmospheric oxygen concentration inferred by the GEOCARBSULF model is also consistent with monotonic oxygen production during and immediately following glaciation.

ACKNOWLEDGMENTS

T. D. R. was supported by a National Science Foundation Graduate Fellowship, and we gratefully acknowledge support from the Agouron Institute and the National Aeronautics and Space Administration Exobiology program.

REFERENCES

Beerling, D. J., and R. A. Berner. 2005. Feedbacks and the coevolution of plants and atmospheric CO2. *Proceedings of the National Academy of Sciences U.S.A.* 102(5):1302-1305.

Bekker, A., S. Ono, and D. Rumble. 2005. Low atmospheric pO_2 in the aftermath of the oldest Paleoproterozoic glaciation. *Astrobiology* 5(2):244.

Berner, R. A. 2006. GEOCARBSULF: A combined model for Phanerozoic atmospheric O_2 and CO_2. *Geochimica et Cosmochimica Acta* 70(23):5653-5664.

Bowring, S., P. Myrow, E. Landing, and J. Ramezani. 2003. Geochronological constraints on terminal Neoproterozoic events and the rise of metazoans. *Geophysical Research Abstracts* 5:13219.

Calver, C. R., and M. R. Walter. 2000. The late Neoproterozoic Grassy Group of King Island, Tasmania: Correlation and palaeogeographic significance. *Precambrian Research* 100(1-3):299-312.

Calver, C. R., L. P. Black, J. L. Everard, and D. B. Seymour. 2004. U-Pb zircon age constraints on late Neoproterozoic glaciation in Tasmania. *Geology* 32(10):893-896.

Canfield, D. E., S. W. Poulton, and G. M. Narbonne. 2006. Late-Neoproterozoic deep-ocean oxygenation and the rise of animal life. *Science* 315(5808):92-95.

Carlson, R. W., M. S. Anderson, R. E. Johnson, W. D. Smythe, A. R. Hendrix, C. A. Barth, L. A. Soderblom, G. B. Hansen, T. B. McCord, J. B. Dalton, R. N. Clark, J. H. Shirley, A. C. Ocampo, and D. L. Matson. 1999. Hydrogen peroxide on the surface of Europa. *Science* 283(5410):2062-2064.

Condon, D., M. Zhu, S. A. Bowring, W. Wang, A. Yang, and Y. Jin. 2005. U-Pb ages from the neoproterozoic Doushantuo Formation, China. *Science* 308(5718):95-98.

Corkeron, M. 2007. "Cap carbonates" and Neoproterozoic glacigenic successions from the Kimberley region, north-west Australia. *Sedimentology* 54:871-903.

Cornell, D. H., S. S. Schutte, and B. L. Eglington. 1996. The Ongeluk basaltic andesite formation in Griqualand West, South-Africa—submarine alteration in a 2222 Ma Proterozoic sea. *Precambrian Research* 79(1-2):101-123.

Dorland, H. 2004. Provenance, Ages, and Timing of Sedimentation of Selected Neoarchean and Paleoproterozoic Successions on the Kaapvaal Craton. D.Phil. thesis. Rand Afrikaans University, Johannesburg.

Embleton, B. J. J., and G. E. Williams. 1986. Low paleolatitude of deposition for late Precambrian periglacial varvites in South Australia—implications for Paleoclimatology. *Earth and Planetary Science Letters* 79(3-4):419-430.

Evans, D. A. 1998. True polar wander, a supercontinental legacy. *Earth and Planetary Science Letters* 157(1-2):1-8.

Evans, D. A. D. 2000. Stratigraphic, geochronological, and paleomagnetic constraints upon the Neoproterozoic climatic paradox. *American Journal of Science* 300:347-433.

Evans, D. A. D. 2003. A fundamental Precambrian-Phanerozoic shift in Earth's glacial style? *Tectonophysics* 375(1-4):353-385.

Evans, D. A. D. 2006. Proterozoic low orbital obliquity and axial-dipolar geomagnetic field from evaporite palaeolatitudes. *Nature* 444(7115):51-55.

Evans, D. A., N. J. Beukes, and J. L. Kirschvink. 1997. Low-latitude glaciation in the Paleoproterozoic. *Nature* 386(6622):262-266.

Evans, D. A. D., J. Gutzmer, N. J. Beukes, and J. L. Kirschvink. 2001. Paleomagnetic constraints on ages of mineralization in the Kalahari manganese field, South Africa. *Economic Geology* 96:621-631.

Fike, D. A., J. P. Grotzinger, L. M. Pratt, and R. E. Summons. 2006. Oxidation of the Ediacaran Ocean. *Nature* 444(7120):744-747.

Frey, M. M., R. W. Stewart, J. R. McConnell, and R. C. Bales. 2005. Atmospheric hydroperoxides in West Antarctica: Links to stratospheric ozone and atmospheric oxidation capacity. *Journal of Geophysical Research, Atmospheres* 110(D23).

Frey, M. M., R. C. Bales, and J. R. McConnell. 2006. Climate sensitivity of the century-scale hydrogen peroxide (H_2O_2) record preserved in 23 ice cores from West Antarctica. *Journal of Geophysical Research, Atmospheres* 111(D21301).

Gutzmer, J. N. Nhleko, N. J. Beukes, A. Pickard, and M. E. Barley. 1999. Geochemistry and ion microprobe (SHRIMP) age of a quartz porphyry sill in the Mozaan Group of the Pongola Supergroup: Implications for the Pongola and Witwatersrand supergroups. *South African Journal of Geology* 102(2):139-146.

Halverson, G. P., A. C. Maloof, and P. F. Hoffman. 2004. The Marinoan glaciation (Neoproterozoic) in northeast Svalbard. *Basin Research* 16(3):297-324.

Hambrey, M. J., and W. B. Harland. 1981. *Earth's Pre-Pleistocene Glacial Record*. Cambridge: Cambridge University Press.

Hannah, J. L., A. Bekker, H. J. Stein, R. J. Markey, and H. D. Holland. 2004. Primitive Os and 2316 Ma age for marine shale: Implications for Paleoproterozoic glacial events and the rise of atmospheric oxygen. *Earth and Planetary Science Letters* 225(1-2):43-52.

Hegner, E., A. Kröner, and P. Hunt. 1994. A precise U-Pb zircon age for the Archean Pongola Supergroup volcanics in Swaziland. *Journal of African Earth Sciences* 18:339-341.

Hilburn, I. A., J. L. Kirschvink, E. Tajika, R. Tada, Y. Hamano, and S. Yamamoto. 2005. A negative fold test on the Lorrain Formation of the Huronian Supergroup: Uncertainty on the paleolatitude of the Paleoproterozoic Gowganda glaciation and implications for the great oxygenation event. *Earth and Planetary Science Letters* 232:315-332.

Hoffman, P. F. 2007. Comment on "Snowball Earth on Trial." *EOS, Transactions of the American Geophysical Union* 88(9):110.

Hoffman, P. F., and D. P. Schrag. 2002. The Snowball Earth hypothesis: Testing the limits of global change. *Terra Nova* 14(3):129-155.

Hoffman, P. F., A. J. Kaufman, G. P. Halverson, and D. P. Schrag. 1998. A Neoproterozoic Snowball Earth. *Science* 281(5381):1342-1346.

Hoffman, P. F., G. P. Halverson, E. W. Domack, J. M. Husson, J. A.Higgins, and D. P. Schrag. 2007. Are basal Ediacaran (635 Ma) post-glacial "cap dolostones" diachronous? *Earth and Planetary Science Letters* 258(1-2):114-131.

Hoffmann, K. H., D. Condon, S. A. Bowring, and J. L. Crowley. 2004. U-Pb zircon date from the Neoproterozoic Ghaub Formation, Namibia: Constraints on Marinoan glaciation. *Geology* 32(9):817-820.

Huey, R. B., and P. D. Ward. 2005. Hypoxia, global warming, and terrestrial Late Permian extinctions. *Science* 308:398-401.

Hurtgen, M. T., M. A. Arthur, and G. P. Halverson. 2005. Neoproterozoic sulfur isotopes, the evolution of microbial sulfur species, and the burial efficiency of sulfide as sedimentary pyrite. *Geology* 33(1):41-44.

Huston, D. L., and G. A. Logan. 2004. Barite, BIFs and bugs: Evidence for the evolution of the Earth's early hydrosphere. *Earth and Planetary Science Letters* 220(1-2):41-55.

Hutterli, M. A., J. R. McConnell, R. W. Stewart, H. W. Jacobi, and R. C. Bales. 2001. Impact of temperature-driven cycling of hydrogen peroxide (H_2O_2) between air and snow on the planetary boundary layer. *Journal of Geophysical Research, Atmospheres* 106(D14):15395-15404.

Hutterli, M. A., J. R. McConnell, G. Chen, R. C. Bales, D. D. Davis, and D. H. Lenschow. 2004. Formaldehyde and hydrogen peroxide in air, snow and interstitial air at South Pole. *Atmospheric Environment* 38(32):5439-5450.

Kirschvink, J. L. 1992. Late Proterozoic low-latitude global glaciation: The Snowball Earth. Section 2.3 in *The Proterozoic Biosphere: A Multidisciplinary Study*, eds. J. W. Schopf et al., pp. 51-52. Cambridge: Cambridge University Press.

Kirschvink, J. L., and T. D. Raub. 2003. A methane fuse for the Cambrian explosion: Carbon cycles and true polar wander. *Comptes Rendus Geoscience* 335(1):65-78.

Kirschvink, J. L., E. J. Gaidos, L. E. Bertani, N. J. Beukes, J. Gutzmer, L. N. Maepa, and R. E. Steinberger. 2000. Paleoproterozoic Snowball Earth: Extreme climatic and geochemical global change and its biological consequences. *Proceedings of the National Academy of Sciences U.S.A.* 97(4):1400-1405.

Klein, C., and N. J. Beukes. 1993. Sedimentology and geochemistry of the glaciogenic Late Proterozoic Rapitan iron-formation in Canada. *Economic Geology and the Bulletin of the Society of Economic Geologists* 88(3):542-565.

Knoll, A. H., M. R. Walter, G. M. Narbonne, and N. Christie-Blick. 2006. The Ediacaran Period: A new addition to the geologic time scale. *Lethaia* 39(1):13-30.

Kopp, R. E., J. L. Kirschvink, I. A. Hilburn, and C. Z. Nash. 2005. The paleoproterozoic snowball Earth: A climate disaster triggered by the evolution of oxygenic photosynthesis. *Proceedings of the National Academy of Sciences U.S.A.* 102:11131-11136.

Le Guerroue, E., P. A. Allen, A. Cozzi, J. L. Etienne, and M. Fanning. 2006. 50 Myr recovery from the largest negative delta C-13 excursion in the Ediacaran ocean. *Terra Nova* 18(2):147-153.

Liang, M. C., H. Hartman, R. E. Kopp, J. L. Kirschvink, and Y. L. Yung. 2006. Production of hydrogen peroxide in the atmosphere of a Snowball Earth and the origin of oxygenic photosynthesis. *Proceedings of the National Academy of Sciences U.S.A.* 103(50):18896-18899.

Lovelock, J. E. 2006. *The Revenge of Gaia: Earth's Climate Crisis and the Fate of Humanity.* New York: BasicBooks.

McElhinny, M. W. 1964. Statistical significance of the fold test in paleomagnetism. *Geophysical Journal of the Royal Astronomical Society* 8:338-340.

Myrow, P. M., and A. J. Kaufman. 1999. A newly discovered cap carbonate above Varanger-age glacial deposits in Newfoundland, Canada. *Journal of Sedimentary Research* 69(3):784-793.

Nhelko, N. 2004. The Pongola Supergroup in Swaziland. D.Phil. thesis. Rand Afrikaans University, Johannesburg.

Ohmoto, H., Y. Watanabe, H. Ikemi, S. R. Poulson, and B. E. Taylor. 2006. Sulphur isotope evidence for an oxic Archaean atmosphere. *Nature* 442(7105):908-911.

Peterson, K. J., M. A. McPeek, and D. A. D. Evans. 2005. Tempo and mode of early animal evolution: Inferences from rocks, Hox, and molecular clocks. *Paleobiology* 31(2):36-55.

Rainbird, R. H., M. A. Hamilton, and G. M. Young. 1990. Formation and diagenesis of a sub-Huronian saprolith—comparison with a modern weathering profile. *Journal of Geology* 98(6):801-822.

Raub, T. D., J. L. Kirschvink, and D. A. D. Evans. 2007. True polar wander: Linking deep and shallow geodynamics to hydro- and bio-spheric hypotheses. In *Treatise on Geophysics*, vol. 5, ch. 14, eds. M. Kono and G. Schubert. Washington, D.C.: American Geophysical Union.

Sohl, L. E., D. V. Kent, and N. Christie-Blick. 1999. Paleomagnetic polarity reversals in Marinoan (ca 600 Ma) glacial deposits of Australia: Implications for the duration of low-latitude glaciation in neoproterozoic time. *Geological Society of America Bulletin* 111(8):1120-1139.

Sumner, D. Y., J. L. Kirschvink, and B. N. Runnegar. 1987. Soft-sediment paleomagnetic field tests of late Precambrian Glaciogenic Sediments. *Transactions of the American Geophysical Union* 68:1251.

Walker, J. C. G., P. B. Hays, and J. F. Kasting. 1981. A negative feedback mechanism for the long-term stabilization of Earth's surface temperature. *Journal of Geophysical Research, Oceans and Atmospheres* 86(NC10):9776-9782.

Williams, G. E., P. W. Schmidt, and B. J. J. Embleton. 1995. The Neoproterozoic (1000-540 Ma) Glacial Intervals—no more Snowball Earth. Comment. *Earth and Planetary Science Letters* 131(1-2):115-122.

Young, G. M., D. G. F. Young, C. M. Fredo, and H. W. Nesbitt. 1998. Earth's oldest reported glaciation: Physical and chemical evidence from the Archean Mozaan Group (similar to 2.9 Ga) of South Africa. *Journal of Geology* 106(5):523-538.

Cooper, A. K., P. J. Barrett, H. Stagg, B. Storey, E. Stump, W. Wise, and the 10th ISAES editorial team, eds. (2008). *Antarctica: A Keystone in a Changing World.* Proceedings of the 10th International Symposium on Antarctic Earth Sciences. Washington, DC: The National Academies Press.

Tectonics of the West Antarctic Rift System: New Light on the History and Dynamics of Distributed Intracontinental Extension

C. S. Siddoway[1]

ABSTRACT

The West Antarctic rift system (WARS) is the product of multiple stages of intracontinental deformation from Jurassic to Present. The Cretaceous rifting phase accomplished >100 percent extension across the Ross Sea and central West Antarctica, and is widely perceived as a product of pure shear extension orthogonal to the Transantarctic Mountains that led to breakup and opening of the Southern Ocean between West Antarctica and New Zealand. New structural, petrological, and geochronological data from Marie Byrd Land reveal aspects of the kinematics, thermal history, and chronology of the Cretaceous intracontinental extension phase that cannot be readily explained by a single progressive event. Elevated temperatures in "Lachlan-type" crust caused extensive crustal melting and mid-crustal flow within a dextral transcurrent strain environment, leading to rapid extension and locally to exhumation and rapid cooling of a migmatite dome and detachment footwall structures. Peak metamorphism and onset of crustal flow that brought about WARS extension between 105 Ma and 90 Ma is kinematically, temporally, and spatially linked to the active convergent margin system of East Gondwana. West Antarctica-New Zealand breakup is distinguished as a separate event at 83-70 Ma, from the standpoint of kinematics and thermal evolution.

INTRODUCTION

Heightened interest in West Antarctica (WANT) and the West Antarctic rift system (WARS) comes from new determinations of the mantle thermal profile (Lawrence et al., 2006) and the context for active volcanism (Behrendt et al., 1994, 1996) arising at a time of instability of the West Antarctic ice sheet, when information is sought about the influence of underlying crustal structures on glaciological and glacial-marine systems (e.g., Holt et al., 2006; Lowe and Anderson, 2002; Vaughan et al., 2006). The question of heat flux arising from warm mantle beneath thinned crust is of obvious consequence for ice-sheet dynamics (Maule et al., 2005). The area of thin crust corresponding to the WARS (Figure 1) includes the Ross Sea and Ross Ice Shelf, the West Antarctic ice sheet (WAIS), and Marie Byrd Land (Behrendt et al., 1991; Storey et al., 1999; Fitzgerald, 2002; Siddoway et al., 2005).

In the geological record the WARS has distinctive but differing expressions in both Cenozoic and Mesozoic time. By far the better-known rift phase is the mid-Cenozoic to Present interval. Widespread basaltic volcanism (Behrendt et al., 1994, 1996; Finn et al., 2005; Rocchi et al., 2005), slow mantle seismic velocities (Danesi and Morelli, 2001; Ritzwoller et al., 2001; Sieminski et al., 2003), and anomalous elevation of thinned continental crust (LeMasurier and Landis, 1996; LeMasurier, 2008) are the hallmarks of the Cenozoic rift. The Victoria Land Basin and Terror rift, on the western limit of the WARS, record modest extension on the order of 150 km in Eocene-Oligocene time (Stock and Cande, 2002; Davey and DeSantis, 2006). The dramatic relief of the Transantarctic Mountains (TAM) developed in the Eocene (ten Brink et al., 1997; Fitzgerald, 2002), and voluminous basin sedimentation commenced (Hamilton et al., 1998; Cape Roberts Science Team, 2000; Luyendyk et al., 2001; Karner et al., 2005), considerably later than major extension in the WARS. Not surprising in light of the dominantly Eocene activation of the TAM boundary (ten Brink et al., 1997; Fitzgerald, 2002), on-land structures attributable to preceding Cretaceous events are few in the TAM (Wilson, 1992).

[1]Department of Geology Colorado College, Colorado Springs, CO 80903, USA (csiddoway@coloradocollege.edu).

FIGURE 1 The Cretaceous West Antarctic rift system at ca. 90 Ma, illustrating the positions of Marie Byrd Land and New Zealand/Campbell Plateau along the East Gondwana margin. The rifted margin corresponds to the −1500 m contours (dashed-line pattern). The tight fit of the reconstruction, the linear to curvilinear continental margin, and the pronounced depth increase suggest fault control and steep fault geometry. The diagram is based on the reconstructions of Lawver and Gahagan (1994) and Sutherland (1999). The present-day position of the Transantarctic Mountains, as labeled, corresponds to the western tectonic boundary of the West Antarctic Rift System. FM = Fosdick Mountains; EP = Edward VII Peninsula; 270 = DSDP site 270.

The lesser-known phase of extension and lithospheric thinning that brought about formation of the vast rift system ($\sim 1.2 \times 10^6$ km^2) did not occur in Cenozoic but in Mesozoic time (Tessensohn and Wörner, 1991; Lawver and Gahagan, 1994; Luyendyk, 1995). Although the origins of the WARS may be linked to Weddell Sea opening and Ferrar magmatism in the Jurassic (Grunow et al., 1991; Wilson, 1993; Jokat et al., 2003; Elliot and Fleming, 2004), dramatic intracontinental extension occurred in Cretaceous time. Much of the basis of knowledge about the Ross Sea sector of the rift comes from ocean bottom seismograph, multichannel seismic reflection, and gravity surveys that revealed a N-S structure of elongate basins marked by a positive gravity anomaly and high seismic velocities in the lower crust and 1-4 km of inferred Mesozoic sedimentary fill (Cooper and Davey, 1985; Cooper et al., 1997; Trey et al., 1997). Paradoxically, major sedimentary infilling of basins with material postdat-

ing regional unconformity RU6 evidently was delayed until the Eocene to Miocene (Hamilton et al., 1998; Wilson et al., 1998; Cape Roberts Science Team, 2000; Luyendyk et al., 2001; Karner et al., 2005). This is despite the rapidity of the large magnitude extension, on the order of 600 km in the south, up to >1000 km in the north achieved in as little as 20 m.y. (DiVenere et al., 1996; Luyendyk et al., 1996). A second paradox is that breakup between WANT and New Zealand (NZ) did not exploit rift structures but rather cut at a high angle across basement highs and basins of the Ross Sea (Tessensohn and Wörner, 1991; Lawver and Gahagan, 1994; Sutherland, 1999). Wrench deformation and the presence of strike slip transfer systems was postulated (Grindley and Davey, 1982) but not substantiated from exposures on land.

New perspective on intracontinental extension in the WARS comes from geological and geophysical research that investigates the exposed bedrock of WANT, NZ, and the Tasman Sea region (Figure 1). WANT, NZ, and submarine plateaus formed a contiguous segment of the convergent margin of East Gondwana in Early Cretaceous time, with arc magmatism recorded in Marie Byrd Land-NZ. Transtension–extension occurred in a back arc to inboard setting, forming the intracontinental West Antarctic rift system and Great South Basin-Campbell Plateau extensional province (Figure 1).

Since 1990, data acquired from geological investigations on land and from airborne and marine geophysical surveys in the region of Marie Byrd Land have dramatically increased the understanding of the eastern WARS, with consequences for our conception of the Cretaceous to Present multistage evolution of the West Antarctic rift system as a whole. The aim of this paper is to summarize the tectonic evolution of western Marie Byrd Land (MBL) (Figure 2) and of neighboring segments of the proto-Pacific margin of East Gondwana (Figure 3). Little affected by Cenozoic events (cf. Fitzgerald, 2002; Stock and Cande, 2002), the eastern Ross Sea region preserves a clear record of the kinematics, magmatism, and thermal history of the Early Cretaceous large-scale opening of the WARS.

Knowledge of the Cretaceous tectonic evolution of the WARS-NZ-Tasman Sea region provides an important foundation for contemporary research in WANT, including studies of the Cretaceous to present landscape evolution (LeMasurier and Landis, 1996; LeMasurier, 2008) involving a postulated orogenic plateau (Bialas et al., 2007; Huerta, 2007), the origins of the Southern Ocean's diffuse alkaline magmatism (Finn et al., 2005; Rocchi et al., 2005), the causes for Cenozoic structural reactivation (e.g., Salvini et al., 1997; Rossetti et al., 2003a,b) and seismicity (Winberry and Anandakrishnan, 2004), and the affects of inherited structures upon ice-bedrock interactions of the dynamic WAIS (Lowe and Anderson, 2002; Holt et al., 2006; Vaughan et al., 2006; Sorlien et al., 2007).

Extent of the West Antarctic Rift System (WARS) and Character of WARS Crust

The region corresponding to the Cretaceous WARS includes the Ross Sea and Ross Ice Shelf, the area of the WAIS, and Marie Byrd Land (Behrendt, 1991, 1999; Behrendt et al., 1991; Cooper et al., 1991a,b; Storey et al., 1999; Trey et al., 1999). Measured orthogonal to the TAM, the WARS widens from 600 km in the south (Storey et al., 1999) to 1200 km across the northern Ross Sea (Luyendyk et al., 2003). Thickness of continental crust ranges from 17-19 km for the Central and Eastern Basins of the Ross Sea to 23-24 km beneath basement highs (Cooper et al., 1991b, 1997; Davey and Brancolini, 1995; Luyendyk et al., 2001). There is a similar range beneath central West Antarctica (Behrendt et al., 1994; Bell et al., 1998). The crust underlying western MBL is ca. 23 km thick, based on airborne geophysics (Figure 2) (Ferraccioli et al., 2002; Luyendyk et al., 2003). This provides evidence that the region of western MBL that is above sea level is part of the WARS province.

The western margin of the WARS extensional province coincides with the TAM, at the long-standing tectonic boundary of the East Antarctica (EANT) craton that initiated as a Neoproterozoic rift margin (Dalziel, 1997), underwent convergence during the Ross Orogeny (Stump, 1995), and was reactivated during the initial two-plate phase of Gondwana breakup in the Jurassic Era (Dalziel et al., 1987). Tholeiitic Ferrar magmatism (Elliot et al., 1999; Elliot and Fleming, 2004) and modest extension to transtension initiated in the WARS at this time (Storey, 1991, 1992; Wilson, 1993). The Cretaceous WARS has been inferred to be an asymmetrical extensional system with the TAM forming the structural upper plate and the WARS, the lower plate (e.g., Fitzgerald et al., 1986; Stern and ten Brink, 1989; Fitzgerald and Baldwin, 1997; compare Lister et al., 1991).

Beneath the Ross Sea, gravity and marine seismic data delineate a crustal structure of N-S grabens, marked by high-density material in the axial regions, separated by basement highs. A large positive gravity anomaly in the basin axes is interpreted as mafic igneous material emplaced into the lower crust (Cooper et al., 1997; Trey et al., 1997). Sedimentary fill in the deep basins is cut by faults and overlapped by a regional unconformity, RU6, that predates thick glacial sediments. The thickness of Mesozoic to early Tertiary sediments is comparatively modest, reaching 4 km in the Eastern Basin, diminishing toward the coast of MBL (Luyendyk et al., 2001).

In a region of thinned crust but exposed above sea level, the Ford Ranges and Edward VII Peninsula (Figure 2) are key locations for examining the crust that constitutes the WARS and observing the structures responsible for the Cretaceous extension. The oldest rocks exposed are Swanson Formation metagreywacke and Ford Granodiorite of Paleozoic age (Bradshaw et al., 1983; Weaver et al., 1991).

These are intruded by Cretaceous alkalic plutonic rocks (Figure 4) that are genetically linked to the WARS (Weaver et al., 1992, 1994). Lower Paleozoic Swanson Formation represents one of the packages of voluminous quartz-rich turbidites deposited in regionally extensive clastic fans shed from the Ross-Delamerian Orogen (Fergusson and Coney, 1992) or distant Transgondwana orogen (Squire et al., 2006). The rock assemblages that were contiguous along the East Gondwana margin (Figure 3) include the Swanson Formation in MBL (Bradshaw et al., 1983), the western Lachlan Belt in Australia (Glen, 2005), the Robertson Bay group in north Victoria Land (NVL) (Rossetti et al., 2006), and the Greenland Group in NZ (Cooper and Tulloch, 1992; Adams et al., 1995, 2005; Gibson and Ireland, 1996; Adams, 2004; Bradshaw, 2007).

The Swanson Formation was deformed and metamorphosed to low greenschist grade (Adams, 1986) prior to emplacement of latest Devonian to Carboniferous calc-alkaline plutons of the Ford Granodiorite (Figure 3) (Adams, 1987; Weaver et al., 1991). Ford Granodiorite represents the first in a succession of convergent margin arcs developed upon the East Gondwana margin from Ordovician through Early Cretaceous time (Pankhurst et al., 1998), and it has correlatives in New Zealand (Muir et al., 1996). Both Swanson Formation and Ford Granodiorite were affected by high-temperature (HT) metamorphism and their high-grade equivalents are exposed in the Fosdick Mountains migmatite gneiss dome (Siddoway et al., 2004b; Saito et al., 2007) (Figure 4). Temperatures in excess of 800°C, sufficient to cause voluminous melting, were attained two to three times in the history of the dome (Siddoway et al., 2006; Korhonen et al., 2007a,b). The most recent migmatization phase coincided with alkalic plutonism in MBL marked by Byrd Coast granite and mafic dikes (Weaver et al., 1992, 1994; Adams et al. 1995; Siddoway et al., 2005).

Pankhurst et al. (1998) introduced the term "Ross Province" for the Swanson-Ford association in western MBL (Figure 2). Correlatives of the Ross Province exist throughout the former Gondwana margin, including a number of culminations of HT metamorphic rocks derived from Paleozoic protoliths (Tulloch and Kimbrough, 1989; Morand, 1990; Ireland and Gibson, 1998; Vernon and Johnson, 2000; Richards and Collins, 2002; Hollis et al., 2004) (Figure 4). It is probable that Ross-Delamerian orogenic sediments and the intermediate plutonic rocks that intrude them constitute the majority of the crust within the Ross Sea sector of the WARS (Bradshaw, 2007), its continuation into New Zealand (e.g., Cook et al., 1999), and into the submerged extended continental crust bordering the Tasman Sea. Within the WARS crust, there is also sparse evidence of "Ross-aged" basement rocks with an affinity to the TAM (Fitzgerald and Baldwin, 1997; Pankhurst et al., 1998; Bradshaw, 2007).

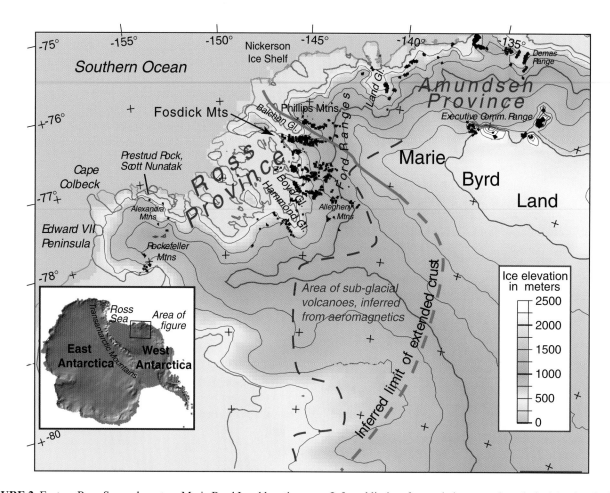

FIGURE 2 Eastern Ross Sea and western Marie Byrd Land location map. Inferred limits of extended crust and a subglacial volcanic field, determined from airborne geophysics (Luyendyk et al., 2003) are indicated. The Ross Province (Pankhurst et al., 1998) of the Ford Ranges comprises lower Paleozoic sedimentary rocks intruded by Devono-Carboniferous intermediate plutons. The rock exposures east of Land Glacier are dominated by intermediate to mafic arc-related plutonic rocks, with subsidiary, younger alkalic intrusions; an association termed the "Amundsen Province" by Pankhurst et al. (1998). Base map by D. Wilson.

THE ACTIVE MARGIN OF EAST GONDWANA AND FORMATION OF THE MESOZOIC WEST ANTARCTIC RIFT SYSTEM

Convergent Margin Plutonism

Mesozoic convergent tectonism with intermittent subduction-related plutonism and terrane accretion is recorded in West Antarctica (Vaughan and Livermore, 2005) and contiguous parts of NZ (e.g., Bradshaw et al., 1997). The calc-alkaline, I-type Median Batholith was emplaced in NZ between 145 Ma and 120 Ma, with some ages older, to 170 Ma (Muir et al., 1998; Mortimer et al., 1999a,b; Tulloch and Kimbrough, 2003); and tectonic reconstructions show continuity of the magmatic arc, together with associated tectonic terranes, into MBL-Thurston Island (Figure 3) (e.g., Bradshaw et al., 1997). The arc province in MBL, termed the "Amundsen Province" (Pankhurst et al., 1998), was the site

of intermediate plutonism spanning the interval 124 to 96 Ma (Pankhurst et al., 1998; Mukasa and Dalziel, 2000).

The timing of HT metamorphism in NZ is determined by U-Pb ages on metamorphic zircon or titanite sampled from gneisses at sites distributed along the convergent margin. These include the Paparoa range at 119-109 Ma (Kimbrough and Tulloch, 1989; Ireland and Gibson, 1998; Spell et al., 2000); and Fiordland at 126-110 Ma (Ireland and Gibson, 1998; Hollis et al., 2004; Scott and Cooper, 2006). Granulite metamorphism documented in Fiordland at 108 ± 3 Ma (Gibson and Ireland, 1995) gives an indication that elevated and compressed crustal isotherms developed during convergent tectonism (Figure 5).

In the Fosdick Mountains gneiss dome in MBL, new U-Pb SHRIMP ages of 115 ± 1 Ma have been acquired for igneous zircon within K-feldspar leucogranite equated with anatectic leucosome, that has been sampled at deepest structural levels (Siddoway et al., 2006). Nd isotope data indicate

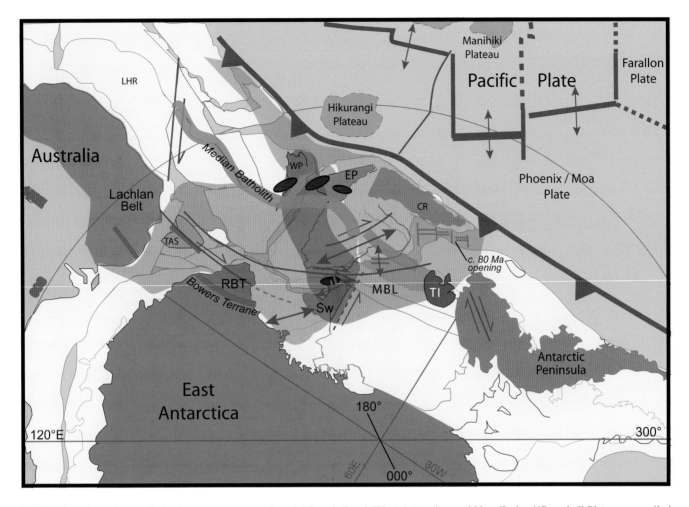

FIGURE 3 Tectonic correlation between terranes of north Victoria Land, West Antarctica, and New Zealand/Campbell Plateau, compiled from Bradshaw (1989) and Bradshaw et al. (1997). Reconstruction of the Cretaceous East Gondwana margin is based on Gaina et al. (1998) and Kula et al. (2007), with oceanic plates configuration based on Sutherland and Hollis (2001) and Larter et al. (2002). Representation of oceanic plateaus is based on Taylor (2006) and Hoernle et al. (2004). Lower Paleozoic orogenic sediments are shown in olive green and tan. Belt of Cretaceous magmatism is shown in violet. Paparoa metamorphic core complex in the Western Province (WP), Fosdick gneiss dome in Marie Byrd Land (MBL), and detachment systems are marked by ellipses. TAS = Tasmania; RBT = Robertson Bay terrane; EP = Eastern Province; Sw = Swanson Formation; LHR = Lord Howe Rise; CR = Chatham Rise; TI = Thurston Island terrane.

a Ford Granodiorite source for the leucogranites (Saito et al., 2007) at T, P conditions of 820-870°C and 6.5-7.5 kbar determined from mineral equilibria modeling (Korhonen et al., 2007a,b). There is evidence of metamorphic zircon growth as early as ca. 140 Ma. A summary of U-Pb SHRIMP analyses of igneous and metamorphic zircon from Fosdick Mountains migmatites (Figure 6) reveals that there is a bimodal distribution of ages. Whereas HT metamorphism and zircon growth is recorded as early as 150 Ma, a majority of points analyzed thus far fall within the interval of 120-100 Ma. Anatectic leucosomes from sites in MBL's Amundsen Province, the Demas Range (Figure 2) yield ages of 128 Ma to 113 Ma for igneous zircon (Mukasa and Dalziel, 2000). The MBL data fall within the 126-107 Ma age range of the youngest arc-related intrusions identifid in NZ by Muir et al. (1997,

1998), those of the Separation Point and Rahu suites in the Western Province and the deeper level Fiordland Orthogneiss in Fiordland. The Rahu suite granites are interpreted to derive from crustal melting of preexisting rocks (Ireland and Gibson, 1998). Thus, the conditions for HT metamorphism and granite genesis in the Fosdick Mountains were attained and overlapped in time with arc plutonism in the Median Batholith and in the Amundsen Province.

By contrast, the alkaline plutonism attributed to back-arc extension occurred in eastern MBL (Figure 5) at 105-102 Ma (Weaver et al., 1992, 1994; Mukasa and Dalziel, 2000), distinctly later than onset of high temperature metamorphism. In western MBL the Ford Ranges experienced alkalic plutonism at 105-103 Ma and ca. 99 Ma (Richard et al., 1994) and in Edward VII Peninsula at 103-98 Ma (Mukasa and Dalziel,

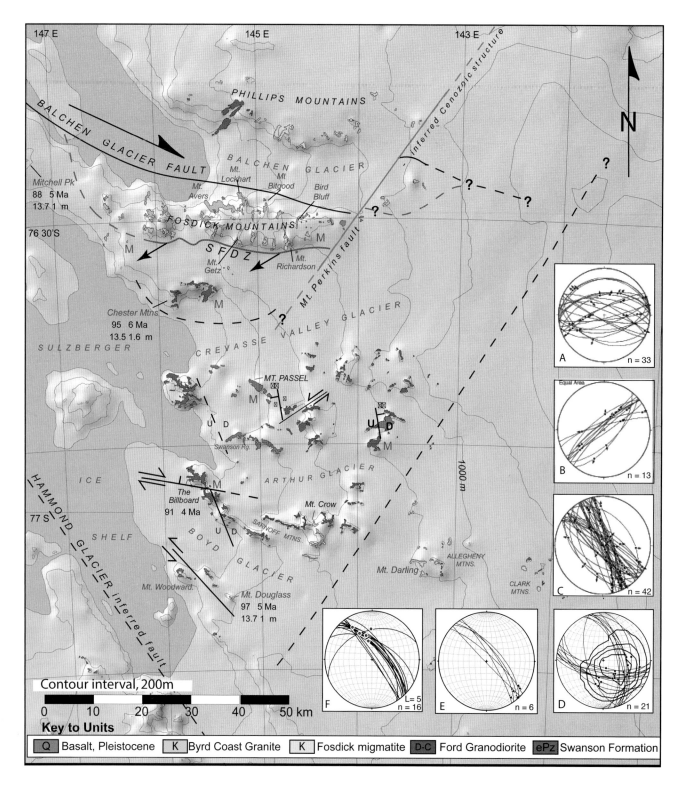

FIGURE 4 Structural-geological map of the Ford Ranges, western Marie Byrd Land. Inferred faults that are concealed by ice are mapped on the basis of contrasts in metamorphic grade between ranges, geophysical lineaments or boundaries, and zones of penetrative brittle deformation in rock exposures. AFT cooling ages and track lengths (Richard et al., 1994; Lisker and Olesch, 1998) are indicated for selected sites. Brittle mesoscopic fault data are shown in stereographic plots (insets). Sites and kinematic sense are as follows: (A) southern Ford Ranges, normal oblique; (B) Mt. Darling, sinistral (Cenozoic); (C) southern Ford Ranges, sinistral oblique; (D) Sarnoff Range, normal oblique; (E) Mt. Crow, sinistral; (F) Mt. Woodward, sinistral oblique. The label "M" indicates sites of glacial deposits examined for clast provenance. Stereonet v. 6.3.3 and FaultKin 4.3.5, by R. Allmendinger 1989-2004, were used for plotting stereographic diagrams. Shaded relief ice topography and base map prepared from Antarctic Digital Database by G. Balco.

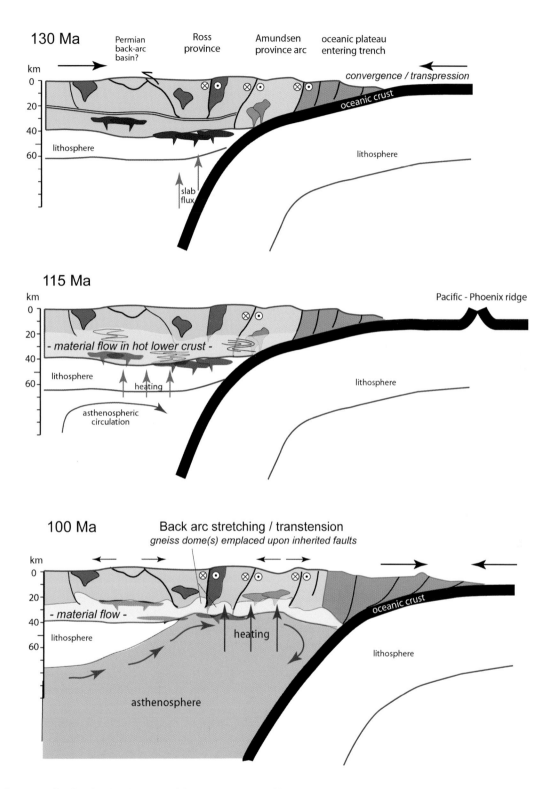

FIGURE 5 Conceptualization for development of the West Antarctic rift system inboard of the Mesozoic convergent margin during oblique plate convergence and subduction of young oceanic lithosphere, including oceanic plateaus. *Top:* Crustal thickening and advective heating during development of Amundsen province magmatic arc; active margin undergoing transpression due to oblique convergence. *Middle:* Heating of the lower crust causes partial melting and lateral flow in the middle and lower crust; thermal gradient is increased. Thickening of the crust continues but the lower lithosphere thins. Upper crust undergoes brittle faulting. *Bottom:* Change to transtension, with oblique opening across preexisting high-angle faults. Lateral flow of hot, weak, partially molten lower crust is accompanied by brittle deformation in shallow upper crust. Strain perturbation along faults allows localized gravity-driven vertical flow of lower-density migmatite-diatexite and formation of gneiss dome(s). Isotherms are elevated, tectonic exhumation and cooling are enhanced, next to transcurrent faults.

2000; Siddoway et al., 2004a). The dominant plutonic rock is Byrd Coast Granite (Figure 4). Mafic alkalic dikes and syeno-granites were emplaced over a wide region. A dolerite dike swarm at 107 + 5 Ma was followed closely by 102-95 Ma syenite and alkalic granite in the Amundsen province (Storey et al., 1999). A wider a range of dike ages, 142 to 96 Ma, comes from the Ross Province (Siddoway et al., 2005). In the Ross province an early phase of Byrd Coast alkalic granite from the Allegheny Mountains is ca. 142 Ma, and another at Mt. Corey is 131 Ma (Figure 2) (Adams, 1987). Subduction ceased in New Zealand at 105 ± 5 Ma (Muir et al., 1994, 1995, 1997, 1998).

Alkalic magmatism in NZ coincided with extension, development of a regional unconformity, and dramatic sedimentation, including thick deposits of sedimentary breccia (Laird and Bradshaw, 2004). Metamorphic core complexes developed in South Island (Figure 3) (Tulloch and Kimbrough, 1989; Forster and Lister, 2003), together with deep level shear zones that were active in Fiordland (Gibson et al., 1988; Scott and Cooper, 2006).

The Lachlan belt on continental Australia and Robertson Bay terrane in NVL occupied an inboard position in middle Cretaceous time and did not experience tectonism related to the active margin (Figure 3), although Australia-EANT breakup was in its initial stages (Li and Powell, 2001, and references cited). Remnants of the active margin are the submarine plateaus that border the Tasman Sea and provide a sparse geological record of the change in plate dynamics (Tulloch et al., 1991; Mortimer et al., 1999a, 2006; Mortimer, 2004).

GEOLOGICAL STRUCTURE OF THE FORD RANGES, MARIE BYRD LAND: DATA BEARING ON TECTONISM IN THE WEST ANTARCTIC RIFT SYSTEM

In MBL the general absence of dynamic fabrics in plutonic units and the elusive nature of crustal-scale faults in a region with extensive ice cover have long hindered the understanding of the strain evolution. New progress has been made in the region through tectonic investigations focused on the structure and metamorphic petrology of the Fosdick Mountains gneiss dome (Siddoway et al., 2004b; Korhonen et al., 2007a; McFadden et al., 2007), the configuration of mafic dikes representing a regional tensile array (Siddoway et al., 2005), and kinematic analysis of mesoscopic brittle faults (Luyendyk et al., 2003). Airborne geophysical data over the Ford Ranges (Luyendyk et al., 2003), and Edward VII Peninsula (Ferraccioli et al., 2002) delineate regional-scale faults. Geochemical investigations reveal the granite petrogenesis (Pankhurst et al., 1998; Mukasa and Dalziel, 2000; Saito et al., 2007) and U-Pb geochronology provide critical age control (Siddoway et al., 2004a,b, 2006; McFadden et al., 2007). A review of the recent findings from MBL in the next section will begin with the lower crustal exposure provided by the Fosdick Mountains, where migmatites provide

a view of crustal rheology, kinematics, and dynamics of Cretaceous tectonism that pertain to the West Antarctic rift system as a whole.

The Fosdick Mountains Gneiss Dome

The Fosdick Mountains form an elongate migmatite gneiss dome (Wilbanks, 1972; Siddoway et al., 2004b) delimited by a S-dipping, dextral-oblique detachment zone on the south (McFadden et al., 2007) and by an inferred steep dextral strike-slip zone on the north, the Balchen Glacier fault (Siddoway et al., 2004b, 2005). From lower to higher structural levels, gneisses that exhibit features indicative of melt-present ductile flow give way to mylonitic rocks exhibiting mixed ductile-brittle deformation textures, indicative of solid-state deformation (McFadden et al., 2007). Kinematic axes calculated from nappe-scale folds and subsidiary folds, mineral lineation, and anisotropy of magnetic susceptibility (AMS) fabrics within the 15 × 80 km dome are subhorizontal, 065 to 072. The orientation is oblique to the long axis of the dome and to the Balchen Glacier fault.

The migmatite gneisses forming the core of the Fosdick Mountains dome reached temperatures (T) and pressures (P) of the upper amphibolite to granulite facies (Siddoway et al., 2004b; Korhonen et al., 2007a). Granite formed by biotite breakdown (Saito et al., 2007) forms sheets, stocks, and extensive interconnected networks on a scale of hundreds of meters. Leucogranite occupies structural sites—within foliation-parallel sheets, shear bands, and interboudin necks—suggestive of deformation-enhanced migration and coalescence of melt products (Sawyer, 2001). Concordant layers of leucogranite may exceed 10 m in thickness. New U-Pb SHRIMP zircon studies (Siddoway et al., 2004b, 2006; McFadden et al., 2007) and isotope geochemistry (Saito et al., 2007) aid in the task of determining the extent and distribution of anatectic granites formed in Late Cretaceous time, coincident with development of the WARS.

At intervals along the Fosdick Mountains dome, leucogranite sills form vertical sequences that reach 1000 m in thickness ("leucogranite sheeted complex" of McFadden et al., 2007). The thin layers (<1-3 m) of para- and orthogneiss that separate the sills contain microstructures indicative of the former presence of melt and of deformation mechanisms dominated by melt-assisted grain boundary diffusion creep. Kinematic data obtained from the horizons of shallowly dipping paragneiss and orthogneiss include fold axes of symmetrical, tight to isoclinal recumbent folds that trend 062-242 (n = 118), and sparse mineral lineation aligned 072-252 (n = 38) (Siddoway et al., 2004b). The consistent linear data suggest ENE-WSW stretching under suprasolidus conditions when the leucogranite sheets were emplaced. The layers of paragneiss and orthogneiss in this setting are "diatexitic" (Brown, 1973; Milord et al., 2001), in that the color distinction between leucosome and melanosome is subdued, with melanosome a light grey color, and boundaries between the light- and darker-colored portions, indistinct. Diatexite textures indicate a high degree of chemical interaction of leucogranite melt with host gneisses under suprasolidus conditions (Sawyer, 1998, 2004). Conventional thermobarometry carried out on the diatexitic paragneisses yielded P = 4 kbar to 6 kbar and T = 680°C to 780°C (Smith, 1992, 1997; Siddoway et al., 2004b). New results of comparative thermobarometry using THERMOCALC indicate considerably higher conditions of 820-870°C and 6.5-7.5 kbar for the Cretaceous peak (Korhonen et al., 2007a,b).

The leucogranites exhibit compositional layering and igneous microstructures, such as euhedral grains and tiling of large feldspars; together with evidence of magmatic solid-state deformation (Blumenfeld and Bouchez, 1988; Weinberg, 2006) such as mechanical kinking at grain-to-grain contacts in coarse-grained phases. The textures suggest that the interstices between solid phases represented a permeable melt network through which melt flowed. The migmatite structures of the Fosdick Mountains suggest that sills and leucosome networks are remnants of a melt transfer system that allowed magma flux through a zone of anatexis (e.g., Olsen et al., 2005; Weinberg, 2006) within the middle and lower crust (e.g., Brown and Pressley, 1999; Brown, 2007). Deformation aided melt-migration and melt enhanced deformation in a mutually complementary process.

The extensive Cretaceous leucogranites within the gneiss dome contain a dominant population of prismatic igneous zircons. The bipyramidal, elongate zircon grains exhibit oscillatory zoning and lack inheritance, suggesting that they crystallized from a melt. U-Pb SHRIMP ages determined for the zircons are 115-101 Ma (Figure 6), suggesting that elevated temperatures were attained and that melt transfer and crustal flow initiated during middle Early Cretaceous (oblique?) plate convergence, then continued during the transition to extension/transtension in the WARS.

South Fosdick Detachment Zone

The leucogranitic sheeted complex passes upward into metatexite at highest structural levels on the southern flank of the range. Metatexite (Brown, 1973; Milord et al., 2001; Sawyer, 2004) is a migmatite type that consists of mesoscopic, cm- to dm-scale compositional layering with a sharp color distinction between light-colored quartzofeldspathic leucosomes and dark, biotite-rich melanosomes. Leucosomes that are volumetrically minor represent a mobile portion, or metatect, and melanosomes, a nonmobilized, depleted component (Brown, 1973; Milord et al., 2001; Sawyer, 2004). Sills and interconnected networks of leucosome that would be indicative of melt transfer and coalescence are poorly developed to absent in the melanosome.

Solid-state deformation is indicated by pervasive mylonitic microstructures indicative of plane strain-simple shear, including C-S fabrics and asymmetric porphyroclasts with tails (McFadden et al., 2007). Kinematic criteria show

dextral normal oblique shear sense, with top-to-the-SW transport along azimuth 240. Foliation dips steepen from west toward east, and give way to strong subhorizontal L-tectonite fabrics trending 070-075 within Ford-phase granodiorite at Mt. Richardson (Figure 4). U-Pb SHRIMP zircon data bracket the time of deformation upon the South Fosdick detachment zone between 107 Ma and 96 Ma (McFadden et al., 2007).

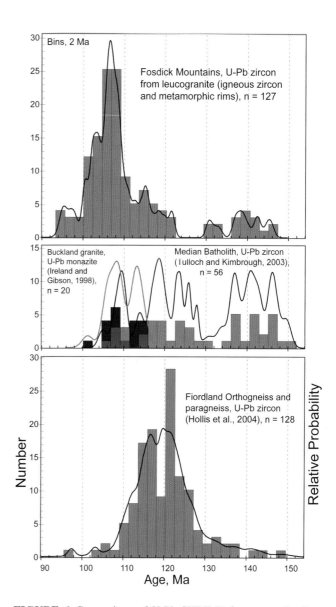

FIGURE 6 Comparison of U-Pb SHRIMP frequency distribution for Fosdick Mountains leucogranites, with New Zealand data including from the Fiordland Orthogneiss (Hollis et al., 2004), Median Batholith (Tulloch and Kimbrough, 2003), and Buckland Granite (Ireland and Gibson, 1998). Fosdick Mountains data were acquired on SHRIMP II at Australian National University under the direction of C. M. Fanning. The relative probability plots with stacked histograms of 206Pb/238U ages (207Pb corrected) were calculated using ISOPLOT/EX by K. Ludwig.

Structural Analysis of Mesoscopic Brittle Fault Arrays Throughout the Ford Ranges

Geometrical and kinematic data have been gathered for systematic mesoscopic brittle structures, including dikes, faults, shear fractures, and joints that cut the isotropic plutonic units, Ford Granodiorite, and Byrd Coast Granite. Structures that cut Byrd Coast Granite or mafic dikes, the majority of which fall in the age range 104-96 Ma, are known to be middle Cretaceous or younger. Mesoscopic faults are striated planes accommodating >2 m offset or zones of cataclasis exceeding 15 cm thickness. The term "shear fractures" refers to slickenside surfaces, sometimes mineralized but generally lacking gouge, and rarely associated with geological markers that allow quantification of offset. In most instances, therefore, brittle criteria are used for interpretation of shear sense (Marrett and Allmendinger, 1990). Few data come from Swanson Formation, because brittle shear planes typically reactivated bedding or preexisting cleavage, making the kinematic significance uncertain.

Outside of the Fosdick Mountains migmatite dome most exposures of crystalline rocks in the Ford Ranges lack dynamic fabrics (Siddoway et al., 2005) and mylonitic zones are found only rarely. Four sites hosting ductile shear zones are situated near locations for which thermochronology data are now available (Table 1). Narrow mylonitic shear zones (1-5 m wide) cut Ford Granodiorite at Mt. Crow (Figure 4, inset E) and Mt. Cooper (Siddoway et al., 2005). High-temperature shear zones exist along the present-day Ross Sea coast, at Mt. Woodward (60 m exposed width) (Figure 4, inset F) and at Prestrud Rock (30 m minimum width). Each of the sites is adjacent to an inferred crustal-scale strike-slip zone that is concealed by ice (Ferraccioli et al., 2002; Luyendyk et al., 2003; Siddoway et al., 2005).

Mafic dikes provide a very valuable kinematic dataset due to their regional distribution and very consistent regional orientation of azimuth 344, subvertical, throughout the Ford Ranges. Tensile opening perpendicular to the dike margins is a reflection of ENE stretching at the time of emplacement. The dike array cuts Byrd Coast granite of 102-98 Ma age, and the range of $^{40}Ar/^{39}Ar$ ages for the majority of the mafic dikes is 104-96 Ma (groundmass concentrates on microcrystalline dikes; Siddoway et al., 2005). U-Pb titanite and $^{40}Ar/^{39}Ar$ hornblende ages for discordant mafic dikes within the Fosdick range are 99-96 Ma (Richard et al., 1994; Siddoway et al., 2006). Mutually crosscutting relationships between mafic dikes and faults indicate that they are contemporaneous.

Brittle fault data offer the most tenuous data to interpret due to the paucity of offset markers and the need to use brittle criteria for kinematic shear sense. Nonetheless, consistent fault and shear fracture arrays are identified. In the central and southern Ford Ranges a well-defined ~NW-SE-oriented conjugate fault array hosts moderately oblique, SE-plunging striae on both SSW and NE dipping planes (Figure 4, inset D). The array is expressed both as minor faults and shear fractures with kinematic criteria indicative of oblique slip, with top-to-the-ESE translation (Luyendyk et al., 2003). The dominant orientation in this array is ESE-striking, with dextral normal oblique kinematic sense. A second generation of brittle structures in the Sarnoff and Denfield ranges consists of NNW-striking, normal- to oblique-slip shear fractures (Figure 4, inset C). The widespread NNW-oriented mesoscopic structures have strikes parallel to the regional mafic dike array, and to the prevalent fault orientations offshore in the easternmost Ross Sea (Luyendyk et al., 2001; Decesari et al., 2003). A late ENE-oriented array is strongly expressed in the Chester Mountains (Figure 4), forming chloritic brittle shear zones up to 15 cm in width, and chloritic and oxidized shear fractures. Brittle criteria on the planes oriented N75E (mean) indicate normal dextral oblique slip upon SSW- and NNE-trending striae (Figure 4, inset A).

NE-oriented, sinistral strike-slip shears are spatially associated with inferred Cenozoic faults that trend NE-SW and offset the Fosdick Mountains gneiss dome (Figure 4, inset B). Pleistocene mafic lavas erupted from small volcanic centers along the trend (Gaffney and Siddoway, 2007) (Figure 4). A brittle fault data set comes from the series of outcrops forming the easternmost exposures in the Ford Ranges, which is situated near a prominent NE-trending escarpment imaged in the bedrock topography (Luyendyk et al., 2003) that corresponds with a geophysical anomaly arising from inferred sub-ice volcanic centers (Figure 7). These are NE-SW shear fractures with strike slip striae, with consistent sinistral-sense offset from brittle kinematic criteria (Figure 4, inset B).

With respect to timing, regional deformation caused by mid-crustal flow arising from melt accumulation in the lower and middle crust (Figure 8) was under way as early as 115 Ma, based on the ages determined for melt-present deformation in the lower crustal exposures in the Fosdick Mountains rocks. The older limit on the time of brittle deformation and formation of mylonitic zones in the upper crustal rocks of the Ford Ranges is provided by Cretaceous plutonic rocks of 104-96 Ma age (Byrd Coast granite and mafic alkalic dikes) that are cut by brittle faults. The cooling history of the block-faulted mountain ranges constrained by $^{40}Ar/^{39}Ar$ and apatite fission track thermochronology (summarized below) provides a younger age limit on regional tectonism.

Crustal Structure from Airborne Geophysics

Airborne gravity and radar soundings over western MBL indicate that the crustal thickness beneath the Ford Ranges is 22-25 km, increasing to the north and inland by 8-9 km for central MBL (Figure 2) (Luyendyk et al., 2003). The inferred steep gradient in crustal thickness coincides spatially with the linear northern front of the Fosdick Mountains, where migmatites were exhumed from mid-crustal

TABLE 1 Summary of 40Ar/39Ar and AFT Thermochronology Data for Sites in the Central and Eastern West Antarctic Rift System

Location	Feature	Age (Ma)	AFT Track Length (mm)	Method	Source of Data	Field Association	Kinematics
Ford Ranges	Mafic dikes	104-96	n.a.	^{40}Ar/^{39}Ar groundmass	Siddoway et al., 2005	Tabular, vertical to sub-vertical dikes throughout the Ford Ranges	Tensile, 074-254
Mt. Cooper	Mylonite zone	96.92 ± 0.34	n.a.	^{40}Ar/^{39}Ar biotite	Siddoway et al., 2005	3- to 5-m-wide zone cutting Ford Granodiorite	Normal sense, down to East
Prestrud Rock	Gneiss	91 ± 4	13.1 ± 0.2	AFT	Lisker and Olesch, 1998; Smith, 1996	Contrast in grade and fabrics; strong lineation suggest shear zone	Strike oblique; kinematic sense not determined
The Billboard	Unfoliated Ford granodiorite	91 ± 4	n.a.	AFT	Lisker and Olesch, 1998	Ford granodiorite bounded by an inferred east-west fault; borders inferred east-west fault	Strike normal oblique, inferred dextral
Mt. Douglass	Unfoliated Byrd Coast granite	97 ± 5	13.7 ± 0.2	AFT	Lisker and Olesch, 1998	Unfoliated Byrd Coast granite, located 6 km from shear zone in calcsilicate gneisses at Mt. Woodward	Sinistral oblique sense; shear zone >100 m wide
Chester Mountains	Unfoliated Ford granodiorite	95 ± 6	13.5	AFT	Lisker and Olesch, 1998	Ford granodiorite	Hanging wall of South Fosdick detachment
Mitchell Peak	Migmatite gneiss	88 ± 5	14.2 ± 0.1	AFT	Lisker and Olesch, 1998	Migmatite gneiss	Hanging wall of South Fosdick detachment
DSDP 270	Calc-silicate gneiss	90 ± 6	n.a.	AFT	Fitzgerald and Baldwin, 1997	Cataclasite/breccia in detachment zone	Not possible to determine
Colbeck Trough	Mylonite	86 ± 5	13.8 ± 1.4	AFT	Siddoway et al., 2004a	Dredged material from submarine escarpment	Not possible to determine
	Mylonite	71 ± 5	14.1 ± 1.3	AFT			
	Mylonite	98-95	n.a.	^{40}Ar/^{39}Ar K-feldspar, biotite			

NOTE: n.a. = not applicable; AFT = Apatite fission track; DSDP = Deep Sea Drilling Project. Kinematic determinations from Siddoway et al., 2005, or Siddoway, unpublished.

depths. A steep gradient also is observed in the magnetics, and there is a well-defined lineament in the bedrock topography beneath Balchen Glacier (Figures 2 and 7). Present-day bed topography for much of the surveyed area defines distinct NNW-SSE and NE-SW-oriented lineaments that are oblique to the density distributions. They are generally parallel to the NNW-oriented, normal-sense and NE-striking, sinistral-sense, second generation shear fractures measured throughout the Ford Ranges and to the mafic dike array. High-gradient magnetic anomalies inferred to be subglacial volcanoes of Cenozoic age are mapped to the east (Figures 2 and 4) (Luyendyk et al., 2003), and are spatially associated

with a postulated NE-SW strike slip fault of Cenozoic age (Siddoway et al., 2005) (Figure 4).

Linear magnetic anomalies on Edward VII Peninsula have three dominant trends. They trend ~E-W, NNW, and NE (Ferraccioli et al., 2002). Outlet glacier troughs in the southern Ford Ranges and Edward VII Peninsula appear to be controlled by the NNW trend (Luyendyk et al., 2001; Wilson and Luyendyk, 2006; Sorlien et al., 2007), and have a narrow, deep linear morphology that suggests an association with the regional normal fault array.

Magnetic and gravity anomalies together with bedrock topography calculated from airborne geophysics data

(Luyendyk et al., 2003) provide the means to assess the regional significance of structures identified from outcrop study. Conversely, information from geological structures may clarify the kinematic history of subglacial features inferred from geophysics. A pattern of generally east-west-oriented gravity anomalies, interpreted as pronounced density variations in the bedrock, is identified in the Ford Ranges and further south (Figure 7). Their trends generally correspond to the approximately east-west orientation of mountain ranges and outlet glaciers in the central and northern Ford Ranges, which have long been considered to be structurally controlled (e.g., Luyendyk et al., 1994).

The margins of the low-density areas are distinct and fairly linear over tens of kilometers, and are parallel to each other, with a nearly E-W trend (Figure 7). Their orientation is parallel or subparallel to the elongate, ESE-trending, low-gradient magnetic anomaly over the Fosdick dome (Luyendyk et al., 2003), and to the known crustal-scale faults bounding the Fosdick Mountains (Siddoway et al., 2004b, 2005; McFadden et al., 2007). Kinematic data within the dome and from neighboring exposures that represent the brittle upper crust show that the dome-bounding faults accommodated dextral to dextral oblique motion. The correspondence of the limits of linear gravity anomalies with the north and south faults bounding the Fosdick migmatite dome suggest that the pattern of generally E-W gravity anomalies over the Ford Ranges and eastern MBL, more broadly, are controlled by dextral oblique faults that dropped cover rocks down with respect to crystalline rocks.

A possible interpretation of the E-W-trending, regular variations in density (Figure 7) is that they correspond to sedimentary basins and intervening highs that originated during Late Cretaceous development of the WARS. A number of observations suggest that this is not the case. First of all, sharply defined features in the sub-ice topography do not correspond to the boundaries separating regions of contrasting density (Luyendyk et al., 2003), as might be expected if comparatively young, poorly indurated clastic sediments existed at depth. Furthermore, no Late Cretaceous-Eocene sedimentary or volcanic strata crop out in Marie Byrd Land (see Pankhurst et al., 1998), and the makeup of glacier-transported clasts within bouldery till upon bedrock—determined from careful searches at six sites bordering three outlet glaciers (labeled "M" in Figure 4)—is dominated by greenschist-grade Swanson Formation, with a small proportion of Byrd Coast Granite and Ford Granodiorite clasts. The prevalence of Swanson Formation clasts and absence or low abundance of younger clastic rocks favors the hypothesis that Swanson Formation constitutes the bedrock beneath the glacier drainages that traverse three of the low-density regions within the Ford Ranges (Figure 7).

The apparent absence or low abundance of Late Cretaceous-Paleocene sedimentary deposits on land is enigmatic, in light of the WARS extension of >100 percent and the subsequent breakup between MBL and NZ. Possibly

the region experienced an interval of low sediment supply (Karner et al., 2005), or did not develop significant relief above or subside below sea level (e.g., LeMasurier and Landis, 1996; Luyendyk et al., 2001).

To summarize, the findings from aerogeophysical surveys and structural geology studies in the region suggest that the brittle crustal architecture along the eastern margin of the Ross Sea rift developed through NE-SW regional transtension upon high-angle faults, with extensional strain accommodated upon NNW-striking mafic dikes and normal faults, and wrench deformation occurring upon a conjugate array of approximately E-W-oriented dextral strike slip zones, and NE-SW sinistral strike slip faults (Figure 4), affecting Cretaceous and older rock units (Ferraccioli et al., 2002; Luyendyk et al., 2003). Generally E-W-oriented crustal-scale structures in the northern Ford Ranges may have originated as contractional faults during prior NNE-SSW convergence along the East Gondwana margin in the Mesozoic Era or before. The correspondence to inferred geological boundaries between supracrustal and crystalline rocks is a possible indication that Paleozoic structures were reactivated during opening of the WARS.

Thermochronology and Cooling History of the WARS

^{40}Ar/^{39}Ar and apatite fission track (AFT) data for sites distributed across the West Antarctic rift system are beginning to reveal the regional cooling pattern for the WARS. Sample sites (Figures 1 and 2) include DSDP 270 on the Central High (Fitzgerald and Baldwin, 1997), Colbeck Trough on the eastern margin of Ross Sea (Siddoway et al., 2004a), Edward VII Peninsula and the southern Ford Ranges (Adams et al., 1995; Lisker and Olesch, 1998), and the Fosdick Mountains (Richard et al., 1994). ^{40}Ar/^{39}Ar K-feldspar data available from three sites record cooling at 98 Ma to 94 Ma (Table 1) (Figure 4). The AFT data record rapid cooling across the region by ca. 90 Ma, but also show complexity in the cooling histories that cannot be explained by a single event. AFT data fall within two age populations of 97-88 Ma and 80-70 Ma, each characterized by long track lengths indicative of rapid cooling (Table 1).

Diachronous cooling was first noted by Richard et al. (1994) for the Fosdick Mountains. Migmatites rapidly cooled through ^{40}Ar/^{39}Ar closure temperatures for four mineral phases, having equilibrated at T >700°C at ca. 101 Ma, then cooled to T <165°C by ca. 94 Ma. The AFT study determined long track lengths indicative of rapid cooling, but the AFT ages are considerably younger at 76-67 Ma. Thus the thermochronology data suggest two pulses of cooling due to tectonic exhumation of the Fosdick Mountains gneiss dome, separated by a quiescent interval. Additional AFT data over the broader region confirm the bimodal distribution of ages of 97-88 Ma and 80-70 Ma in the Ford Ranges, Edward VII Peninsula (Adams et al., 1995; Lisker and Olesch, 1998), and central Ross Sea (Fitzgerald and Baldwin, 1997).

New mapping and structural geology research reveals that seven of the sites recording the older AFT or $^{40}Ar/^{39}Ar$ ages (Table 1) are situated near regional-scale faults or within shear zones. These include Prestrud Rock, bordering the Ross Sea at the edge of the Alexandra Mountains (Figure 2) (Smith, 1996; Ferracioli et al., 2002); the Colbeck trough on the margin of Edward VII Peninsula (Figure 2) (Siddoway et al., 2004a); and DSDP site 270 on the Ross Sea Central High (Figures 1 and 9) (Fitzgerald and Baldwin, 1997). Adams et al. (1995) attributed a pronounced contrast in cooling history between the Alexandra and Rockefeller Mountains to relative fault motion. The two off-shore sites correspond to inferred detachment faults submerged along the eastern margin and Central High of the Ross Sea (Fitzgerald and Baldwin, 1997; Siddoway et al., 2004a). The tectonites from both offshore sites exhibit ductile fabrics overprinted by brittle cataclasis, suggestive of translation from deeper to shallower crustal levels. No geometrical or kinematic

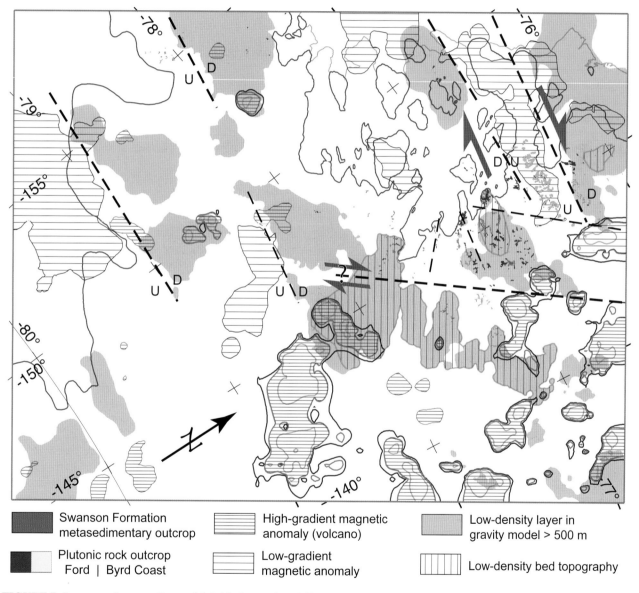

FIGURE 7 Summary diagram of potential fields data and modeling (Luyendyk et al., 2003) for the region surveyed by the Support Office for Aerogeophysical Research in 1998. North direction is toward the upper right. High-gradient positive magnetic anomalies delimited by the +50 nT contour are interpreted as subglacial volcanic centers. Broader, lower-gradient positive magnetic anomalies are attributed to crustal magnetization contrasts. Gray-shaded areas correspond to low-density regions (2300 kg/m³) exceeding 500 m thickness, determined from gravity modeling (Luyendyk et al., 2003). Small polygons filled with solid color are rock exposures. Dashed straight lines are inferred faults between crust with low-density bodies and crust with sources of low-gradient magnetic anomalies. Modeling carried out by D. Wilson, University of California, Santa Barbara.

information was obtained for the structures since sample retrieval was by drill core (Ford and Barrett, 1975; Hayes and Davey, 1975) and dredge (Luyendyk et al., 2001). DSDP 270 yielded a multicomponent AFT sample from a few grains of apatite extracted from calcsilicate gneiss (n = 16) (Fitzgerald and Baldwin, 1997). The dominant component is 90 ± 6 Ma in age.

A recently discovered shear zone site that yields critical kinematic data is at Mt. Woodward, bordering a pronounced lineament along the Haines Glacier (Figure 4, inset F). The steep high-strain zone developed in high-temperature calcsilicate gneisses is oriented 160-340 and exceeds 100 m in width. Asymmetrical folds indicate sinistral shear sense (Siddoway, unpublished). The thermochronology data obtained from Mt. Douglass, 6 km away, yield the region's oldest AFT cooling age of 97 ± 5 Ma on Byrd Coast granite (Lisker and Olesch, 1998). Northwest-southeast-oriented bedrock faults are inferred to control the Sarnoff Range trend, where pronounced narrow troughs, oriented 150-330, are evident in the bedrock topography (Luyendyk et al., 2003). A 3-m-wide mylonitic shear zone at Mt. Crow (Figure 4, inset E) parallels this trend and offers kinematic information that possibly is representative of the concealed fault. The Mt. Crow shear zone exhibits shallow-plunging sinistral-sense stretching lineation oriented 20, 138, on steep foliation. An AFT age of 91 ± 4 Ma (Lisker and Olesch, 1998) came from Ford Granodiorite at The Billboard in the Sarnoff Range (Figure 4).

The remaining sites with AFT data in the older age range (Table 1) are associated with the hanging wall of the South Fosdick detachment zone. They are Mitchell Peak, the isolated nunatak forming the westernmost outcrop in the Fosdick range, and the Chester Mountains, south of the Fosdick range (Figure 4). There is a pronounced contrast in cooling age across the South Fosdick detachment zone (Figure 4), with 95 Ma to 88 Ma AFT ages obtained from sites in the hanging wall, and 76 Ma to 67 Ma from sites in the gneiss dome core (Table 1).

Remaining AFT localities in western MBL record moderate to slow cooling between 83 Ma and 67 Ma. The ages correspond to the time of initiation of seafloor spreading between Campbell Plateau (NZ) and WANT (Figure 1) at 83-79 Ma (chron 33r) (McAdoo and Laxon, 1997; Larter et al., 2002; Stock and Cande, 2002; Eagles et al., 2004), suggesting that the second AFT cooling pulse was triggered by the lithospheric separation (Figure 9c) (Siddoway et al., 2004a; Kula et al., 2007). There is a good correspondence in timing and tectonic history of MBL localities with detachment structures in New Zealand (Kula et al., 2007).

In summary, examination of AFT data together with mapped structures shows that the early stage of rapid cooling in western Marie Byrd Land at 95-85 Ma was localized upon high-angle conjugate wrench zones. The timing of fault activity determined from U-Pb zircon geochronology, [40]Ar/[39]Ar, and AFT thermochronology corresponds with the

time of development of the eastern WARS. The younger phase of rapid cooling at ~75 Ma reflects regional uplift and cooling, coincident in time with and attributable to modest denudation in response to onset of seafloor spreading and separation between WANT and New Zealand, upon a new divergent plate boundary that continued in to the Tasman Sea (Figure 3a) (Gaina et al., 1998; Sutherland, 1999; Kula et al., 2007). The observation that rapid cooling occurred first upon discrete WARS fault zones (101-92 Ma) suggests a localized landscape response, reflected in the thermochronology cooling histories data. The affected area covers 250,000 km² of western Marie Byrd Land and the neighboring Ross Sea.

DISCUSSION

The Role of a Hot Middle Crust in the Regional Structural Evolution of the WARS

The Fosdick Mountains gneiss dome is a structure of vast complexity pervaded by sills and discordant networks of leucogranite. Crosscutting relationships and varying degrees of deformation suggest multiple cycles of melt migration and emplacement within structurally controlled, dilatant sites. Thick sills of leucogranite containing microstructures indicative of horizontal magmatic flow are interlayered with thin sheets of diatexitic gneisses that exhibit consistent ENE-WSW kinematic sense, leading to the interpretation that the Fosdick gneiss dome represents an exposure of deep middle crust that underwent directional viscous, magma-like flow (Figure 8).

Relationships in the Fosdick Mountains dome suggest that partial melting and rheological weakening of the crust in MBL was a consequence of crustal heating during orogenesis, affecting "Lachlan"-type sedimentary rocks and middle Paleozoic intermediate plutons. Argillaceous rocks of Lachlan type are chemically fertile (e.g., Thompson, 1996) and may generate substantial quantities of melt. Subjected to a differential stress in a convergent orogen or to gravity forces in the region of thickened crust at the convergent margin, viscous flow commenced (Figures 5 and 8). The localization of strain at the interface between the region of hot versus cold crust caused detachment structures to initiate (Figure 8b) (e.g., Teyssier et al., 2005) and/or reactivated preexisting faults (Siddoway et al., 2004b, 2005), leading to gneiss dome emplacement (Figure 8c).

The development of thermal perturbations of this type in a convergent margin setting has been noted as a characteristic of hot accretionary orogens (Collins, 2002a). The Lachlan belt exemplifies this type of orogen as it has undergone multiple cycles of contractional orogeny and extensional collapse involving HT metamorphism (Foster et al., 1999; Collins, 2002a,b; Gray and Foster, 2004; Fergusson et al., 2007). In MBL the elevated heat flow into the base of the continental crust may have arisen during subduction of hot oceanic lithosphere newly formed at the Phoenix-Pacific

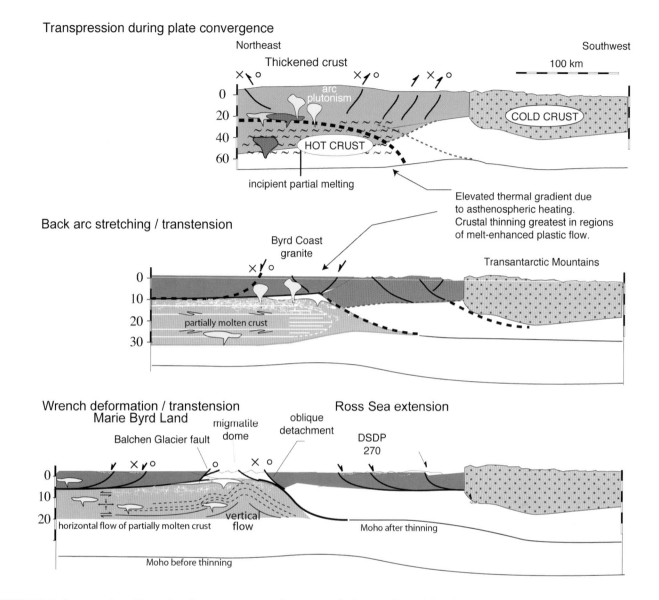

FIGURE 8 Cross-sections illustrating the consequences of presence of a hot, weak, partial melt-rich horizon in the lower crust. Portrays the geometry and kinematics of brittle structures in the upper crust and propagation of viscous flow region within the middle to lower crust. *Top:* Crustal heating and crustal thickening within convergent margin setting, wrench faults active. *Middle:* Partially molten crust flows laterally, advecting heat into new regions and weakening the crust. *Bottom:* Transtension affects a large region of warm crust near the active margin. Sites of melt transfer and accumulation exhibit vertical translation if focused upon a fault. X, O symbols indicate motion out of and in to the plane of the profile.

ridge (Figures 4 and 5b) (Bradshaw, 1989; Luyendyk, 1995), or due to back arc extension and lithospheric thinning (Figure 5b and 5c) (Weaver et al., 1991, 1994; Mukasa and Dalziel, 2000); or from basal heating in the presence of a postulated mantle plume (Weaver et al., 1994). Singly or collectively these factors could promote magmatic underplating and advection of heat into the crust, with corresponding effects on crustal rheology (Regenauer-Lieb et al., 2006). Further effects could arise from infiltration of fluids into the overriding plate due to dehydration of the newly subducted slab, or to radiogenic heat production in the East Gondwana crust

that had been thickened during Mesozoic convergence (e.g., Collins and Hobbs, 2001; McKenzie and Priestly, 2007), augmented by heat advection by fluids. The rheological partitioning interpreted to exist in the Cenozoic WARS, with a brittle to ductile gradient across the Ross Sea (Salvini et al., 1997) may have been established at this time.

The question of whether the substantial volumes of leucogranite derived from constituent gneisses of the Fosdick dome (phases of Ford Granodiorite, Swanson Formation) or were produced within other parts of a crustal zone of magma flux (e.g., Olsen et al., 2005) then migrated into the dome

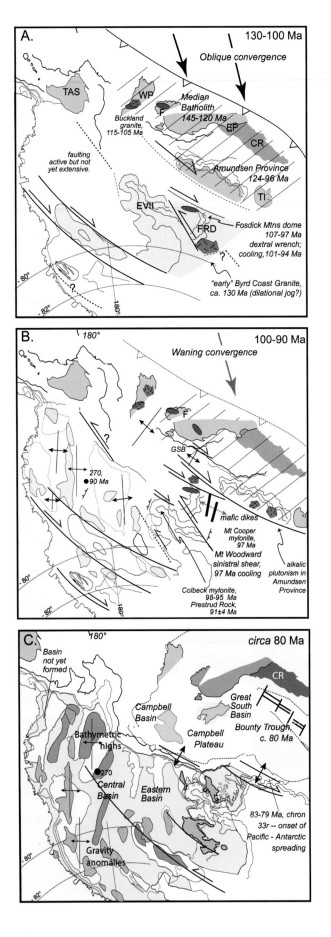

FIGURE 9 Development of the WARS in three hypothesized stages, based on the observations of strain accommodated upon wrench and extensional fault systems in Marie Byrd Land, a part of the WARS that resides above sea level. Basement graben configuration based on Cooper et al. (1991b) and Trey et al. (1997).

(A) 130 Ma to 100 Ma, interpreted configuration of the East Gondwana margin. HT metamorphism and crustal melting (ellipse symbol) is contemporaneous with arc magmatism at inboard sites including Fiordland (F); Paparoa complex in Western Province, NZ (WP); and Fosdick Mountains gneiss dome in MBL. Diachroneity of calcalkaline plutonism in the Median Batholith and Amundsen province may be a reflection of subducted slab geometry or the configuration of the Phoenix-Pacific spreading ridge offshore (Figures 3 and 5). Sites of alkalic magmatism may be an expression of dilational jogs in wrench zones or within detachment structures. TAS = Tasmania; EP = Eastern Province, NZ; CR = Chatham Rise; TI = Thurston Island.

(B) 100-90 Ma, major phase of intracontinental deformation in the WARS. Tensile dike arrays and alkalic plutons were emplaced across the back-arc region. Blue shaded areas are gravity anomalies corresponding with high-density material along basin axes (Cooper et al., 1991b). Differential movement upon steep wrench zones is recorded by ^{40}Ar/^{39}Ar and AFT data that record rapid cooling between 97 Ma and 90 Ma for fault zone samples (Table 1). Based on the available data there appears to be an age progression from northeast toward southwest, with site DSDP 270 in the Ross Sea recording the youngest of the older subset of cooling ages, if the dominant population is accepted as the AFT cooling age (Fitzgerald and Baldwin, 1997). Dominant wrench deformation is documented in Marie Byrd Land, and prevalent normal faulting is inferred in the Ross Sea. The regional strain variation may be due to contrasts in competency of the pre-Mesozoic continental lithosphere, contrasting thermal conditions arising from lithospheric thinning or the dynamics of the convergent plate boundary (e.g., Figure 8), or the transition from a region undergoing oblique subduction of young continental lithosphere (eastern WARS) to one experiencing slow rifting between mature continental crust of Antarctica-Australia (western WARS).

(C) Ca. 80 Ma, the time of breakup between WANT and NZ-Campbell Plateau. Continental extension across the WARS and Campbell Plateau exceeded 100 percent and was completed prior to onset of seafloor spreading. Blue shaded areas are gravity anomalies corresponding with high-density material along basin axes; red areas are bathymetric (and basement) highs (Cooper et al., 1991b). Extension direction for breakup was nearly orthogonal to that for WARS opening and the rifted margin cuts at a high angle across Ross Sea basins (Lawver and Gahagan, 1994). It is probable that a preexisting wrench fault structure was reactivated at the time of breakup. This would explain the exceptionally abrupt ocean-continent boundary along the coast of Marie Byrd Land (Gohl, 2008, this volume).

upon structurally controlled pathways, is being addressed by isotopic and geochemical investigation. A petrogenetic link between Ford Granodiorite (source) and Byrd Coast granite (product) has been demonstrated in the region (Weaver et al., 1991; Pankhurst et al., 1998). New Nd isotope data from the Fosdick Mountains strengthen this interpretation (Saito et al., 2007). Therefore it is plausible that regional melting of Ford Granodiorite contributed to anatectic granite magmas that were capable of vertical or lateral migration during oblique convergence (Weaver et al., 1995) to transtension (Siddoway et al., 2005) along the Early Cretaceous plate margin.

HT metamorphism and exhumation of deep crustal rocks on detachment structures are documented over a wide region proximal to the Gondwana margin arc in New Zealand and MBL (Kimbrough and Tulloch, 1989; Fitzgerald and Baldwin, 1997; Forster and Lister, 2003; Siddoway et al., 2004b; Kula et al., 2007), suggesting pervasive middle to lower crustal flow and advection of magmatic heat (Ehlers, 2005) over a large region (Figures 8 and 9). The existing geochronological and thermochronological data reviewed above suggest a comparatively short and dynamic development of the Fosdick gneiss dome and other strike slip shear zones active in the eastern WARS in Cretaceous time. U-Pb SHRIMP zircon data indicate that HT metamorphism was in effect by 140 Ma and growth of new igneous zircon within the anatectic granites was under way by 115 Ma (Figure 6) (Siddoway et al., 2006); a possible indication that crustal melting and conditions favorable for viscous flow arose during convergent tectonism and crustal thickening along the East Gondwana active margin.

In the Fosdick Mountains gneiss dome syn- to post-tectonic granite intrusions of 107 Ma to 96 Ma age delimit the duration of detachment tectonics and exhumation, with upward translation through ductile to brittle conditions and development of a mylonitic shear zone at the transition (McFadden et al., 2007). Following the emplacement of the dome, mid-crustal magmatism and flow ceased and dome rocks cooled rapidly from >700°C to <200°C at rates as high as 70°C/m.y. (Richard et al., 1994). Overprinting textures of cordierite possibly record decompression of footwall rocks due to translation upon the detachment structure; and the late-tectonic melt-filled, normal-sense shear bands that cut all older structures may be an indication of a small volume of late leucogranite formed by decompression-induced melting (Siddoway et al., 2004b; Korhonen et al., 2007b). The very high rates of cooling recorded by $^{40}Ar/^{39}Ar$ mineral cooling data are comparable to rates of advective heat loss that arise from pluton emplacement in cold country rock (Fayon et al., 2004). These observations are possible indications of a component of upward, gravity-driven flow (diapirism) during emplacement of the dome (e.g., Teyssier and Whitney, 2002). The association of the Fosdick Mountains gneiss dome with the Balchen Glacier fault, which is known to be an inherited Paleozoic structure (Richard et al., 1994; Siddoway et al.,

2004b), suggests that a crustal to lithospheric-scale discontinuity has a role in gneiss dome emplacement.

The predominance of subhorizontal fabrics (rather than vertical geometry expected for strike slip faults in the brittle upper crust) is considered to be either (1) an expression of coupling between crustal layers of contrasting compentency (e.g., metatexite versus diatexite plus leucogranite) that accommodates strain by different mechanisms (e.g., Tikoff et al., 2002), or (2) accentuation of vertical shortening at the "melt propagation front" for melt-rich diatexite-leucogranite as melt-rich material migrated upward and was arrested at the thermal or permeability boundary (e.g., Sawyer, 2001) represented by metatexite (Figure 8) or (3) a change in orientation of the shortening axis of strain due to unroofing, to coincide with direction of gravitational load.

Overview of the West Antarctic Rift System

Structural and geochronological data from sites throughout the eastern WARS show a broad compatibility with respect to ENE coordinates for principal finite strain and timing of crustal thinning deformation at ca. 105-95 Ma. The kinematic compatibility between structures of the brittle upper and viscous lower crust is a great aid to interpretation of the mechanisms of formation of the WARS. The best-exposed, crustal-scale structure with lateral extent in Marie Byrd Land is the South Fosdick detachment fault, which accommodated dextral normal oblique translation of a pervasively brittlely deformed hanging wall block to the SW and WSW kinematic sense (dip dependent) along a mean direction of 240° (McFadden et al., 2007). The transport direction agrees with stretching axes at deeper levels within the Fosdick dome, determined to be 060-240 to 070-250 (Siddoway et al., 2004b).

Structural data that support an ENE direction for the maximum principle finite strain axis for the Ford Ranges (Figure 4) include the regional mafic dike array (Siddoway et al., 2005); mapped and inferred NW-SE dip-slip normal faults in the southern Ford Ranges; and brittle kinematic criteria on ESE-striking dextral oblique minor faults and on NE-striking sinistral shear fractures. Among mesoscopic brittle structures on land, strike slip faults are prevalent, forming populations that accommodated both dextral and sinistral motion (Figure 4). Mutually crosscutting relationships with alkalic dikes indicate that the strike slip faults were active during the 070-250-directed opening that is recorded by the mafic dike array. Regional mapping indicates that Byrd Coast plutons are spatially associated with inferred major faults and may occupy releasing bends (Figures 4 and 9).

Consequently, the dextral and sinistral regional-scale faults in western Marie Byrd Land are viewed as contemporaneous conjugate structures whose motion aided ENE-WSW dextral transtension in the eastern WARS (Figure 4). Kinematics of normal faults mapped offshore of Edward VII Peninsula (Luyendyk et al., 2001) are consistent, as is the direction of margin-parallel divergence across a short-lived MBL-Bellingshausen plate boundary further east (Heinemann et al., 1999). The stretching direction for the eastern WARS in MBL is generally parallel to that predicted from the orientation and geometry of basement grabens in the Ross Sea (Cooper et al., 1991a, 1997; Davey and Brancolini, 1995; Trey et al., 1997). The documentation of important wrench deformation in MBL supports past interpretations of wrench and transfer faults within the WARS (e.g., Grindley and Davey, 1982) and for the first time determines their orientation and kinematics.

The apparent change from prevalent normal faults in the Ross Sea portion of the rift (Cooper et al., 1991a,b; Tessensohn and Wörner, 1991) to dextral strike slip in the eastern WARS implies a rotation of principal stress axes from σ_1 vertical in the west to σ_2 vertical in the east. The spatial variation in kinematics and dynamics across the WARS probably is related to the geometry of subducted lithosphere at the active margin (e.g., Bradshaw, 1989; Luyendyk, 1995), or is an expression of a regional strain gradient between the East Gondwana convergent boundary and the Australia-Antarctica boundary, undergoing slow divergence since ca. 125 Ma (Cande and Mutter, 1982; Tessensohn and Wörner, 1991).

The conjugate wrench zones active in MBL between 107 Ma and 97 Ma indicate a vertical orientation for the intermediate axis of principle finite strain, with axis of minimum finite strain oriented NW-SE in the plane of the earth. Transcurrent strain in MBL is consistent with the oblique convergence vector (Figures 9a and 9b) determined for Late Cretaceous time (Vaughan and Storey, 2000; Sutherland and Hollis, 2001; Vaughan et al., 2002a). Therefore, the postulated tectonic boundary separating the Ross continental province from the Amundsen arc province (Pankhurst et al., 1998) probably corresponds to an intracontinental dextral transform fault. The Amundsen province boundary is inferred on paleomagnetic grounds to have an approximately E-W trend in Marie Byrd Land (Figure 9b) (DiVenere et al., 1996). Restoration of dextral motion across an E-W transform fault potentially would place the Amundsen province magmatic arc outboard of the Ford Ranges (Figure 9a). Such a reconstruction helps explain the extent and degree of regional heating throughout the Ross province that elevated crustal isotherms (Figure 5b and 5c), induced extensive mid-crustal flow of the type documented in the Fosdick Mountains (Figure 8c), promoted rapid intracontinental extension across the WARS, and prevented development of orogenic topography. Dynamic subduction (e.g., Giunchi et al., 1996) or a postulated mantle plume (Weaver et al., 1994; Storey et al., 1999) may have been responsible for preventing dramatic subsidence and voluminous infilling of sedimentary basins of the Ross Sea (Wilson et al., 1998; Luyendyk et al., 2001; Karner et al., 2005).

CONCLUSIONS

The determination of dextral transtensional strain in the eastern WARS in Marie Byrd Land is consistent with the current picture of tectonic plate interactions at the Phoenix-East Gondwana (Pacific sector) boundary, with the final stages of subduction marked by oblique convergence of young oceanic crust (Bradshaw, 1989; Luyendyk et al., 1995; Sutherland and Hollis, 2001; Wandres and Bradshaw, 2005), as far east as Palmer Land (Figure 3) (Vaughan and Storey, 2000; Vaughan et al., 2002b). It is now clear that strike slip fault systems, thought to be in existence during oblique convergence at the Early Cretaceous margin (Weaver et al., 1995), remained active and accommodated intracontinental extension in the WARS until ca. 90 Ma. The structural and thermochronology record from MBL and the eastern Ross Sea indicates that the intracontinental extension between EANT and WANT that brought about opening the WARS by 90 Ma is distinct from the NZ-WANT breakup at 83 Ma and later. There is compelling evidence that the sharp continent-ocean boundary that distinguishes the MBL margin from the other gradational continent boundaries of the Antarctic Plate (Gohl, 2008, this volume) is controlled by a wrench zone formed during opening of the WARS.

It may be that relict subvertical transcurrent zones penetrating to the base of the crust provide a deep-seated conduit for magmatism in the linear volcanic mountain ranges of the MBL volcanic province (LeMasurier and Rex, 1989), or control the deep narrow lineaments in the subglacial topography beneath the Pine Island and Thwaites ice streams, 950 km to the east (Holt et al., 2006; Vaughan et al., 2006). In this way the lithospheric-scale structures formed during development of the West Antarctic rift system continue to exert fundamental influences on the long-term continental evolution of West Antarctica.

ACKNOWLEDGMENTS

Sincere thanks are extended to J. D. Bradshaw, F. J. Davey, and F. Tessensohn who provided reviews; to A. K. Cooper for his dedicated service as editor; and to the ISAES 2007 program committee for inviting this contribution. M. Brown and N. Mortimer provided input on a prior manuscript. Interpretations represented here have arisen through collaborations with B. P. Luyendyk, C. M. Fanning, R. McFadden, C. Teyssier, D. L. Whitney, C. A. Ricci, D. Wilson, F. J. Korhonen, and S. Saito. Other contributors include the Support Office for Aerogeophysical Research (SOAR) (1998), A. Whitehead, L. Sass III, S. Kruckenberg, J. Haywood, S. Fadrhonc, and M. Siddoway. Mike Roberts (polar guide), Raytheon Polar Services, the 109th Air National Guard, and Kenn Borek Air provided logistical support over several years. Warm thanks are also due to GANOVEX VII (1992-1993); Spedizione X, Italiantartide (1996-1997); J. Müller; and members of the South Pacific Rim Tectonics Expedition (SPRITE). S. Borg and J. Palais are thanked for program leadership. The U.S. National Science Foundation provided support through grants OPP-0338279, 0443543, 9702161, and 9615282. The content of the article is the work of the author and does not necessarily reflect the views of the National Science Foundation. Colorado College faculty research awards provided further support.

REFERENCES

Adams, C. J. 1986. Geochronological studies of the Swanson Formation of Marie Byrd Land, West Antarctica, and correlation with northern Victoria Land, East Antarctica and the South Island, New Zealand. *New Zealand Journal of Geology and Geophysics* 29:345-358.

Adams, C. J. 1987. Geochronology of granite terranes in the Ford Ranges, Marie Byrd Land, West Antarctica. *New Zealand Journal of Geology and Geophysics* 30:51-72.

Adams, C. J. 2004. Rb-Sr age and strontium isotopic characteristics of the Greenland Group, Buller terrane, New Zealand, and correlations at the East Gondwana margin. *New Zealand Journal of Geology and Geophysics* 47:189-200.

Adams, C. J., D. Seward, and S. D. Weaver. 1995. Geochronology of Cretaceous granites and metasedimentary basement on Edward VII Peninsula, Marie Byrd Land, West Antarctica. *Antarctic Science* 7:265-277.

Adams, C. J., R. J. Pankhurst, R. Maas, and I. L. Millar. 2005. Nd and Sr isotopic signatures of metasedimentary rocks around the South Pacific margin and implications for their provenance. *Geological Society of London Special Publication* 246:113-141.

Behrendt, J. C. 1999. Crustal and lithospheric structure of the West Antarctic Rift System from geophysical investigations—a review. *Global and Planetary Change* 23:25-44.

Behrendt, J. C., W. E. LeMasurier, A. K., Cooper, F., Tessensohn, A. Trehu, and D. Damaske. 1991. Geophysical Studies of the West Antarctic Rift System. *Tectonics* 10:1257-1273.

Behrendt, J. C., D. D. Blankenship, C. A. Finn, R. E. Bell, R. E. Sweeney, S. M. Hodge, and J. M. Brozena. 1994. Casertz aeromagnetic data reveal Late Cenozoic flood basalts (?) in the West Antarctic Rift System. *Geology* 22:527-530.

Behrendt, J. C., R. Saltus, D. Damaske, A. McCafferty, C. A. Finn, D. Blankenship, and R. E. Bell. 1996. Patterns of late Cenozoic volcanic and tectonic activity in the West Antarctic rift system revealed by aeromagnetic surveys. *Tectonics* 15:660-676.

Bell, R. E., D. D. Blankenship, C. A. Finn, D. L. Morse, T. A. Scambos, J. M. Brozena, and S. M. Hodge. 1998. Influence of subglacial geology on the onset of a West Antarctic ice stream from aerogeophysical observations. *Nature* 394:58-62.

Bialas, R. W., W. R. Buck, M. Studinger, and P. G. Fitzgerald. 2007. Plateau collapse model for the Transantarctic Mountains-West Antarctic Rift System: Insights from numerical experiments. *Geology* 35:687-690.

Blumenfeld, P., and J.-L. Bouchez. 1988. Shear criteria in granite and migmatite deformed in the magmatic and solid state. *Journal of Structural Geology* 10:361-372.

Bradshaw, J. D. 1989. Cretaceous geotectonic patterns in the New Zealand region. *Tectonics* 8:803-820.

Bradshaw, J. D. 2007. The Ross Orogen and Lachlan Fold Belt in Marie Byrd Land, Northern Victoria Land and New Zealand: Implication for the tectonic setting of the Lachlan Fold Belt. In *Antarctica: A Keystone in a Changing World—Online Proceedings for the Tenth International Symposium on Antarctic Earth Sciences*, eds. Cooper, A. K., C. R. Raymond et al., USGS Open-File Report 2007-1047. Short Research Paper 059, doi:10.3133/of2007-1047.srp059.

Bradshaw, J. D., B. Andrew, and B. D. Field. 1983. Swanson Formation and related rocks of Marie Byrd Land and a comparison with the Robertson Bay Group of northern Victoria Land. In *Antarctic Earth Science*, eds. R. L. Oliver, P. R. James, J. B. Jago, pp. 274-279. Canberra: Australian Academy of Science.

Bradshaw, J. D., R. J. Pankhurst, S. D. Weaver, B. C. Storey, R. J. Muir, and T. R. Ireland. 1997. New Zealand superterranes recognized in Marie Byrd Land and Thurston Island. In *The Antarctic Region, Geological Evolution and Processes*, ed. C. A. Ricci, pp. 429-436. Siena: *Terra Antartica* Publication.

Brown, M. 1973. The definition of metatexis, diatexis, and migmatite. *Proceedings of the Geologists' Association* 84:371-382.

Brown, M. 2007. Crustal melting and melt extraction, ascent and emplacement in orogens: Mechanisms and consequences. *Journal of the Geological Society of London* 164:709-730.

Brown, M., and R. A. Pressley. 1999. Crustal melting in nature: Prosecuting source processes. *Physics and Chemistry of the Earth, Part A: Solid Earth and Geodesy* 24:305-316.

Cande, S. C., and J. C. Mutter. 1982. A revised identification of the oldest sea-floor spreading anomalies between Australia and New Zealand. *Earth and Planetary Science Letters* 58(2):151-160.

Cape Roberts Science Team. 2000. Studies from the Cape Roberts Project; Ross Sea, Antarctica, Initial Report on CRP-3. *Terra Antartica* 7(1-2):185-209.

Collins, W. J. 2002a. Nature of extensional accretionary orogens. *Tectonics* 21(4), doi:10.1029/2000TC001272.

Collins, W. J. 2002b. Hot orogens, tectonic switching, and creation of continental crust. *Geology* 30(6): 535-538.

Collins, W. J., and B. E. Hobbs. 2001. What caused the Early Silurian change from mafic to silicic (S-type) magmatism in the eastern Lachlan Fold Belt? *Australian Journal of Earth Sciences* 48:25-41.

Cook, R. A., R. Sutherland, and H. Zhu. 1999. *Cretaceous-Cenozoic Geology and Petroleum Systems of the Great South Basin, New Zealand.* Lower Hutt, New Zealand: Institute of Geological and Nuclear Sciences.

Cooper, A. K., and F. J. Davey. 1985. Episodic rifting of Phanerozoic rocks in the Victoria Land Basin, western Ross Sea, Antarctica. *Science* 229:1085-1087.

Cooper, A. K., F. J. Davey, and J. C. Behrendt. 1991a. Structural and depositional controls on Cenozoic and (?)Mesozoic strata beneath the western Ross Sea. In *Geological Evolution of Antarctica*, eds. M. R. A. Thomson, J. A. Crame, J. W. Thomson, pp. 279-283. Cambridge: Cambridge University Press International.

Cooper, A. K., F. J. Davey, and K. Hinz. 1991b. Crustal extension and origin of sedimentary basins beneath the Ross Sea and Ross Ice Shelf, Antarctica. In *Geological Evolution of Antarctica*, eds. M. R. A. Thomson, J. A. Crame, and J. W. Thomson, pp. 285-291. Cambridge: Cambridge University Press.

Cooper, A. K., H. Trey, G. Pellis, G. Cochrane, F. Egloff, M. Busetti, and ACRUP Working Group. 1997. Crustal structure of the Southern Central Trough, Western Ross Sea. In *The Antarctic Region: Geological Evolution and Processes*, ed. C. A. Ricci, pp. 637-642. Siena: *Terra Antartica* Publication.

Cooper, R. A., and A. Tulloch. 1992. Early Paleozoic terranes in New Zealand and their relationship to the Lachlan Fold Belt. *Tectonophysics* 214:129-144.

Dalziel, I. W. D. 1997. Neoproterozoic-Paleozoic geography and tectonics: Review, hypothesis, environmental speculation. *Geological Society of America Bulletin* 109:16-42.

Dalziel, I. W. D., B. C. Storey, S. W. Garrett, A. M. Grunow, L. D. B. Herrod, and R. J. Pankhurst. 1987. Extensional tectonics and the fragmentation of Gondwanaland. In *Continental Extensional Tectonics*, eds. M. P. Coward, J. F. Dewey, and P. L. Hancock, *Geological Society Special Publication* 28:433-441.

Danesi, S., and A. Morelli. 2001. Structure of the upper mantle under the Antarctic Plate from surface wave tomography. *Geophysical Research Letters* 28:4395-4398.

Davey, F. J., and G. Brancolini. 1995. The Late Mesozoic and Cenozoic structural setting of the Ross Sea region. In *Geology and Seismic Stratigraphy of the Antarctic Margin,* eds. A. K. Cooper, P. F. Barker, and G. Brancolini, *Antarctic Research Series* 68:167-182. Washington, D.C.: American Geophysical Union.

Davey, F. J., and L. De Santis. 2006. A multi-phase rifting model for the Victoria Land Basin, western Ross Sea, in *Antarctica, Contributions to Global Earth Sciences*, eds. D. K. Fütterer, D. Damaske, G. Kleinschmidt, H. Miller, and F. Tessensohn, pp. 303-308. Berlin: Springer.

Decesari, R. C., C. C. Sorlien, B. P. Luyendyk, L. R. Bartek, J. B. Diebold, and D. S. Wilson. 2003. Tectonic evolution of the Coulman High and Central Trough along the Ross Ice Shelf, Southwestern Ross Sea, Antarctica. *Eos Transactions American Geophysical Union,* Fall Meeting Supplement, Abstract T12A-0447.

DiVenere, J., D. V. Kent, and I. W. D. Dalziel. 1996. Summary of palaeomagnetic results from West Antarctica; implications for the tectonic evolution of the Pacific margin of Gondwana during the Mesozoic. In *Weddell Sea Tectonics and Gondwana Break-up*, eds. B. C. Storey, E. C. King and R. A. Livermore, *Geological Society Special Publication* 108:31-43.

Eagles, G., K. Gohl, and R. D. Larter. 2004. High-resolution animated tectonic reconstruction of the South Pacific and West Antarctic Margin. *Geochemistry Geophysics Geosystems* 5:Q07002, doi:10.1029/2003GC000657.

Ehlers, T. A. 2005. Crustal thermal processes and the interpretation of thermochronometer data. *Reviews in Mineralogy and Geochemistry* 58:315-350.

Elliot, D. H., and T. H. Fleming. 2004. Occurrence and dispersal of magmas in the Jurassic Ferrar large igneous province, Antarctica. *Gondwana Research* 7(1):223-237.

Elliot, D. H., T. H. Fleming, P. Kyle, and K. A. Foland. 1999. Long-distance transport of magmas in the Jurassic Ferrar large igneous province, Antarctica. *Earth and Planetary Science Letters* 167:89-104.

Fayon, A. K., D. L. Whitney, and C. Teyssier. 2004. Exhumation of orogenic crust: Diapiric ascent versus low-angle normal faulting. In *Gneiss Domes in Orogeny*, eds. D. L. Whitney, C. Teyssier, and C. S. Siddoway, *Geological Society of America Special Paper* 380:129-139.

Fergusson, C. L., and J. Coney. 1992. Implications of a Bengal fan-type deposit in the Paleozoic Lachlan fold belt of southeastern Australia. *Geology* 20:1047-1049.

Fergusson, C. L., R. A. Henderson, I. Withnall, C. M. Fanning, D. Phillips, and K. Lewthwaite. 2007. Structural, metamorphic, and geochronological constraints on alternating compression and extension in the Early Paleozoic Gondwanan Pacific margin, northeastern Australia. *Tectonics* 26(3):TC300.

Ferraccioli, F., E. Bozzo, and D. Damaske. 2002. Aeromagnetic signatures over western Marie Byrd Land provide insight into magmatic arc basement, mafic magmatism and structure of the eastern Ross Sea rift flank. *Tectonophysics* 347:139-165.

Finn, C. A., R. D. Müller, and K. S. Panter. 2005. A Cenozoic diffuse alkaline magmatic province (DAMP) in the southwest Pacific without rift or plume origin. *Geochemistry Geophysics Geosystems* 6, doi:10.1029/2004GC000723.

Fitzgerald, P. G. 2002. Tectonics and landscape evolution of the Antarctic plate since the breakup of Gondwana, with an emphasis on the West Antarctic Rift System and the Transantarctic Mountains. In *Antarctica at the Close of a Millennium, eds. J. A. Gamble, D. N. B. Skinner, and S. Henry, pp. 453-469. Wellington: Royal Society of New Zealand.

Fitzgerald, P. G., and S. L. Baldwin. 1997. Detachment fault model for the evolution of the Ross Embayment. In *The Antarctic Region: Geological Evolution and Processes*, ed. C. A. Ricci, pp. 555-564. Siena: *Terra Antartica* Publication.

Fitzgerald, P. G., M. Sandiford, P. J. Barrett, and A. J. W. Gleadow. 1986. Asymmetric extension associated with uplift and subsidence in the Transantarctic Mountains and Ross Embayment. *Earth and Planetary Science Letters* 81:67-78.

Ford, A. B., and P. J. Barrett. 1975. Basement rocks of the south-central Ross Sea, Site 270, DSDP Leg 28. In *Initial Reports of the Deep Sea Drilling Project*, eds. D. E. Hayes, I. A. Frakes et al. Washington, D.C.: U.S. Government Printing Office, doi:10.2973/dsdp.proc.28.131.1975.

Forster, M. A., and G. S. Lister. 2003. Cretaceous metamorphic core complexes in the Otago Schist, New Zealand. *Australian Journal of Earth Sciences* 50:181-198.

Foster, D. A., D. R. Gray, and Bucher, M. 1999. Chronology of deformation within the turbidite-dominated Lachlan orogen: Implications for the tectonic evolution of eastern Australia and Gondwana. *Tectonics* 18, doi:10.1029/1998TC900031.

Gaffney, A. M., and C. S. Siddoway. 2007. Heterogeneous sources for Pleistocene lavas of Marie Byrd Land, Antarctica: New Data from the SW Pacific Diffuse Alkaline Magmatic Province. In *Antarctica: A Keystone in a Changing World—Online Proceedings for the Tenth International Symposium on Antarctic Earth Sciences,* eds. Cooper, A. K., C. R. Raymond et al., USGS Open-File Report 2007-1047, Extended abstract 063, http://pubs.usgs.gov/of/2007/1047/.

Gaina, C., R. D. Müller, J.-Y. Royer, J. Stock, J. Hardebeck, and A. Symonds. 1998. The tectonic history of the Tasman Sea: A puzzle with 13 pieces. *Journal of Geophysical Research* 103:12413-12433.

Gibson, G. M., and T. R. Ireland. 1995. Granulite formation during continental extension in Fiordland, New Zealand. *Nature* 375:479-482.

Gibson, G. M., and T. R. Ireland. 1996. Extension of Delamerian (Ross) Orogen into western New Zealand; evidence from zircon ages and implications for crustal growth along the Pacific margin of Gondwana. *Geology* 24:1087-1090.

Gibson, G. M., I. McDougall, and T. R. Ireland. 1988. Age constraints on metamorphism and the development of a metamorphic core complex in Fiordland, southern New Zealand. *Geology* 16:405-408.

Giunchi, C., R. Sabadini, E. Boschi, and P. Gasperini. 1996. Dynamic models of subduction—Geophysical and geological evidence in the Tyrrhenian Sea. *Geophysical Journal International* 126:555-578.

Glen, R. A. 2005. The Tasmanides of eastern Australia. In *Terrane Processes at the Margin of Gondwana*, eds. A. Vaughan, P. Leat, and R. J. Pankhurst, *Geological Society of London Special Publication* 246:23-96.

Gohl, K. 2008, this volume. Antarctica's continent-ocean boundaries: Consequences for tectonic reconstructions. In *Antarctica: A Keystone in a Changing World,* eds. A. K. Cooper et al. Washington, D.C.: The National Academies Press.

Gray, D. R., and D. A. Foster. 2004. Tectonic evolution of the Lachlan Orogen, southeast Australia: Historical review, data synthesis and modern perspectives. *Australian Journal of Earth Sciences* 51(6):773-817.

Grindley, G. W., and F. J. Davey. 1982. The reconstruction of New Zealand, Australia and Antarctica. In *Antarctic Geoscience*, ed. C. Craddock, pp. 15-26. Madison: University of Wisconsin Press.

Grunow, A. M., D. V. Kent, and I. W. D. Dalziel. 1991. New paleomagnetic data from Thurston Island: Implications for the tectonics of West Antarctica and Weddell Sea opening. *Journal of Geophysical Research* 96:17935-17954.

Hamilton, R. J., C. C. Sorlien, B. P. Luyendyk, L. R. Bartek, and S. A. Henrys. 1998. Tectonic regimes and structural trends off Cape Roberts, Antarctica. *Terra Antartica* 5(3):261-272.

Hayes, D. E., and F. J. Davey. 1975. A geophysical study of the Ross Sea, Antarctica. In *Initial Reports of the Deep Sea Drilling Project*, eds. D. E. Hayes, L. A. Frakes et al. Washington, D.C.: U.S. Government Printing Office, doi:10.2973/dsdp.proc.28.134.1975.

Heinemann, J., J. Stock, K. Clayton, S. Hafner, S. Cande, and C. Raymond. 1999. Constraints on the proposed Marie Byrd Land-Bellingshausen plate boundary from seismic reflection data. *Journal of Geophysical Research* 104:25321-25330.

Hoernle, K., F. Hauff, R. Werner, N. Mortimer, P. van den Bogaard, J. Geldmacher, and D. Garbe-Schoenberg. 2004. New insights into the origin and evolution of the Hikurangi oceanic plateau. *Eos Transactions of the American Geophysical Union* 85:401-406, doi:10.1029/2004EO410001.

Hollis, J. A., G. L. Clarke, K. A. Klepeis, N. R. Daczko, and T. R. Ireland. 2004. The regional significance of Cretaceous magmatism and metamorphism in Fiordland, New Zealand, from U-Pb zircon geochronology. *Journal of Metamorphic Geology* 22:607-627.

Holt, J. W., D. D. Blankenship, D. L. Morse, D. A. Young, M. E. Peters, S. D. Kempf, T. G. Richter, D. G. Vaughan, and H. F. J. Corr. 2006. New boundary conditions for the West Antarctic Ice Sheet: Subglacial topography of the Thwaites and Smith glacier catchments. *Geophysical Research Letters* 33:L09502, doi:10.1029/2005GL025561.

Huerta, A. 2007. Byrd drainage system: Evidence of a Mesozoic West Antarctic Plateau. In *Antarctica: A Keystone in a Changing World—Online Proceedings for the Tenth International Symposium on Antarctic Earth Sciences,* eds. Cooper, A. K., C. R. Raymond et al., USGS Open-File Report 2007-1047, Extended Abstract 091, http://pubs.usgs.gov/of/2007/1047/.

Ireland, T. R., and G. M. Gibson. 1998. SHRIMP monazite and zircon geochronology of high-grade metamorphism in New Zealand. *Journal of Metamorphic Geology* 16:149-167.

Jokat, W., T. Boebel, M. König, and M. Uwe. 2003. Timing and geometry of early Gondwana breakup. *Journal of Geophysical Research* 108:2428.

Karner, G. D., M. Studinger, and R. E. Bell. 2005. Gravity anomalies of sedimentary basins and their mechanical implications: Application to the Ross Sea basins, West Antarctica. *Earth and Planetary Science Letters* 235:577-596.

Kimbrough, D., and A. J. Tulloch. 1989. Early Cretaceous age of orthogneiss from the Charleston Metamorphic Group, New Zealand. *Earth and Planetary Science Letters* 95:130-140.

Korhonen, F. J., M. Brown, and C. S. Siddoway. 2007a. Petrologic and geochronological constraints on the polymetamorphic evolution of the Fosdick Migmatite Dome, Marie Byrd Land, West Antarctica. In *Antarctica: A Keystone in a Changing World—Online Proceedings for the Tenth International Symposium on Antarctic Earth Sciences,* eds. Cooper, A. K., C. R. Raymond et al., USGS Open-File Report 2007-1047, Extended abstract 049, http://pubs.usgs.gov/of/2007/1047/.

Korhonen, F. J., M. Brown, and C. S. Siddoway. 2007b. Unraveling polyphase high-grade metamorphism and anatexis in the Fosdick migmatite dome, West Antarctica, using mineral equilibria modeling and in situ monazite geochronology. Eos Transactions of American Geophysical Union, Fall Meeting Supplement 88, Abstract V04-4536.

Kula, J., A. J. Tulloch, T. L. Spell, and M. L. Wells. 2007. Two-stage rifting of Zealandia-Australia-Antarctica: Evidence from $^{40}Ar/^{39}Ar$ thermochronometry of the Sisters shear zone, Stewart Island, New Zealand. *Geology* 35:411-414, doi:10.1130/G23432A.1.

Laird, M. G., and J. D. Bradshaw. 2004. The break-up of a long-term relationship: The Cretaceous separation of New Zealand from Gondwana. *Gondwana Research* 7:273-286.

Larter, R. D., A. P. Cunningham, F. Barker, K. Gohl, and F. O. Nitsche. 2002. Tectonic evolution of the Pacific margin of Antarctica. 1. Late Cretaceous tectonic reconstructions. *Journal of Geophysical Research* 107:2345.

Lawrence, J. F., D. A. Wiens, A. A. Nyblade, S. Anandakrishnan, J. Shore, and D. Voigt. 2006. Upper mantle thermal variations beneath the Transantarctic Mountains inferred from teleseismic S-wave attenuation. *Geophysical Research Letters* 33:L03303, doi:10.1029/2005GL024516.

Lawver, L. A., and L. M. Gahagan. 1994. Constraints on the timing of extension in the Ross Sea region. *Terra Antartica* 1:545-552.

LeMasurier, W. E., and C. A. Landis. 1996. Mantle-plume activity recorded by low relief erosion surfaces in West Antarctica and New Zealand. *Bulletin of the Geological Society of America* 108:1450-1466.

LeMasurier, W. E. 2008. Neogene extension and basin deepening in the West Antarctic rift inferred from comparisons with the East African rift and other analogs. *Geology* 36(3):247-250, doi:10.1130/G24363A.1.

LeMasurier, W. E., and D. C. Rex. 1989. Evolution of linear volcanic ranges in Marie Byrd Land, West Antarctica. *Journal of Geophysical Research, B* 94:7223-7236.

Li, Z. X., and C. M. Powell. 2001. An outline of the palaeogeographic evolution of the Australasian region since the beginning of the Neoproterozoic. *Earth-Science Reviews* 53.

Lisker, F., and M. Olesch. 1998. Cooling and denudation history of western Marie Byrd Land, Antarctica, based on apatite fission-tracks. In *Advances in Fission-Track Geochronology,* eds. P. Van den haute and F. De Corte, pp. 225-240. Dordrecht: Kluwer Academic Publishers.

Lister, G. S., M. Etheridge, and P. Symonds. 1991. Detachment models for the formation of passive continental margins. *Tectonics* 10:1038-1064.

Lowe, A. L., and J. B. Anderson. 2002. Reconstruction of the West Antarctic ice sheet in Pine Island Bay during the Last Glacial Maximum and its subsequent retreat history. *Quaternary Science Reviews* 21:1879-1897.

Luyendyk, B. 1995. Hypothesis for Cretaceous rifting of East Gondwana caused by subducted slab capture. *Geology* 23:373-376.

Luyendyk, B. P., C. H. Smith, and F. M. van der Wateren. 1994. Glaciation, block faulting, and volcanism in western Marie Byrd Land. *Terra Antartica* 1:541-543.

Luyendyk, B., S. Cisowski, C. H. Smith, S. Richard, and D. Kimbrough. 1996. Paleomagnetic study of the northern Ford Ranges, western Marie Byrd Land, West Antarctica: Motion between West and East Antarctica. *Tectonics* 15:122-141.

Luyendyk, B. P., C. C. Sorlien, D. S. Wilson, L. R. Bartek, and C. S. Siddoway. 2001. Structural and tectonic evolution of the Ross Sea Rift in the Cape Colbeck region, eastern Ross Sea, Antarctica. *Tectonics* 20:933-958.

Luyendyk, B. P., D. S. Wilson, and C. S. Siddoway. 2003. The eastern margin of the Ross Sea Rift in western Marie Byrd Land: Crustal structure and tectonic development. *Geochemistry Geophysics Geosystems,* doi:10.1029/2002GC000462.

Marrett, R., and R. W. Allmendinger. 1990. Kinematic analysis of fault-slip data. *Journal of Structural Geology* 12:973-986.

Maule, C. F., M. E. Purucker, N. Olsen, and K. Mosegaard. 2005. Heat flux anomalies in Antarctica revealed by satellite magnetic data. *Science* 309:464-467.

McAdoo, D., and S. Laxon. 1997. Antarctic tectonics: Constraints from an ERS-1 satellite marine gravity field. *Science* 276:556-561.

McFadden, R., C. S. Siddoway, C. Teyssier, C. M. Fanning, and S. C. Kruckenberg. 2007. Cretaceous oblique detachment tectonics in the Fosdick Mountains, Marie Byrd Land, Antarctica. In *Antarctica: A Keystone in a Changing World—Online Proceedings for the Tenth International Symposium on Antarctic Earth Sciences,* eds. Cooper, A. K., C. R. Raymond et al., USGS Open-File Report 2007-1047. Short Research Paper 046, doi:10.3133/of2007-1047.srp046.

McKenzie, D., and K. Priestley. 2007. The influence of lithospheric thickness variations on continental evolution. *Lithos,* doi:10.1016/j.lithos.2007.05.005.

Milord, I., E. W. Sawyer, and M. Brown. 2001. Formation of diatexite migmatite and granite magma during anatexis of semi-pelitic metasedimentary rocks: An example from St. Malo, France. *Journal of Petrology* 42:487-505.

Morand, V. J. 1990. Low-pressure regional metamorphism in the Omeo Metamorphic Complex, Victoria, Australia. *Journal of Metamorphic Geology* 8:1-12.

Mortimer, N. 2004. Basement gabbro from the Lord Howe Rise. *New Zealand Journal of Geology and Geophysics* 47:501-507.

Mortimer, N., A. J. Tulloch, R. N. Spark, N. W. Walker, E. Ladley, A. Allibone, and D. L. Kimbrough. 1999a. Overview of the Median Batholith, New Zealand: A new interpretation of the geology of the Median Tectonic Zone and adjacent rocks. *Journal of African Earth Sciences* 29:257-268.

Mortimer, N., B. Gans, A. Calvert, and N. W. Walker. 1999b. Geology and geochronometry of the east edge of the Median Batholith (Median Tectonic Zone): A new perspective on Permian to Cretaceous crustal growth of New Zealand. *The Island Arc* 8:404-425.

Mortimer, N., K. Hoernle, F. Hauff, J. M. Palin, W. J. Dunlap, R. Werner, and K. Faure. 2006. New constraints on the age and evolution of the Wishbone Ridge, southwest Pacific Cretaceous microplates, and Zealandia-West Antarctica breakup. *Geology* 3:185-188.

Muir, R., T. R. Ireland, S. D. Weaver, and J. D. Bradshaw. 1994. Ion microprobe U-Pb zircon geochronology of granitic magmatism in the Western Province of the South Island, New Zealand. *Chemical Geology* 113:171-189.

Muir, R. J., S. D. Weaver, J. D. Bradshaw, G. N. Eby, and J. A. Evans. 1995. The Cretaceous Separation Point Batholith, New Zealand; granitoid magmas formed by melting of mafic lithosphere. *Journal of Geological Society of London* 152:689-701.

Muir, R. J., T. R. Ireland, S. D. Weaver, and J. D. Bradshaw. 1996. Ion microprobe dating of Paleozoic granitoids: Devonian magmatism in New Zealand and correlations with Australia and Antarctica. *Chemical Geology* 127:191-210.

Muir, R. J., T. R. Ireland, S. D. Weaver, J. D. Bradshaw, T. E. Waight, R. Jongens, and G. N. Eby. 1997. SHRIMP U-Pb geochronology of Cretaceous magmatism in northwest Nelson-Westland, South Island, New Zealand. *New Zealand Journal of Geology and Geophysics* 40:453-463.

Muir, R., T. R. Ireland, S. D. Weaver, J. D. Bradshaw, J. A. Evans, G. N. Eby, and D. Shelley. 1998. Geochronology and geochemistry of a Mesozoic magmatic arc system, Fiordland, New Zealand. *Journal of the Geological Society of London* 155:1037-1053.

Mukasa, S. B., and I. W. D. Dalziel. 2000. Marie Byrd Land, West Antarctica: Evolution of Gondwana's Pacific margin constrained by zircon U-Pb geochronology and feldspar common-Pb isotopic compositions. *Bulletin of the Geological Society of America* 112:611-627.

Olsen, S. N., B. D. Marsh, and L. P. Baumgartner. 2005. Modelling mid-crustal migmatite terrains as feeder zones for granite plutons: The competing dynamics of melt transfer by bulk versus porous flow. *Transactions of the Royal Society of Edinburgh, Earth Sciences* 95:49-58.

Pankhurst, R. J., S. D. Weaver, J. D. Bradshaw, B. C. Storey, and T. R. Ireland. 1998. Geochronology and geochemistry of pre-Jurassic superterranes in Marie Byrd Land, Antarctica. *Journal of Geophysical Research* 103:2529-2547.

Regenauer-Lieb, K., R. F. Wienberg, and G. Rosenbaum. 2006. The effects of energy feedbacks on continental strength. *Nature* 442:67-70.

Richard, S. M., C. H. Smith, D. K. Kimbrough, G. Fitzgerald, B. P. Luyendyk, and M. O. McWilliams. 1994. Cooling history of the northern Ford Ranges, Marie Byrd Land, West Antarctica. *Tectonics* 13:837-857.

Richards, S. W., and W. J. Collins. 2002. The Cooma Metamorphic Complex, a low-P, high-T (LPHT) regional aureole beneath the Murrumbidgee Batholith. *Journal of Metamorphic Geology* 20(1):119-134, doi:10.1046/j.0263-4929.2001.00360.x.

Ritzwoller, M. H., N. M. Shapiro, A. L. Levshin, and G. M. Leahy. 2001. Crustal and upper mantle structure beneath Antarctica and surrounding oceans. *Journal of Geophysical Research B* 106:30645-30670.

Rocchi, S., P. Armienti, and G. Di Vincenzo. 2005. No plume, no rift magmatism in the West Antarctic Rift. In *Plates, Plumes, and Paradigms,* eds. G. R. Foulger, J. Natland, D. Presnall and D. L. Anderson, *Geological Society of America Special Paper* 388:435-447.

Rossetti, F., G. DiVincenzo, A. L. Läufer, F. Lisker, S. Rocchi, and F. Storti. 2003a. Cenozoic right-lateral strike-slip faulting in North Victoria Land: An integrated structural, AFT and ⁴⁰Ar-³⁹Ar study. In *9th International Symposium on Antarctic Earth Sciences—Antarctic Contributions to Global Earth Science,* eds. D. K. Fütterer, D. Damaske, G. Kleinschmidt, H. Miller and F. Tessensohn, pp. 283-284. Potsdam: *Terra Nostra.*

Rossetti, F., F. Lisker, F. Storti, and A. L. Läufer. 2003b. Tectonic and denudational history of the Rennick Graben, North Victoria Land: Implications for the evolution of rifting between East and West Antarctica. *Tectonics* 22(2):1016, doi:10.1029/2002TC001416, 2003.

Rossetti, F., F. Tecce, L. Aldega, M. Brilli, and C. Faccenna. 2006. Deformation and fluid flow during orogeny at the palaeo-Pacific active margin of Gondwana: The Early Palaeozoic Robertson Bay accretionary complex (north Victoria Land, Antarctica). *Journal of Metamorphic Geology* 24:33-53.

Saito, S., F. Korhonen, M. Brown, and C. S. Siddoway. 2007. Petrogenesis of granites in the Fosdick migmatite dome, Marie Byrd Land, West Antarctica. In *Antarctica: A Keystone in a Changing World—Online Proceedings for the Tenth International Symposium on Antarctic Earth Sciences,* eds. Cooper, A. K., C. R. Raymond et al., USGS Open-File Report 2007-1047. Extended Abstract 105, http://pubs.usgs.gov/of/2007/1047/.

Salvini, F., G. Brancolini, M. Busetti, F. Storti, F. Mazzarini, and F. Coren. 1997. Cenozoic geodynamics of the Ross Sea region, Antarctica: Crustal extension, intraplate strike-slip faulting, and tectonic inheritance. *Journal of Geophysical Research* 102:24669-24696.

Sawyer, E. W. 1998. Formation and evolution of granite magmas during crustal reworking: The significance of diatexites. *Journal of Petrology* 39:1147-1167.

Sawyer, E. W. 2001. Melt segregation in the continental crust: Distribution and movement of melt in anatectic rocks. *Journal of Metamorphic Geology* 19:291-309.

Sawyer, E. W. 2004. Morphological variations in migmatites: importance of melt fraction and syn-migmatization strain. In *Scientific Sessions: Abstracts (part 1), 32nd International Geological Congress, Florence, Italy, 2004*, Abstract 86-9, p. 423.

Scott, J. M., and A. F. Cooper. 2006. Early Cretaceous extensional exhumation of the lower crust of a magmatic arc: Evidence from the Mount Irene Shear Zone, Fiordland, New Zealand. *Tectonics* 25, doi.10.1029/2005TC001890.

Siddoway, C. S., S. Baldwin, G. Fitzgerald, C. M. Fanning, and B. P. Luyendyk. 2004a. Ross Sea mylonites and the timing of intracontinental extension within the West Antarctic rift system. *Geology* 32:57-60.

Siddoway, C. S., S. M. Richard, C. M. Fanning, and B. P. Luyendyk. 2004b. Origin and emplacement of a middle Cretaceous gneiss dome, Fosdick Mountains, West Antarctica. In *Gneiss Domes in Orogeny*, eds. D. L. Whitney, C. Teyssier, and C. S. Siddoway, *Geological Society of America Special Paper* 380:267-294.

Siddoway, C. S., L. C. Sass III, and R. Esser. 2005. Kinematic history of Marie Byrd Land terrane, West Antarctica: Direct evidence from Cretaceous mafic dykes. In *Terrane Processes at the Margin of Gondwana*, eds. A. Vaughan, P. Leat and R. J. Pankhurst, *Geological Society of London Special Publication* 246:417-438.

Siddoway, C. S., C. M. Fanning, S. C. Kruckenberg, and S. C. Fadrhonc. 2006. U-Pb SHRIMP investigation of the timing and duration of melt production and migration in a Pacific margin gneiss dome, Fosdick Mountains, Antarctica. *Eos Transactions of the American Geophysical Union*, Fall Meeting Supplement 87, Abstract V23D-0661.

Sieminski, A., E. Debayle, and J. J. Leveque. 2003. Seismic evidence for deep low-velocity anomalies in the transition zone beneath West Antarctica. *Earth and Planetary Science Letters* 216:645-661.

Smith, C. H. 1992. Metapelitic migmatites at the granulite facies transition, Fosdick Mountains, Antarctica. In *Recent Progress in Antarctic Earth Science*, eds. Y. Yoshida, K. Kaminuma, and K. Shiraishi, pp. 295-301. Tokyo: Terra Publications.

Smith, C. H. 1996. Migmatites of the Alexandra Mountains, West Antarctica: Pressure-temperature conditions of formation and regional context. *Geologisches Jahrbuch B* 52:169-178.

Smith, C. H. 1997. Mid-crustal processes during Cretaceous rifting, Fosdick Mountains, Marie Byrd Land. In *The Antarctic Region: Geological Evolution and Processes*, ed. C. A. Ricci, pp. 313-320. Siena: *Terra Antartica* Publication.

Sorlien, C. C., B. P. Luyendyk, D. S. Wilson, R. C. Decesari, L. R. Bartek, and J. B. Diebold. 2007. Oligocene development of the West Antarctic Ice Sheet recorded in eastern Ross Sea strata. *Geology* 35:467-470.

Spell, T. L., I. McDougall, and A. J. Tulloch. 2000. Thermochronological constraints on the breakup of the Pacific Gondwana margin: The Paparoa metamorphic core complex, South Island, New Zealand. *Tectonics* 19:433-451.

Squire, R. J., I. H. Campbell, C. M. Allen, and C. J. L. Wilson. 2006. Did the Transgondwanan Supermountain trigger the explosive radiation of animals on Earth? *Earth and Planetary Science Letters* 250:116-133.

Stern, T., and U. S. ten Brink. 1989. Flexural uplift of the Transantarctic mountains. *Journal of Geophysical Research* 94:10315-10330.

Stock, J. M., and S. C. Cande. 2002. Tectonic history of Antarctic seafloor in the Australia-New Zealand-South Pacific sector: Implications for Antarctic continental tectonics. In *Antarctica at the Close of a Millennium*, eds. J. A. Gamble, D. N. B. Skinner and S. Henry, pp. 251-259. Welllington: Royal Society of New Zealand.

Storey, B. C. 1991. The crustal blocks of West Antarctica within Gondwana: Reconstruction and break-up model. In *Geological Evolution of Antarctica*, eds. M. R. A. Thomson, J. A. Crame, and J. W. Thomson, pp. 587-592. Cambridge: Cambridge University Press.

Storey, B. C. 1992. Role of subduction-plate boundary forces during the initial stages of Gondwana break-up: Evidence from the proto-Pacific margin of Antarctica. In *Magmatism and the Causes of Continental Breakup*, eds. B. C. Storey, T. Alabaster, and R. J. Pankhurst, *Geological Society Special Publication* 68:165-184.

Storey, B., T. Leat, S. D. Weaver, R. J. Pankhurst, J. D. Bradshaw, and S. Kelley. 1999. Mantle plumes and Antarctica-New Zealand rifting: Evidence from mid-Cretaceous mafic dykes. *Journal of the Geological Society of London* 156:659-671.

Stump, E. 1995. *The Ross Orogen of the Transantarctic Mountains*: New York: Cambridge University Press.

Sutherland, R. 1999. Basement geology and tectonic development of the greater New Zealand region: An interpretation from regional magnetic data. *Tectonophysics* 308:341-362.

Sutherland, R., and C. Hollis. 2001. Cretaceous demise of the Moa plate and strike-slip motion at the Gondwana margin. *Geology* 29:279-282.

Taylor, B. 2006. The single largest oceanic plateau: Ontong Java-Manihiki-Hikurangi. *Earth and Planetary Science Letters* 241:372-380.

ten Brink, U. S., R. I. Hackney, S. Bannister, T. A. Stern, and Y. Makovsky. 1997. Uplift of the Transantarctic Mountains and the bedrock beneath the East Antarctic ice sheet. *Journal of Geophysical Research, Solid Earth* 102:27603-27621.

Tessensohn, F., and G. Wörner. 1991. The Ross Sea Rift System: Structure, evolution and analogues. In *Geological Evolution of Antarctica*, eds. M. R. A. Thomson, J. A. Crame, and J. W. Thomson., pp. 273-277. Cambridge: Cambridge University Press.

Teyssier, C., and D. L. Whitney. 2002. Gneiss domes and orogeny. *Geology* 30:1139-1142.

Teyssier, C., E. Ferré, D. L. Whitney, B. Norlander, O. Vanderhaeghe, and D. Parkinson. 2005. Flow of partially molten crust and origin of detachments during collapse of the Cordilleran orogen. In *High-Strain Zones: Structure and Physical Properties*, eds. D. Bruhn and L. Burlini. *Geological Society of London Special Publication* 245:39-64.

Thompson, A. B. 1996. Fertility of crustal rocks during anatexis. *Transactions of the Royal Society of Edinburgh* 87:1-10.

Tikoff, B., C. Teyssier, and C. Waters. 2002. Clutch tectonics and the partial attachment of lithospheric layers. *EGU Stephan Mueller Special Publication Series* 1:57-73.

Trey, H., J. Makris, G. Brancolini, and A. K. Cooper. 1997. The Eastern Basin Crustal model from wide-angle reflection data, Ross Sea, Antarctica Structure of the Southern Central Trough, Western Ross Sea. In *The Antarctic Region: Geological Evolution and Processes*, ed. C. A. Ricci, pp. 643-648. Siena: *Terra Antartica* Publication.

Trey, H., A. K. Cooper, G. Pellis, B. della Vedova, G. Cochrane, G. Brancolini, and J. Makris. 1999. Transect across the West Antarctic rift system in the Ross Sea, Antarctica. *Tectonophysics* 301:61-74.

Tulloch, A. J., and D. L. Kimbrough. 1989. The Paparoa metamorphic core complex, New Zealand: Cretaceous extension associated with fragmentation of the Pacific margin of Gondwana. *Tectonics* 8:1217-1234.

Tulloch, A. J., and D. L. Kimbrough. 2003. Paired plutonic belts in convergent margins and the development of high Na, Al, Sr, low Y magmatism: The Peninsular Ranges Batholith of California and the Median Batholith of New Zealand. *Geological Society of America Special Publication* 374:275-295.

Tulloch, A. J., D. Kimbrough, D., and R. Wood. 1991. Carboniferous granite basement dredged from a site on the southwest margin of the Challenger Plateau. *New Zealand Journal of Geology and Geophysics* 34:121-126.

Vaughan, A. M., and R. A. Livermore. 2005. Episodicity of Mesozoic terrane accretion along the Pacific margin of Gondwana: Implications for superplume-plate interactions. In *Terrane Processes at the Margin of Gondwana*, eds. A. Vaughan, P. Leat and R. J. Pankhurst, *Geological Society of London Special Publication* 246:143-178.

Vaughan, A. M., and B. C. Storey. 2000. The eastern Palmer Land shear zone: A new terrane accretion model for the Mesozoic development of the Antarctic Peninsula. *Journal of the Geological Society of London* 157:1243-1256.

Vaughan, A. M., S. P. Kelley, and B. C. Storey. 2002a. Age of ductile deformation on the Eastern Palmer Land Shear Zone, Antarctica, and implications for timing of Mesozoic terrane collision. *Geological Magazine* 139:465-471.

Vaughan, A. M., R. J. Pankhurst, and C. M. Fanning. 2002b. A mid-Cretaceous age for the Palmer Land event, Antarctic Peninsula: Implications for terrane accretion timing and Gondwana palaeolatitudes. *Journal of the Geological Society of London* 159:113-116.

Vaughan, D. G., F. Ferraccioli, N. Frearson, H. F. J. Corr, A. O'Hare, D. Mach, J. W. Holt, D. D. Blankenship, D. L. Morse, and D. A. Young. 2006. New boundary conditions for the West Antarctic Ice Sheet: Subglacial topography beneath Pine Island Glacier. *Geophysical Research Letters* 33:L09501, doi:10.1029/2005GL025588.

Vernon, R. H., and S. E. Johnson. 2000. Transition from gneiss to migmatite and the relationship of leucosome to peraluminous granodiorite in the Cooma Complex, SE Australia. *Journal of the Virtual Explorer* 2. http://virtualexplorer.com.au/special/meansvolume/contribs/vernon/index.html.

Wandres, A. M., and J. D. Bradshaw. 2005. New Zealand tectonostratigraphy and implications from conglomeratic rocks for the configuration of the SW Pacific margin of Gondwana. In *Terrane Processes at the Margin of Gondwana*, eds. A. Vaughan, P. Leat and R. J. Pankhurst, *Geological Society of London Special Publication* 246:179-216.

Weaver, S. D., J. D. Bradshaw, and C. J. Adams. 1991. Granitoids of the Ford Ranges, Marie Byrd Land, Antarctica. In *Geological Evolution of Antarctica*, eds. M. R. A. Thomson, J. A. Crame, and J. W. Thomson, pp. 345-351. Cambridge: Cambridge University Press.

Weaver, S. D., C. J. Adams, R. J. Pankhurst, and I. L. Gibson. 1992. Granites of Edward VII Peninsula, Marie Byrd Land: Anorogenic magmatism related to Antarctic-New Zealand rifting. In *Proceedings of the Second Hutton Symposium on the Origin of Granites and Related Rocks,* eds. E. Brown and B. W. Chappell. *Transactions of the Royal Society of Edinburgh, Earth Sciences* 83:281-290.

Weaver, S. D., B. C. Storey, R. J. Pankhurst, S. B. Mukasa, V. Divenere, and J. D. Bradshaw. 1994. Antarctic-New Zealand rifting and Marie Byrd Land lithospheric magmatism linked to ridge subduction and mantle plume activity. *Geology* 22:811-814.

Weaver, S. D., R. J. Pankhurst, B. C. Storey, J. D. Bradshaw, and R. Muir. 1995. Cretaceous magmatism along the SW Pacific Gondwana margin. In *Abstracts—VII International Symposium on Antarctic Earth Sciences*, ed. C. A. Ricci, p. 402. Siena: *Terra Antartica* Publication.

Weinberg, R. F. 2006. Melt segregation structures in granitic plutons. *Geology* 34:305-308.

Wilbanks, J. R. 1972. Geology of the Fosdick Mountains, Marie Byrd Land. In *Antarctic Geology and Geophysics*, ed. R. J. Adie, pp. 277-284. Oslo: Universitetsforlaget

Wilson, D., and B. P. Luyendyk. 2006. Bedrock platforms within the Ross Embayment, West Antarctica: Hypotheses for ice sheet history, wave erosion, Cenozoic extension, and thermal subsidence. *Geochemistry Geophysics Geosystems* 7(2):Q12011, doi:10.1029/2006GC001294.

Wilson, G. S., A. P. Roberts, K. L. Verosub, F. Florindo, and L. Sagnotti. 1998. Magnetobiostratigraphic chronology of the Eocene-Oligocene transition in the CIROS-1 core, Victoria Land margin, Antarctica; implications for Antarctic glacial history. *Bulletin of the Geological Society of America* 110:35-47.

Wilson, T. J. 1992. Mesozoic and Cenozoic kinematic evolution of the Transantarctic Mountains. In *Recent Progress in Antarctic Earth Science*, eds. Y. Yoshida, K. Kaminuma, and K. Shiraishi, pp. 303-314. Tokyo: Terra Scientific Publishing.

Wilson, T. J. 1993. Jurassic faulting and magmatism in the Transantarctic Mountains; implications for Gondwana breakup. In *Assembly, Evolution and Dispersal: Proceedings of the Gondwana Eight Symposium*, eds. R. H. Findlay, R. Unrug, M. R. Banks and J. J. Veevers, pp. 563-572. Hobart, Tasmania: International Gondwana Symposium.

Winberry, J. P., and S. Anandakrishnan. 2004. Crustal structure of the West Antarctic rift system and Marie Byrd Land hotspot. *Geology* 32:977-980.

The Significance of Antarctica for Studies of Global Geodynamics

R. Sutherland[1]

ABSTRACT

Antarctica has geometric significance for global plate kinematic studies, because it links seafloor spreading systems of the African hemisphere (Indian and Atlantic Oceans) with those of the Pacific. Inferences of plate motions back to 44 Ma, around the onset of rapid spreading south of Australia and formation of a new boundary through New Zealand, are consistent with Antarctic rifting and formation of the Adare Basin during 44-26 Ma (i.e., no additional plate motions are required in the South Pacific). The time period 52-44 Ma represents a profound global and South Pacific tectonic change, and significant details remain unresolved. For 74 Ma a significant nonclosure of the South Pacific plate-motion circuit is identified if Antarctic motion is not included. Alternate inferences of motion through Antarctica during the interval 74-44 Ma imply significantly different subduction volumes and directions around the Pacific, and imply different relative motions between hotspots.

INTRODUCTION

The surface of Earth can be divided into hemispheres with distinct tectonic character. The African hemisphere contains spreading ridges in the Indian and Atlantic Oceans that allow relative plate motions to be determined, and the motions are shown by studies of seamount chains to be well approximated by a single hotspot (absolute) reference frame (Muller et al., 1993). The Pacific hemisphere is surrounded by subduction zones with highly uncertain relative motions between subducting and overriding plates (Figures 1 and 2). A full understanding of geodynamics requires global determinations of past relative plate motions and boundary locations, and motions of plates relative to hotspots (e.g., Lithgow-Bertelloni and Richards, 1998).

Most previous calculations of global plate motions assume that hotspots in the Pacific hemisphere, specifically Hawaii and Louisville, have been fixed relative to each other and to African-hemisphere hotspots during Cretaceous-Cenozoic time (Engebretson et al., 1985; Gordon and Jurdy, 1986). However, paleomagnetic data from the Emperor seamount chain in the North Pacific are inconsistent with this assumption, and mantle flow calculations predict significant advection of the rising mantle plume responsible for the Hawaii hotspot (Tarduno et al., 2003; Steinberger et al., 2004). The only way to determine global relative plate motions and, hence, test predictive hotspot models based on mantle flow calculations is to piece together the kinematic evidence that was formed at plate boundaries; this includes seafloor fracture zones and magnetic anomalies, and the records that are preserved within continents and their margins.

Antarctica is significant in the global relative plate-motion circuit because it geometrically connects the African and Pacific hemispheres along a path that can be directly reconstructed at past times from seafloor and continental records (Figures 1 and 2). Therefore, quantification of internal deformation of Antarctica is an essential part of any global relative-plate-motion model. Further, because internal deformation of plates is commonly characterized by local rotation poles (Gordon, 1998), small local displacements may be described by relatively large rotation angles, which propagate (when incorporated into a plate-motion chain) into

[1]GNS Science, PO Box 30368, Lower Hutt 5040, New Zealand (r.sutherland@gns.cri.nz).

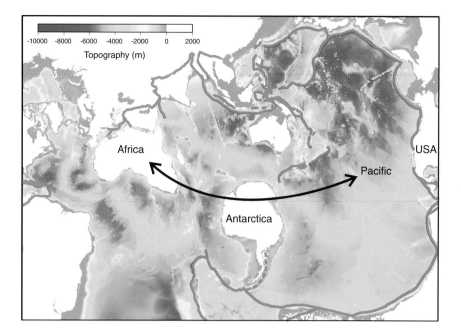

FIGURE 1 Global bathymetry (from Smith and Sandwell, 1997). Convergent plate boundaries shown in orange. Arrow indicates the pathway of kinematic connection between the African and Pacific hemispheres that did not contain destructive boundaries during the Cenozoic era. Oblique Mercator projection.

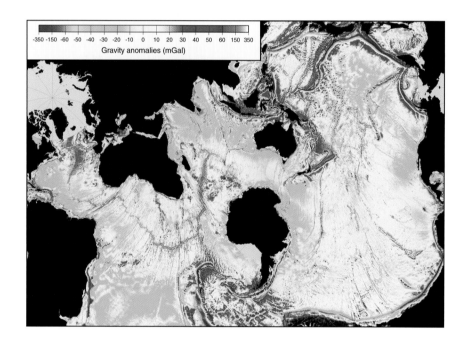

FIGURE 2 Gravity anomalies (from Sandwell and Smith, 1997).

very large predicted plate displacements at greater distances (e.g., at equatorial latitudes). The validity of Antarctic reconstructions must be consistent with other motions in the South Pacific, because Antarctica is part of a closed plate-motion circuit that includes Australia and New Zealand.

This paper presents a model for the block motion of Marie Byrd Land relative to the East Antarctic craton since 74 Ma. The model is simplistic by design, because the primary purpose of this paper is to propose a new and quantitative hypothesis for motion on an intra-Antarctic plate

boundary during the interval 74-50 Ma. It is accepted that refinement to this model will be necessary, to fit observations of crustal strain and the timing of deformation in detail. The hypothesis is presented and then tested against crustal geology of the Antarctic continent, the geometry of seafloor in the South Pacific plate-motion circuit, and the global motions of plates relative to hotspots. A plausible tectonic explanation for why such a model makes physical sense is briefly discussed. Finally, global implications of the hypothesis are explored with regard to subduction budgets since the late

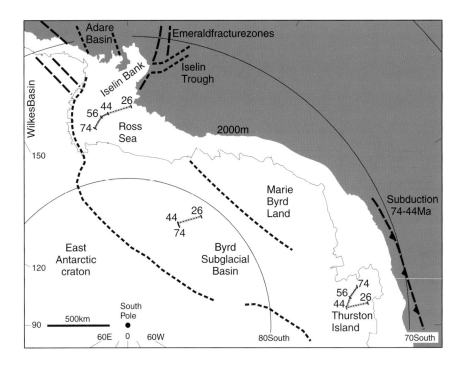

FIGURE 3 Map of the hypothesized intra-Antarctic plate boundary that stretches between the Ross Sea through the Byrd Subglacial Basin to inboard of Thurston Island. Significant tectonic regions are labeled and their margins are shown bold dashed. Arbitrary points within the plate boundary zone show the motion of the Marie Byrd Land plate relocated (pink dotted) relative to the East Antarctic plate using the proposed kinematic model (Table 1).

Cretaceous, because this is of broad international interest for both geodynamics and continental margin geology.

MOTION OF MARIE BYRD LAND RELATIVE TO EAST ANTARCTICA

The plate-motion model (Figure 3) is divided into four phases: (1) 74-56 Ma is a time of slow dextral extension in the Ross Sea and dextral transpression near Thurston Island; (2) 56-44 Ma is a time of accelerated rifting in the Ross Sea and highly oblique dextral transpression near Thurston Island; (3) 44-26 Ma is the time of Adare Basin formation (Cande et al., 2000) and includes rifting inboard of Thurston Island; and (4) there has been no significant motion since 26 Ma (Table 1). The oldest finite rotation corresponds to chron 33y (Cande and Kent, 1995), which is the time of the oldest magnetic anomaly that is widely preserved and recognized in the South Pacific.

Test 1: Antarctic Geology and Geophysical Data

Ross Sea Rift

It has been known for several decades that the Ross Sea has thinned crust, rifted sedimentary basins, and a rift flank uplift called the Transantarctic Mountains, and that Cretaceous-Cenozoic extensional tectonics were implicated (Davey et al., 1982; Behrendt et al., 1991; ten Brink et al., 1993; Fitzgerald, 1994; Cooper et al., 1995). Crustal thickness estimates imply 350-400 km of total extension, which could

TABLE 1 Finite Rotations Describing Marie Byrd Land Relative to East Antarctica

Age (Ma)	Latitude (°N)	Longitude (°E)	Angle (°)
26.6	–18.2	–17.9	0.0
33.6	–18.2	–17.9	0.7
43.8	–18.2	–17.9	1.7
56.0	–70.0	–30.0	4.0
73.6	–80.0	–70.0	8.5

be revised to >400-450 km if crustal addition were accounted for (Behrendt et al., 1991).

The total motion in the Ross Sea that is implied by the model proposed in this paper is ca. 300 km since 74 Ma (Figure 3). Therefore, ca. 70 percent of the total thinning is implied to have occurred after 74 Ma. This estimate could be revised downward if some extension were distributed over a broader region.

The oldest strata in the Ross Sea that have been drilled are late Eocene and Oligocene in age (Hayes et al., 1975; Barrett, 1989; Barrett et al., 1995), and these postdate normal-faulted strata everywhere except the Victoria Land Basin in the western Ross Sea, where faulting continued through Oligocene time (34-24 Ma) (Cooper et al., 1987; Henrys et al., 1998; Hamilton et al., 2001). Hence, a large component of the extension is constrained to be Eocene or older. Models of apatite fission track data from a basement rock sample collected at DSDP site 270, which is sited on a rifted horst in the central Ross Sea, suggest exhumation was completed at some time during the interval 90-50 Ma (Fitzgerald and Baldwin, 1997).

Apatite fission-track and (U-Th)/He thermochronology results from the Transantarctic Mountains record cooling and hence inferred exhumation during the time interval 80-40 Ma (Fitzgerald, 1992, 1994; Stump and Fitzgerald, 1992; Lisker, 2002; Fitzgerald et al., 2006). Age-elevation correlations have been used to suggest an episodic uplift model with an initial phase starting before 80 Ma and a second phase of more rapid exhumation starting at 55-45 Ma (Stump and Fitzgerald, 1992). This model has appeal, because the first phase corresponds to Gondwana breakup (local separation of New Zealand) and the second phase is a time of profound global and South Pacific tectonic change (below). However, close inspection of the data does not provide compelling evidence for a discrete regional (rather than local) event before 80 Ma. Instead, the ages are broadly distributed across the time interval 80-45 Ma (or even older) and thermal model inversions produce results that are consistent with exhumation histories during that interval. The regional increase in cooling rate within the Transantarctic Mountains at ca. 55-45 Ma is compelling, and hence an increase in exhumation rate is inferred.

Byrd Subglacial Basin

South of the Ross Sea the Byrd Subglacial Basin is a region of bedrock elevations below sea level, with some regions being deeper than 1000 m below sea level (Lythe et al., 2001). The basin is interpreted to be of rift origin and has only a thin (<1 km) sedimentary record imaged beneath the ice, but gravity interpretations suggest localized narrow basins with up to 5 km of sediment (Behrendt et al., 1991; Anandakrishnan et al., 1998; Studinger et al., 2001). Fission track data from the southern Transantarctic Mountains (Scott Glacier) yield ages in the range 120-60 Ma (Stump and Fitzgerald, 1992). The Byrd Subglacial Basin has a similar width to the Ross Sea (ca. 800 km) and is entirely ice-covered.

Thurston Island Region

The region of Antarctica from Thurston Island to the Antarctic Peninsula has undergone a complex tectonic history since Jurassic time (Dalziel and Elliot, 1982). Pre-Gondwanaland breakup reconstructions must account for extensional basins, such as the Byrd Subglacial Basin; dextral transpression along the paleosubduction continental margin; and continuity of geological characteristics (Storey and Nell, 1988; Storey, 1991; McCarron and Larter, 1998; Larter et al., 2002). The relative magnitudes, timing, and spatial distribution of dextral transpression and extension remain constrained by only a small number of field observations due to extreme remoteness and ice cover.

Summary of Test 1

The kinematic hypothesis predicts that 70 percent of extension in the Ross Sea occurred after 74 Ma, and that dextral-oblique compression followed by extension occurred in the Thurston Island region. I contend that both predictions are consistent with local observations, but are not necessarily the favored models of local workers. The initial phase is most controversial, because there is so little geological evidence for activity during this time. However, the syn-rift strata have not been sampled and if a heating model were proposed to explain the rift flank uplift (ten Brink et al., 1993), burial rather than exhumation could be expected during the early stages of rifting, which could explain the apatite fission track results.

Test 2: South Pacific Plate Motions

43-0 Ma

Rifted boundaries in the southeast Indian Ocean (East Antarctica Australia) and south of New Zealand (ENZ-WNZ) formed at about chron 20 (43 Ma) (Sutherland, 1995; Tikku and Cande, 2000), after cessation of spreading in the Tasman Sea (Australia-WNZ) (Gaina et al., 1998). Pacific-Marie Byrd Land spreading is constrained by numerous magnetic anomaly and fracture zone picks (Cande et al., 1995; Cande and Stock, 2004). The "missing link" in this plate-motion circuit (Figure 4) is rifting within Antarctica and formation of seafloor in the Adare Basin (Cande et al., 2000). However, the location of the pole of rotation that describes Antarctic rifting is imprecisely constrained by this analysis, and a rotation pole nearer to Antarctica has been suggested (Davey et al., 2006). This paper uses a magnetic anomaly 13 plate-motion circuit inversion and extrapolation to anomaly 20 (Cande et al., 2000); the model (Table 1; Figure 5) produces rifting of only slightly decreased magnitude toward the east (Figure 3).

56-43 Ma

This was a time of profound change throughout the Pacific and Indian Oceans, and uncertainty surrounds the exact nature and timing of these changes. One of the least affected boundaries was that between East New Zealand and Marie Byrd Land, where seafloor spreading continued with only minor changes in direction and rate (Cande et al., 1995). The Tasman Sea stopped opening at chron 24 (52 Ma) (Gaina et al., 1998), but rapid divergence south of New Zealand (WNZ-ENZ) and Australia did not start until ca. 43 Ma (Sutherland, 1995; Wood et al., 1996; Tikku and Cande, 2000). Analysis of plate closure (Figure 5) during this interval is hampered by the relatively short time interval; the possibility of intraplate deformation within New Zealand, which was very close to the instantaneous pole of ENZ-WNZ rotation after 43 Ma;

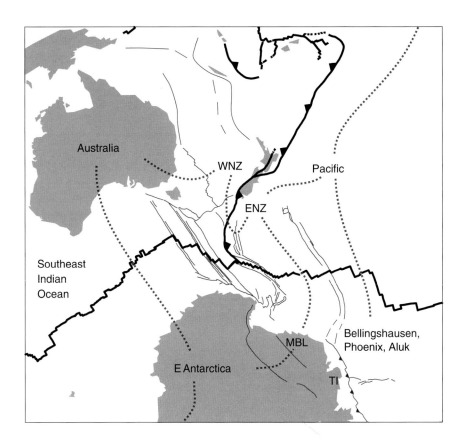

FIGURE 4 The South Pacific plate-motion circuit is the closed loop Australia-East Antarctica-Marie Byrd Land (MBL)-East New Zealand (ENZ)-West New Zealand (WNZ)-Australia. Bold lines show active plate boundaries. Fine blue lines show locations of significant boundaries during the interval 74-0 Ma. TI = Thurston Island.

and the difficulty of interpretation of crust in the southeast Indian Ocean.

74-56 Ma

This was a time of tectonic quiescence within New Zealand, with the only records of tectonic activity coming from the Taranaki Basin of central New Zealand, where very small amounts of rifting are implied (King and Thrasher, 1996). Therefore, the total motion through New Zealand is very similar at 56 Ma and 74 Ma, and is well approximated by fitting rift boundaries (ENZ-WNZ) (Figure 4) south of New Zealand (Sutherland, 1995). Tasman Sea spreading (Australia-WNZ) is quantified by magnetic anomalies (Gaina et al., 1998), as is ENZ-MBL spreading (Cande et al., 1995). It is not clear that magnetic lineations in the southeast Indian Ocean are isochrons, but detailed analyses and published reconstructions exist (Tikku and Cande, 1999, 2000; Muller et al., 2000). It is notable that substantial overlaps between Tasmania and Wilkes Land region have been predicted by some regional analyses (Tikku and Cande, 2000) and are inconsistent with local restoration of seafloor (Royer and Rollet, 1997; Tikku and Cande, 2000); this suggests some internal deformation (extension) of either the Australian or Antarctic plates, or the regional data are open to alternate interpretation (Whittaker et al., 2007). Reconstructions of seafloor older than 74 Ma in the southeast Indian Ocean,

and Australia-Antarctic continental geology, are achieved with only a slightly larger plate movement and the Australia-Antarctica boundary is relatively long, so the implications for global plate motions are relatively small. During this time, oblique convergence occurred adjacent to Thurston Island and a subducting microplate, the Bellingshausen plate, moved independently to Marie Byrd Land until ca. 60 Ma (Stock and Molnar, 1987; Heinemann et al., 1999; Larter et al., 2002). The tectonic model (Table 1; Figure 5) results in acceptable plate closure for this time interval.

Test 3: Global Plate Motions Relative to Hotspots

With a model for intra-Antarctic motion (Table 1), it is possible to compute the relative motion of the Pacific plate relative to those in the African hemisphere by following a path Africa-East Antarctica-Marie Byrd Land-East New Zealand-Pacific (Cande et al., 1995; Nankivell, 1997; Cande and Stock, 2004), assuming the motions of Table 1 and that East New Zealand was fixed to the Pacific during this time. Therefore, it is possible to rotate the African hotspot reference frame to Hawaii and account for predicted movement of the Hawaii hotspot caused by mantle flow (Steinberger et al., 2004). The results (Figure 6) reveal that inclusion of the Antarctic deformation model (Table 1) produces a good fit between observations and predictions.

74.00 Ma

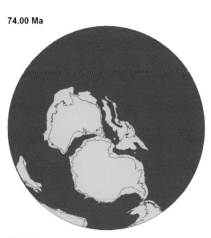

56.00 Ma

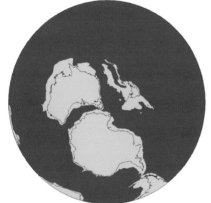

44.00 Ma

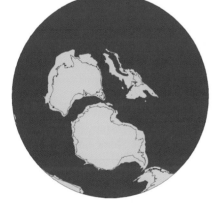

26.00 Ma

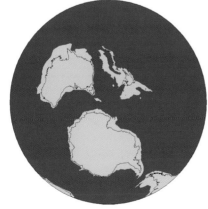

DISCUSSION OF THE ANTARCTIC PLATE-MOTION MODEL

The kinematic model (Table 1) results in a much improved fit to global observations of hotspots and it adequately closes the South Pacific plate-motion circuit. I believe the model to lie within the constraints of local observations from Antarctica, but I accept that fine-tuning will be required as the hypothesis is tested in detail. Specifically, it is likely that the number of continental fragments will be increased and the precise locations of the poles of the relative rotations that describe their movements will be revised.

South Pacific reconstructions (Figure 5) place moderately strong constraints on predicted movements between

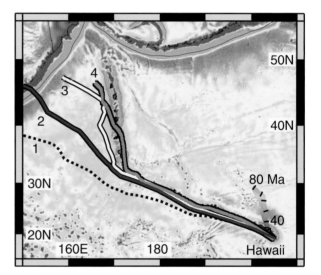

FIGURE 6 Model positions of Emperor-Hawaii seamounts for the time 0-74 Ma, based upon rotation of the African hotspot reference frame (Muller et al., 1993) and predicted movements of the Hawaii hotspot (movement of the hotspot is shown in the Pacific plate frame of reference; it is color-coded green-orange and tick marks are shown at 10 Myr increments; and the two movement lines were computed for models 2 and 3 below) (Steinberger et al., 2004). From east to west the model assumptions are: (1) (dotted), no relative hotspot motion and no intra-Antarctic motion; (2) (red), Adare Basin motion 43-26 Ma (Cande et al., 2000) and hotspot motion; (3) (yellow), hotspot motion and the intra-Antarctic issue avoided by following a path through Australia before 50 Ma; (4) (orange), hotspot motion and the Antarctic deformation model of this paper (Table 1).

FIGURE 5 South Pacific tectonic reconstructions. See discussion in text and animated version at http://www.gns.cri.nz/research/tectonics.

East Antarctica and Marie Byrd Land in the Ross Sea, and to a lesser extent on the angle of rotation between East Antarctica and Marie Byrd Land. Hotspot observations (ages and positions of seamounts) place a moderately strong constraint on the angle of rotation between East Antarctica and Marie Byrd Land, because most hotspot chains are at middle or low latitudes and the likely pole of intra-Antarctic rotation is at high latitude.

Intra-Antarctic motion during the interval 74-56 Ma, which is the most controversial part of the model, is plausible because it connects a complex zone of rifting near the Ross Sea (Figure 3) with a subduction boundary adjacent to Thurston Island and the Antarctic Peninsula. Antarctica was not an isolated continent surrounded by spreading ridges at that time. It is possible that features such as the Iselin Trough and Wilkes Basin also formed during this time interval (Figure 3).

IMPLICATIONS FOR GLOBAL SUBDUCTION BUDGETS

The relative motions of oceanic plates in the Pacific hemisphere can be determined from magnetic anomalies within the Pacific, based upon the assumption of symmetric spreading because conjugate crust has mostly been subducted (Engebretson et al., 1985). The motion of the Pacific plate relative to Africa can be computed from Table 1 and magnetic anomalies in the southwest Indian Ocean (Nankivell, 1997) and South Pacific (Cande et al., 1995; Larter et al., 2002; Cande and Stock, 2004). Hence it is possible to compute the relative positions of Pacific hemisphere oceanic plates with overriding plates that border the African hemisphere: South America (Muller et al., 1993); India (Muller et al., 1993); North America (Klitgord and Schouten, 1986; Muller et al., 1990); and Eurasia (Lawver et al., 1990; Srivastava and Roest, 1996; Rosenbaum et al., 2002). Hence, it is possible to compute subduction histories.

Since the time of the Emperor-Hawaii bend at ca. 50 Ma (Sharp and Clague, 2006), there is fairly good agreement within the published literature with estimates of relative and absolute plate motions (Steinberger et al., 2004). However, there is substantial disagreement before that (Raymond et al., 2000; Tarduno et al., 2003; Steinberger et al., 2004). Consequently, I focus here on the time interval 74-50 Ma. Specifically, the plate pairs examined are Pacific-Eurasia, Kula-North America, Farallon-North America, and Farallon-South America (Figure 7). Two Antarctic models are used (Figure 8): the preferred model is given in Table 1; the second is the same, but assumes no intra-Antarctic motion during the interval 74-44 Ma.

It is clear from Figure 8 that the model of Antarctic motion is significant for the computation of global subduction rates and directions for the interval 74-50 Ma. The preferred model (yellow, Table 1) predicts a hinge point, where Pacific plate subduction rates were very low in southeast Asia; this may have geodynamic significance, because it is close to the southern limit of the plate boundary and may represent a propagating system (perhaps similar to the tectonic setting of the Scotia Sea today). Alternatively, it is

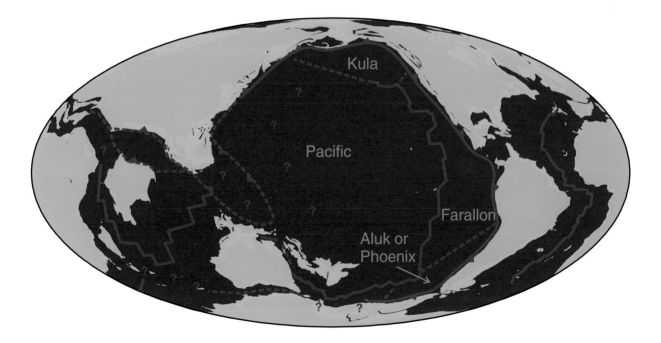

FIGURE 7 Simplified global plate tectonic reconstruction for 55 Ma.

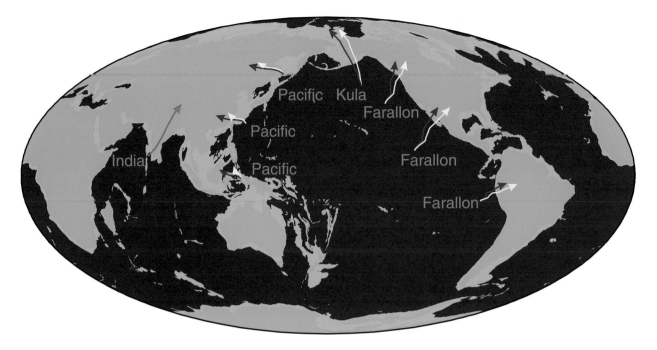

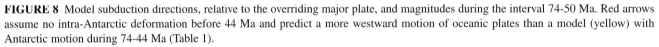

FIGURE 8 Model subduction directions, relative to the overriding major plate, and magnitudes during the interval 74-50 Ma. Red arrows assume no intra-Antarctic deformation before 44 Ma and predict a more westward motion of oceanic plates than a model (yellow) with Antarctic motion during 74-44 Ma (Table 1).

possible that a ridge has since been subducted and that a remanent of the Izanagi plate, rather than Pacific plate, was being subducted beneath Eurasia at that time (Whittaker et al., 2007). Kula and Farallon subduction for the period 74-50 Ma beneath North and South America, when computed using the preferred Antarctic model, is rotated clockwise by 10-15° to an orientation more closely orthogonal to the margin, and has a rate that is ca. 15 percent higher at equatorial latitudes. The increase in rate at equatorial latitudes results from the additional intra-Antarctic rotation about a pole at high latitude.

CONCLUSIONS

An hypothesis for the motion of Marie Byrd Land relative to East Antarctica is presented. The model is shown to be consistent with South Pacific plate motions and it provides an improved global reconciliation of relative versus hotspot-derived plate motions. While the proposed model is broadly consistent with available observations from Antarctica, it is inevitable that a more complex and precise model will emerge as the hypothesis is tested in the future. However, I conclude that South Pacific and global data constrain the motion of Marie Byrd Land relative to East Antarctica to be described by a rotation pole with high latitude and rotation angles similar to those proposed. Such a model is plausible because the proposed intracontinental plate boundary con-

nected active tectonic zones of rifting near the Ross Sea (Figure 3) with a subduction boundary adjacent to Thurston Island and the Antarctic Peninsula.

The key geographic connection that Antarctica provides between subducting oceanic plates of the Pacific with diverging plates of the African hemisphere means that an understanding of Antarctic deformation has significance for both global geodynamics and for subduction-related geology of the entire circum-Pacific.

ACKNOWLEDGMENTS

I thank Bryan Storey, Alan Cooper, and the ISAES 2007 organizing committee for their editorial and facilitation roles, and Carol Finn and Joann Stock for their very helpful reviews. I used GMT software (Wessel and Smith, 1995) and Atlas software of Cambridge Paleomap Services. This research was funded by the New Zealand Foundation for Research, Science and Technology.

REFERENCES

Anandakrishnan, S., D. D. Blankenship, R. B. Alley, and P. L. Stoffa. 1998. Influence of subglacial geology on the position of a West Antarctic ice stream from seismic observations. *Nature* 394:62-65.

Barrett, P. J. E. 1989. Antarctic Cenozoic history from CIROS-1 drillhole, McMurdo Sound. *DSIR Bulletin* 245:1-254.

Barrett, P. J., S. A. Henrys, L. R. Bartek, G. Brancolini, M. Busetti, F. J. Davey, M. J. Hannah, and A. R. Pyne. 1995. Geology of the margin of the Victoria Land Basin off Cape Roberts, southwest Ross Sea. In *Geology and Seismic Stratigraphy of the Antarctic Margin,* eds. A. K. Cooper, P. F. Barker, and G. Brancolini, *Antarctic Research Series* 68:183-207. Washington, D.C.: American Geophysical Union.

Behrendt, J. C., W. E. LeMasurier, A. K. Cooper, F. Tessensohn, A. Trefu, and D. Damaske. 1991. Geophysical studies of the west Antarctic rift arm. *Tectonics* 10:1257-1273.

Cande, S. C., and D. V. Kent. 1995. Revised calibration of the geomagnetic polarity timescale for the Late Cretaceous and Cenozoic. *Journal of Geophysical Research, B, Solid Earth and Planets* 100(4):6093-6095.

Cande, S. C., and J. M. Stock. 2004. Pacific-Antarctic-Australia motion and the formation of the Macquarie Plate. *Geophysical Journal International* 157:399-414.Cande, S. C., C. A. Raymond, J. Stock, and W. F. Haxby. 1995. Geophysics of the Pitman Fracture Zone. *Science* 270:947-953.

Cande, S. C., J. M. Stock, R. D. Mueller, and T. Ishihara. 2000. Cenozoic motion between East and West Antarctica. *Nature* 404(6774):145-150.

Cooper, A. K., F. J. Davey, and J. C. Behrendt. 1987. *Seismic Stratigraphy and Structure of the Victoria Land Basin, Western Ross Sea, Antarctica.* Houston, TX: Circum-Pacific Council for Energy and Mineral Resources.

Cooper, A. K., P. F. Barker, and G. Brancolini. 1995. *Geology and Seismic Stratigraphy of the Antarctic Margin.* Antarctic Research Series 68, Washington, D.C.: American Geophysical Union.

Dalziel, I. W. D., and D. H. Elliot. 1982. West Antarctica: Problem child of Gondwanaland. *Tectonics* 1:3-19.

Davey, F. J., D. J. Bennett, and R. E. Houtz. 1982. Sedimentary basins of the Ross Sea, Antarctica. *New Zealand Journal of Geology and Geophysics* 25(2):245-255.

Davey, F. J., S. C. Cande, and J. M. Stock. 2006. Extension in the western Ross Sea region—Links between Adare Basin and Victoria Land Basin. *Geophysical Research Letters* 33.

Engebretson, D. C., A. Cox, and R. G. Gordon. 1985. Relative motions between oceanic and continental plates in the Pacific basin. *Geological Society of America Special Paper* 206:1-59.

Fitzgerald, P. G. 1992. The Transantarctic Mountains in southern Victoria Land: The application of apatite fission track analysis to a rift-shoulder uplift. *Tectonics* 11:634-662.

Fitzgerald, P. G. 1994. Thermochronologic constraints on post-Paleozoic tectonic evolution of the central Transantarctic Mountains, Antarctica. *Tectonics* 13:818-836.

Fitzgerald, P., and S. Baldwin. 1997. Detachment fault model for the evolution of the Ross Embayment, the Antarctic region, geological evolution and processes. In *Proceedings of the VII International Symposium on Antarctic Earth Sciences,* ed. C. A. Ricci, pp. 555-564. Siena: *Terra Antartica* Publication.

Fitzgerald, P. G., S. L. Baldwin, L. E. Webb, and P. B. O'Sullivan. 2006. Interpretation of (U-Th)/He single grain ages from slowly cooled crustal terranes: A case study from the Transantarctic Mountains of southern Victoria Land. *Chemical Geology* 225:91-120.

Gaina, C., D. R. Mueller, J.-Y. Royer, J. Stock, J. L. Hardebeck, and P. Symonds. 1998. The tectonic history of the Tasman Sea: a puzzle with 13 pieces. *Journal of Geophysical Research* 103(6):12413-12433.

Gordon, R. G. 1998. The plate tectonic approximation: Plate nonrigidity, diffuse plate boundaries, and global plate reconstructions. *Annual Review of Earth and Planetary Sciences* 26:615-642.

Gordon, R. G., and D. M. Jurdy. 1986. Cenozoic global plate motions. *Journal of Geophysical Research* 91:12389-12406.

Hamilton, R. J., B. P. Luyendyk, and C. C. Sorlien. 2001. Cenozoic tectonics of the Cape Roberts rift basin and Transantarctic Mountains front, southwestern Ross Sea, Antarctica. *Tectonics* 20:325-342.

Hayes, D. E., L. A. Frakes, P. J. Barrett, D. A. Burns, P.-H. Chen, A. B. Ford, A. G. Kaneps, E. M. Kemp, D. W. McCollum, D. J. W. Piper, R. E. Wall, and P. N. Webb. 1975. Sites 270, 271, 272. *Initial Reports of the Deep Sea Drilling Project* 28, eds. D. E Hayes, I. A. Frakes et al., pp. 211-334. Washington, D.C.: U.S. Government Printing Office.

Heinemann, J., J. Stock, R. Clayton, K. Hafner, S. Cande, and C. Raymond. 1999. Constraints on the proposed Marie Byrd Land-Bellingshausen plate boundary from seismic reflection data. *Journal of Geophysical Research, B, Solid Earth and Planets* 104(11):25321-25330.

Henrys, S. A., L. R. Bartek, G. Brancolini, B. P. Luyendyk, R. J. Hamilton, C. C. Sorlien, and F. J. Davey. 1998. Seismic stratigraphy of the pre-Quaternary strata off Cape Roberts and their correlation with strata cored in the CIROS-1 drillhole, McMurdo Sound. *Terra Antartica* 5:273-279.

King, P. R., and G. P. Thrasher. 1996. Cretaceous-Cenozoic geology and petroleum systems of the Taranaki Basin, New Zealand. Lower Hutt, N.Z.: Institute of Geological and Nuclear Sciences.

Klitgord, K. D., and H. Schouten. 1986. Plate kinematics of the central Atlantic. In *The Geology of North America,* vol. M, *The Western and North Atlantic Region,* eds. P. R. Vogt and B. E. Tucholke, pp. 351- 378. Boulder, CO: Geological Society of America.

Larter, R. D., A. P. Cunningham, and P. F. Barker. 2002. Tectonic evolution of the Pacific margin of Antarctica. 1. Late Cretaceous reconstructions. *Journal of Geophysical Research* 107:2345.

Lawver, L. A., R. D. Muller, S. P. Srivastava, and E. R. Roest. 1990. The opening of the Arctic Ocean. In *Geological History of the Polar Oceans: Arctic Versus Antarctic,* eds. U. Bleil and J. Thiede, pp. 29-62. Dordrecht: Kluwer Academic.

Lisker, F. 2002. Review of fission track studies in northern Victoria Land, Antarctica-passive margin evolution versus uplift of the Transantarctic Mountains. *Tectonophysics* 349:57-73.

Lithgow-Bertelloni, C., and M. A. Richards. 1998. The dynamics of Cenozoic and Mesozoic plate motions. *Reviews of Geophysics* 36(1):27-78.

Lythe, M. B., D. G. Vaughan, A. Lambrecht, H. Miller, U. Nixdorf, H. Oerter, D. Steinhage, I. F. Allison, M. Craven, I. D. Goodwin, J. Jacka, V. Morgan, A. Ruddell, N. Young, P. Wellman, A. P. R. Cooper, H. F. J. Corr, C. S. M. Doake, R. C. A. Hindmarsh, A. Jenkins, M. R. Johnson, P. Jones, E. C. King, A. M. Smith, J. W. Thomson, M. R. Thorley, K. Jezek, B. Li, H. Liu, H. Mideo, V. Damm, F. Nishio, S. Fujita, P. Skvarca, F. Remy, L. Testut, J. Sievers, A. Kapitsa, Y. Macheret, T. Scambos, I. Filina, V. Masolov, S. Popov, G. Johnstone, B. Jacobel, P. Holmlund, J. Naslund, S. Anandakrishnan, J. L. Bamber, R. Bassford, H. Decleir, P. Huybrechts, A. Rivera, N. Grace, G. Casassa, I. Tabacco, D. Blankenship, D. Morse, H. Conway, T. Gades, N. Nereson, C. R. Bentley, N. Lord, M. Lange, and H. Sanhaeger. 2001. BEDMAP; a new ice thickness and subglacial topographic model of Antarctica. *Journal of Geophysical Research* 106:11335-11351.

McCarron, J. J., and R. D. Larter. 1998. Late Cretaceous to early Tertiary subduction history of the Antarctic Peninsula. *Journal of the Geological Society of London* 155:255-268.

Muller, R. D., D. T. Sandwell, B. E. Tucholke, J. G. Sclater, and P. R. Shaw. 1990. Depth to basement and geoid expression in the Kane Fracture Zone: A comparison. *Marine Geophysical Researches* 13:105-129.

Muller, R. D., J.-Y. Royer, and L. A. Lawver. 1993. Revised plate motions relative to the hotspots from combined Atlantic and Indian Ocean hotspot tracks. *Geology* 21(3):275-278.

Muller, R. D., C. Gaina, A. Tikku, D. Mihut, S. C. Cande, and J. M. Stock. 2000. Mesozoic/Cenozoic tectonic events around Australia. *Geophysical Monograph* 121:161-188.

Nankivell, A. P. 1997. *Tectonic Evolution of the Southern Ocean between Antarctica, South America and Africa over the Past 84 Ma.* Oxford: University of Oxford.

Raymond, C. A., J. M. Stock, and S. C. Cande. 2000. Fast Paleogene motion of the Pacific hotspots from revised global plate circuit constraints. *Geophysical Monograph* 121:359-376.

Rosenbaum, G., G. S. Lister, and C. Duboz. 2002. Relative motions of Africa, Iberia and Europe during Alpine orogeny. *Tectonophysics* 359:117-129.

Royer, J. Y., and N. Rollet. 1997. Plate-tectonic setting of the Tasmanian region. *Australian Journal of Earth Sciences* 44(5):543-560.

Sandwell, D. T., and W. H. F. Smith. 1997. Marine gravity anomaly from Geosat and ERS 1 satellite altimetry. *Journal of Geophysical Research* 102:10039-10054.

Sharp, W. D., and D. A. Clague. 2006. 50-Ma initiation of Hawaiian-Emperor bend records major change in Pacific Plate motion. *Science* 313:1281-1284.

Smith, W. H. F., and D. T. Sandwell. 1997. Global sea floor topography from satellite altimetry and ship depth soundings. *Science* 277:956-1962.

Srivastava, S. P., and W. R. Roest. 1996. Porcupine plate hypothesis: Comment. *Marine Geophysical Researches* 18:89-595.

Steinberger, B., R. Sutherland, and R. J. O'Connell. 2004. Prediction of Emperor-Hawaii seamount locations from a revised model of global plate motion and mantle flow. *Nature* 430:67-173.

Stock, J. M., and P. Molnar. 1987. Revised history of early Tertiary plate motion in the South-west Pacific. *Nature* 325(6104):495-499.

Storey, B. C. 1991. The crustal blocks of West Antarctica within Gondwana: Reconstruction and breakup model. In *Geological Evolution of Antarctica*, eds. M. R. A. Thompson, J. A. Crame, and J. W. Thompson, pp. 587-592. Cambridge: Cambridge University Press.

Storey, B. C., and P. A. R. Nell. 1988. Role of strike-slip faulting in the tectonic evolution of the Antarctic Peninsula. *Journal of the Geological Society of London* 145(2):333-337.

Studinger, M., R. E. Bell, D. D. Blankenship, C. A. Finn, R. A. Arko, D. L. Morse, and I. Joughin. 2001. Subglacial Sediments: A regional geological template for ice flow in West Antarctica. *Geophysical Research Letters* 28:3493-3496.

Stump, E., and P. G. Fitzgerald. 1992. Episodic uplift of the Transantarctic Mountains; with supplemental data 92-08. *Geology* 20(2):161-164.

Sutherland, R. 1995. The Australia-Pacific boundary and Cenozoic plate motions in the SW Pacific; some constraints from Geosat data. *Tectonics* 14(4):819-831.

Tarduno, J. A., R. A. Duncan, D. W. Scholl, R. D. Cottrell, B. Steinberger, T. Thordarson, B. C. Kerr, C. R. Neal, F. A. Frey, M. Torii, and C. Carvallo. 2003. The Emperor Seamounts: Southward motion of the Hawaiian Hotspot plume in Earth's mantle. *Science* 301(5636):1064-1069.

ten Brink, U. S., S. Bannister, B. C. Beaudoin, and T. A. Stern. 1993. Geophysical investigations of the tectonic boundary between east and west Antarctica. *Science* 261:45-50.

Tikku, A. A., and S. C. Cande. 1999. The oldest magnetic anomalies in the Australian-Antarctic Basin; are they isochrons? *Journal of Geophysical Research* 104(1):661-677.

Tikku, A. A., and S. C. Cande. 2000. On the fit of Broken Ridge and Kerguelen Plateau. *Earth and Planetary Science Letters* 180:117-132.

Wessel, P., and W. H. F. Smith. 1995. New version of the generic mapping tools released. *Eos, Transactions of the American Geophysical Union* 76:329.

Whittaker, J. M., R. D. Muller, G. Leitchenkov, H. Stagg, M. Sdrolias, C. Gaina, and A. Goncharov. 2007. Major Australian-Antarctic plate reorganisation at Hawaiian-Emperor bend time. *Science* 318:83-86.

Wood, R. A., G. Lamarche, R. H. Herzer, J. Delteil, and B. Davy. 1996. Paleogene seafloor spreading in the southeast Tasman Sea. *Tectonics* 15:966-975.

Antarctica and Global Paleogeography: From Rodinia, Through Gondwanaland and Pangea, to the Birth of the Southern Ocean and the Opening of Gateways

T. H. Torsvik,[1,2] *C. Gaina,*[1] *and T. F. Redfield*[1]

ABSTRACT

Neoproterozoic Rodinia reconstructions associate East Antarctica (EANT) with cratonic Western Australia. By further linking EANT to both Gondwana and Pangea with relative plate circuits, a Synthetic Apparent Polar Wander (SAPW) path for EANT is calculated. This path predicts that EANT was located at tropical to subtropical southerly latitudes from ca. 1 Ga to 420 Ma. Around 400 Ma and again at 320 Ma, EANT underwent southward drift. Ca. 250 Ma Antarctica voyaged briefly north but headed south again ca. 200 Ma. Since 75 Ma EANT became surrounded by spreading centers and has remained extremely stable. Although paleomagnetic data of the blocks that embrace West Antarctica are sparse, we attempt to model their complex kinematics since the Mesozoic. Together with the SAPW path and a revised circum-Antarctic seafloor spreading history we construct a series of new paleogeographic maps.

INTRODUCTION

Antarctica is the world's last discovered wilderness, still relatively poorly mapped, and the only continent without an indigenous human history. Excepting the very tip of the Antarctic Peninsula, the bulk of the Antarctic land mass lies south of the Antarctic circle, and is covered by ice on a year-round basis. Constrained to isolated nunataks, mountain chains, and coastal exposures, geological studies have been correspondingly limited in scope. Geophysical techniques capable of resolving rock properties beneath the ice cover have proved helpful to delineate the continent's crustal structure, but often fail to shed light on Antarctica's geotectonic evolution. Thus, Antarctica remains the most geologically unexplored continent.

Extending from the Ross to the Weddell Seas, the Transantarctic Mountains (Figure 1) effectively divide Antarctica into two geological provinces: cratonal East Antarctica (EANT) and the collage of tectonic blocks that make up West Antarctica (WANT). Possessed of a long, globetrotting history, portions of EANT can be traced to Rodinia and perhaps even beyond. As a relative newcomer to the paleogeographic parade, WANT comprises discrete tectonic blocks (Figure 1) separated by rifts or topographic depressions.

Today the Antarctic plate is neighbored by six different tectonic plates and almost entirely surrounded by spreading ridges. This tectonic configuration has in part given rise to Antarctica's near-total Cenozoic isolation. Two important hotspots (Kerguelen, Marion) lie within the Antarctic plate. The Bouvet hotspot, which may have been responsible for the catastrophic Karroo igneous outpouring in Jurassic time, is now located near the AFR-SAM-ANT triple junction (Figure 2). In this review we outline the location of the world's fifth largest continent and its neighbors in space and time, and present paleogeographic reconstructions for important periods of assembly and breakup. We give a list of acronyms used in Table 1.

EAST ANTARCTICA IN SPACE AND TIME

EANT comprises Archaean and Proterozoic-Cambrian terranes that amalgamated during Precambrian and Cambrian times (Fitzsimons, 2000; Harley, 2003). Proterozoic base-

[1]Center for Geodynamics, Geological Survey of Norway, Leiv Eirikssons vei 39, NO-7491 Trondheim (Norway) (trond.torsvik@ngu.no).

[2]Also at PGP, University of Oslo, 0316 Oslo (Norway); School of Geosciences, Private Bag 3, University of Witwatersrand, WITS, 2050 (S. Africa).

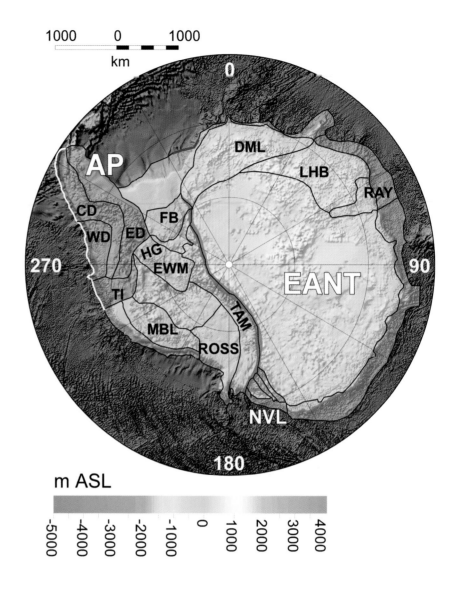

FIGURE 1 Antarctic topography and bathymetry. East Antarctica is subdivided into four provinces (Lythe et al., 2000): DML, LHB, RAY, and a large undivided unit (EANT). West Antarctica consists of five major distinctive terranes: ΛP (comprising Eastern-Central-Western domains: ED-CD-WD), TI, FB, MBL, and EWM. The three northern Victoria Land terranes are grouped together (NVL). ROSS = extended continental crust between MBL and EANT; TAM = Transantarctic Mountains (red and black lines). Most of the circum-Antarctic continent ocean boundaries (outermost polygon boundaries) are the result of nonvolcanic breakup except NW of DML (volcanic; red line) and Western AP (inactive trench, white and red lines) (see Table 1 for more abbreviations).

ment provinces (Fitzsimons, 2003) link EANT and cratonic Western Australia (WAUS, Australia west of the Tasman line), and consequently all reconstructions of Rodinia associate EANT (including the mostly unknown EANT shield) with WAUS in Neoproterozoic time. By further linking EANT to Gondwana at ~550 Ma, Pangea at ~320 Ma, and breakup at ~175 Ma with relative plate circuits we are able to construct a Synthetic Apparent Polar Wander (SAPW) path for EANT (Figure 3) based on Australian paleomagnetic data, Gondwana poles (550-320 Ma) (Torsvik and Van der Voo, 2002), and a global data compilation for the last 320 million years describing Pangea assembly and breakup (Torsvik et al., forthcoming). The SAPW path also includes palaeomagnetic data from EANT (listed in Torsvik and Van der Voo, 2002) whenever considered reliable. Within error, EANT poles match the SAPW path (Figure 3a).

The SAPW path (Table 2) predicts that a given location in EANT (Figure 3b) was located at tropical to subtropical southerly latitudes from about 1 Ga to the Late Ordovician. However, Precambrian data are sparse: This portion of the SAPW path is based only on the ~1070 Ma Bangemall pole (Wingate et al., 2002), the ~755 Ma Mundine pole (Wingate and Giddings, 2000) and several ~600 Ma poles from WAUS. During Late Ordovician and Silurian times (~450-400 Ma), EANT drifted southward (8-11 cm/yr; latitudinal velocity calculated from Figure 3b), followed by another phase of southward drift during the Carboniferous (350-300 Ma; 6-11 cm/yr). The Permo-Triassic (300-200 Ma) was characterized by northerly motion (~5 cm/yr); southerly drift (2-7 cm/yr) again recommenced after 200 Ma. For the last 75 Ma, EANT has remained extremely stable (~0.6 cm/yr) near the South Pole.

EANT was located near the equator in the Early Paleozoic (Figure 3b). Marine invertebrates flourished in tropical seas and the continent hosted a varied range of climates, from deserts to tropical swamps. Excepting a period of

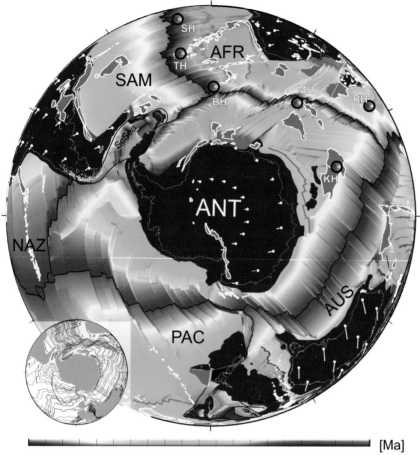

FIGURE 2 Age of the ocean floor surrounding Antarctica. White arrows show "absolute" motion of some tectonic plates based on a moving hotspot frame for the last 5 million years. Red circles denote hotspot locations (BH = Bouvet; KH = Kerguelen; MH = Marion; RH = Reunion; SH = St. Helena; TH = Tristan). Large igneous provinces and other volcanic provinces (including seaward-dipping reflectors) are shown in brown and white. Active plate boundaries shown in black (midocean ridges) and extinct midocean ridges in grey. Inset figure shows isochrons based on present-day magnetic and gravity data. ScSea = Scotia Sea (see Table 1 for more abbreviations).

[Ma]

0 10 20 30 40 50 60 70 80 90 100 110 120 130 140 150 160 170 180 190 200

northerly drift into temperate latitudes near the Triassic-Jurassic boundary (~200 Ma), EANT has remained in polar latitudes for the last 325 myr. Consequently, EANT has commonly been inundated by ice. However, the first recognized Phanerozoic glacial event—the short-lived Late Ordovician Hirnantian (ca. 443 Ma) glacial episode—occurred while EANT occupied temperate latitudes. During this time, NW Africa was located over the South Pole (Cocks and Torsvik, 2002). Conversely, during the Late Paleozoic glacial interval, commencing in the Late Carboniferous and lasting for almost 50 myr, the South Pole was located in EANT (Torsvik and Cocks, 2004). These Permo-Carboniferous glaciations resulted in deposition of widespread tills across South Pangea (Gondwana).

WEST ANTARCTICA AND PALAEOMAGNETIC DATA

In contrast with the great lumbering elephant of continental EANT, WANT comprises several distinct crustal blocks (Dalziel and Elliot, 1982), with independent Mesozoic and Cenozoic geotectonic histories (Figure 1). Because many are inadequately sampled paleomagnetically, it is difficult to portray their latitudinal story with great confidence.

Thus, we restrict our description and paleomagnetic analysis (following the pioneering work of Grunow, Dalziel, and coworkers) to the Antarctic Peninsula (AP), Thurston Island (TI), and the Ellesworth-Whitmore Mountains (EWM). In addition, relative movements between individual blocks or vs. EANT (e.g., Jurassic poles in Figure 4a) are sometimes only slightly greater than the resolving power of the paleomagnetic method.

AP has traditionally been treated as a single Mesozoic-to-Cenozoic continental arc system formed above an eastward-dipping paleo-Pacific subduction zone. Recent studies, however, suggest that AP consists of three fault-bounded terranes (WD, CD, and ED in Figure 1) that amalgamated in Late Cretaceous (Albian) time (Vaughan and Storey, 2000; Vaughan et al., 2002). However, for reasons of simplicity we keep AP blocks together in our reconstructions. Jurassic to Early Cretaceous (between 175 Ma and 140 Ma) paleomagnetic poles from AP differ from the EANT SAPW path while Early Cretaceous (110 Ma) and younger poles overlap within error (Figure 4a). The data are therefore compatible with models that imply that AP moved away from EANT between 175 and 140 Ma while undergoing slow clockwise rotation (Weddell Sea opening at ~5 cm/yr) (Figure 4b) fol-

TABLE 1 Commonly Used Abbreviations

AFR	Africa plate
ANT	Antarctic plate
AP	Antarctic Peninsula (now part of WANT)
AUS	Australia
DML	Dronning Maud Land, includes the Grunehøgna terrane (part of the Kapvaal Archean core of the Kalahari craton, and the Maud orogen and perhaps part of the Coats Land crustal block) (now part of EANT)
EANT	East Antarctica
EWM	Ellesworth-Whitmore Mountains (now part of WANT)
FB	Filchner Block (as defined by Studinger and Miller (1999), partly Coats Land cratonic block and partly extended and intruded "Afar depression like" continental crust (Dalziel and Lawver, 2001) (now part of WANT)
FI	Falkland Island (now part of SAM)
HG	Haag (included in EWM)
LHB	Lützow-Holm Bay (now part of EANT)
KAL	Kalahari (now part of SAFR)
MBL	Marie Byrd Land (now part of WANT)
MEB	Maurice Ewing Bank (now part of SAM)
NAZ	Nazca plate
P	Patagonia (now part of SAM)
PAC	Pacific plate
RAY	Raynor Province (now part of EANT)
SAFR	South Africa
SAM	South American plate
ScSea	Scotia Sea
TI	Thurston Island (now part of WANT)
WANT	West Antarctica
WAUS	Western Australia, Cratonic Australia west of the Tasman line (now part of AUS)

lowed by convergence (Weddell Sea partial subduction) and clockwise rotation (~130-110 Ma) relative EANT (see also Grunow, 1993).

TI has few exposures, but available data indicate that TI was similar morphologically and tectonically to AP (Leat et al., 1993). 110 and 90 Ma poles (Figure 4a) are grossly similar to those from AP and EANT. In our reconstructions (Table 3) (Grunow, 1993), TI follows the overall motion of the AP blocks (see velocity pattern in Figure 4b). However, TI was emplaced in its present position at the southern end of the AP by some ~300 km dextral movement and several degrees of clockwise rotation between 130 Ma and 110 Ma.

EWM is a displaced segment of the cratonic margin (Schopf, 1969) whose past position is constrained by Late Cambrian (e.g., Grunow et al., 1987; Randall and MacNiocaill, 2004) and Jurassic paleomagnetic data (Grunow et al., 1987) (Figure 4a). EWM was probably located in the Natal Embayment of the African plate in Cambrian times and underwent 90° counterclockwise rotation during Pangea breakup. A 175 Ma pole (Figure 4a) from EWM overlaps with a contemporaneous pole from AP (Figure 4a). However, we keep EWM near EANT during Late Jurassic-Early Cretaceous times by interpolating its position with the ~175 Ma pole and its current position fixed to EANT (here modeled to 120 Ma). This implies ~1.3 cm/yr of sinistral movement vs. EANT (Figure 4b).

A PALEOGEOGRAPHIC PARADE

Below we present eight paleogeographic maps from Neoproterozoic to Early Tertiary times. Global reconstructions are based on relative fits and paleomagnetic APW paths (Torsvik and Van der Voo, 2002; Torsvik et al., forthcoming), upgraded with reconstruction parameters for Antarctica and surrounding plates (Table 3). Paleomagnetism yields only latitudes and plate rotations, but longitudinal uncertainty can be minimized if the continent that has moved least in longitude can be identified and is used for reference: Africa is the best candidate (Burke and Torsvik, 2004). In order to reconstruct the continents in the best possible "absolute" manner we here use a hybrid reference frame based on merging an African mantle frame (O'Neil et al., 2005; Torsvik et al., forthcoming) and a paleomagnetic reference frame (>100 Ma) back to the time of Pangea assembly (~320 Ma). The paleomagnetic frame (110-320 Ma), calculated in African coordinates, was adjusted 5° in longitude to correct for the longitudinal motion of Africa during the past 100 Ma inferred in the mantle reference frame. Reconstructions 250 Ma and younger are therefore shown in a absolute sense with paleolongitudes and/or velocity vectors with respect to the Earth's spin axis. While it could prove possible to quantify uncertainties in velocity vectors for the mantle frame (see O'Neil et al., 2005), velocity vector uncertainties in the paleomagnetic frame (110-320 Ma) cannot be quantified since we assume "zero longitude" movement for Africa.

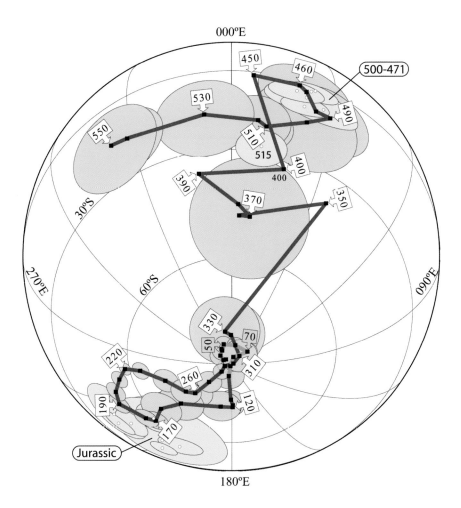

FIGURE 3a Synthetic APW path for EANT (Table 2, see text). Running mean poles (20 Ma window; 10 Ma intervals) are shown with A95 confidence circles (blue color). We also show actual input poles with dp/dm ovals (light brown) from EANT that were included in the calculation of the SAPW path.

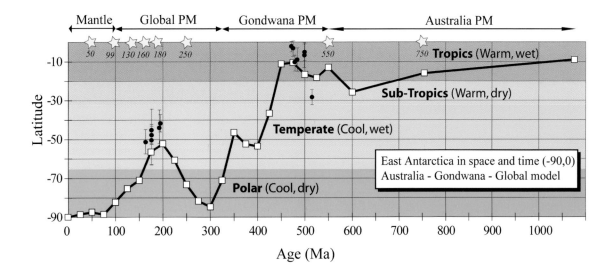

FIGURE 3b Latitude for a geographic location (90°S) in EANT based on the SAPW path in Figure 3a. However, the last 100 Ma is calculated from a moving hotspot frame. We show actual Phanerozoic latitudinal data from EANT (recalculated to 90°S) with error bars. Yellow stars denote the times of reconstructions shown in the paper. PM = Palaeomagnetic data, and where the oldest section is based on paleomagnetic data from cratonic WAUS, then mean data from Gondwana (550-320 Ma), and finally global data until 110 Ma.

TABLE 2 Phanerozoic Synthetic Apparent Polar Wander (SAPW) Path for EANT

Age (Ma)	N	A95	Lat.	Long.
0	18	3.0	−87.8	308.3
10	30	2.5	−87.5	296.8
20	23	2.9	−86.1	307.4
30	18	2.7	−84.9	327.9
40	19	2.8	−85.3	326.6
50	27	2.5	−83.8	338.3
60	30	2.4	−84.4	010.3
70	20	2.6	−84.4	050.7
80	23	2.8	−87.8	010.3
90	27	2.6	−89.2	206.4
100	11	4.2	−86.1	191.6
110	16	3.4	−78.2	181.4
120	24	2.3	−76.1	179.0
130	18	3.1	−75.3	179.1
140	10	5.6	−75.2	191.9
150	16	6.4	−69.6	224.3
160	14	6.0	−62.9	228.1
170	23	3.8	−57.5	222.5
180	26	3.6	−55.3	226.0
190	31	3.5	−50.1	237.0
200	35	3.2	−51.9	243.9
210	32	2.7	−55.4	251.0
220	29	2.0	−58.5	257.8
230	28	2.6	−63.5	257.4
240	35	3.7	−70.1	248.5
250	38	4.3	−73.8	231.9
260	26	4.8	−75.7	224.9
270	28	3.4	−81.4	226.3
280	57	2.4	−86.4	228.9
290	70	1.9	−88.0	251.5
300	39	2.4	−89.6	074.2
310	20	4.8	−86.8	041.7
320	9	9.3	−81.4	358.9
330	4	9.7	−80.2	348.2
340, 350	1	——	−38.2	035.2
360	5	12.5	−48.9	003.1
370	8	16.7	−49.0	007.7
380	5	35.1	−45.8	002.7
390	2	138.6	−36.4	348.7
400	1	——	−33.9	017.6
440, 450	1	——	−1.8	006.1
460	2	99.0	−2.3	019.1
470	5	13.8	−4.4	021.2
480	7	7.7	−11.3	024.4
490	7	12.7	−11.5	028.9
500	7	15.7	−16.6	022.2
510	9	7.6	−21.7	010.4
520	9	9.3	−20.0	007.8
530	8	13.7	−17.9	352.1
540	10	10.7	−17.7	328.6
550	7	14.1	−16.8	323.1

NOTE: N = number of poles; A95 = 95 percent confidence circle; Lat., Long. = mean pole latitude, longitude.

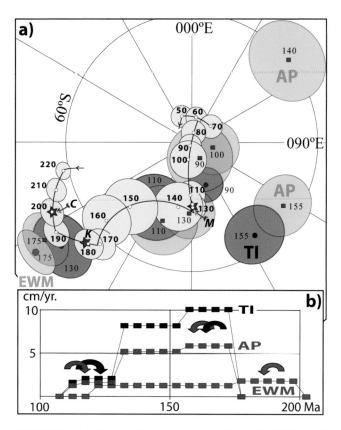

FIGURE 4 (a) Jurassic-Cretaceous poles from AP, EWM, and TI (mean poles with A95 ovals except 175 Ma EWM pole, dp/dm ovals) (Grunow, 1993). We also show the timing of some large igneous province events (red stars) that must have had an effect on Pangea in general (C = Central Atlantic Magmatic Province) or directly affected Antarctica and its margin (K = Karroo; M = Maud Rise/Madagascar Ridge). (b) Velocity for TI (74°S, 248°E), AP (72°S, 290°E) and EWM (81°S, 271°E) relative to a fixed EANT. Colored arrows show sense of rotation relative to EANT. Poles in (a) are compared with the EANT SAPW path (yellow A95 ovals) as in Figure 3a for the last 200 Ma, but fitted with small circles that have RMS values less than 0.6°. Abrupt changes in the balance of forces driving and resisting plate motions should be noticed in the APW paths as cusps.

Rodinia

The identification of 1300-1000 Ma mountain belts (Grenvillian, Sveconorwegian, and Kibaran) presently located on different continents caused geologists of the 1970s to postulate a Precambrian supercontinent (e.g., Dewey and Burke, 1973). Thus, since the early 1990s, Precambrian reconstructions have consistently incorporated a vaguely resolved Neoproterozoic supercontinent, Rodinia (Hoffman, 1991; Dalziel, 1992, 1997; Torsvik et al., 1996) (Figure 5), postulated to have amalgamated about 1.0 Ga and to have disintegrated at around 850-800 Ma (Torsvik,

TABLE 3 Important Relative Reconstruction Parameters Discussed in the Text and Shown in Figures 5–9

Age	Continents	Lat.	Long.	Ang.
750-250	DML–KAL	10.5	148.8	−58.2
750-130	EANT-WAUS	11.1	−137.2	−29.7
750-250	EWM-KAL	56.8	−086.1	92.7
750-250	FI-KAL	31.6	164.8	−119.5
750-250	MEB-KAL	44.3	−032.6	58.9
750	MBL-EANT	68.4	023.9	57.1
250, 180	AP-EANT	72.3	086.8	−35.5
250-99	MBL-EANT	47.2	146.2	−3.0
250	EWM/FB-EANT	81.9	134.1	97.6
250	P-SAFR	47.5	−033.2	63
250, 180	TI-EANT	73.6	089.6	−49.2
180	EWM/FB-EANT	33.1	078.7	−9.0
180	FI-SAFR	32.2	164.0	−119.3
180	MEB-SAFR	44.9	−032.9	58.7
180	P-SAFR	47.5	−033.1	61.3
160	AP-EANT	23.9	−027.0	−11.9
160	EWM/FB-EANT	69.9	093.7	−23.5
160	P/FI/MEB-SAFR	47.5	−033.3	58.0
160	TI-EANT	60.3	−004.6	−20.6
130	AP-EANT	77.9	079.7	−16.4
130	EWM/FB-EANT	69.9	093.7	−5.9
130	P/FI/MEB-SAFR	48.5	−033.4	56.1
130	TI-EANT	74.6	102.1	−31.1
99	EANT-AUS	5.7	034.6	27.8
99	P/FI (SAM)-SAFR	56.0	−034.8	42.7
50	EANT-AUS	13.0	032.9	24.7
50	MBL-EANT	18.2	162.1	−1.7
50	SAM-SAFR	58.2	−031.2	20.5

NOTE: Lat., Long., Ang. = Euler pole latitude, longitude, angle. EWM, FB, TI, and AP fixed to EANT from ca. 110 Ma. Fits derived from this study, Dalziel (1997) (MBL), or interpolated from Grunow (1993) and Torsvik et al. (forthcoming).

2003). However, despite exhaustive research for more than 15 years, including new paleomagnetic studies as well as dating of mobile belts and rift sequences associated with Rodinia's breakup, the details of Rodinia remain obscure. The paleolatitudes of only a few of Rodinia's constituent continents are known at any given time, and in addition to nonconstrained longitudes, the hemispheric position of the individual continents is uncertain.

In Rodinia times WAUS and EANT were clearly linked as illustrated by the two-stage Albany-Fraser-Wilkes orogen (1350-1260 Ma and 1210-1140 Ma) (red shading in Figure 5) and also the older Mawson Craton (Fitzsimons, 2003). We do not associate India with WAUS-EANT as portrayed in many classic Rodinia reconstructions; rather, we consider them to have amalgamated during Gondwana assembly (Fitzsimons, 2000; Torsvik et al., 2001; Torsvik, 2003; Meert, 2003; Collins and Pisarevsky, 2005). The Napier Complex, LHR, and RAY (currently part of EANT) probably belonged to India prior to Pan-African collision (Figure 6).

Rodinia probably formed between 1100 Ma and 1000 Ma , and breakup probably occurred before 750 Ma. Rupture may have commenced with the opening of an equatorial ocean between western Laurentia and WAUS-EANT. Dronning Maud Land (DML; including the Grunehøgna terrane), and EWM (currently part of EANT and WANT) were probably linked to Kalahari (South Africa) during the Neoproterozoic (Figure 5).

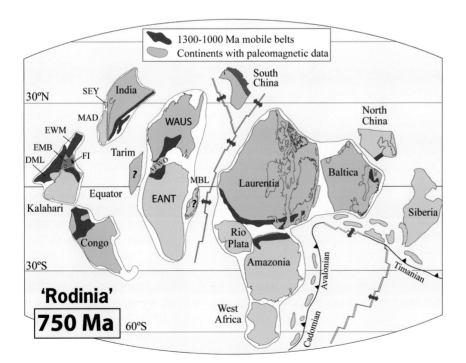

FIGURE 5 750 million year reconstruction of Rodinia just after breakup ("Rodinia New" of Torsvik, 2003). Kalahari (no paleomagnetic data) has been modified to include DML, FI, MEB, and EWM. Outline of the Albany-Fraser-Wilkes Orogen (AFWO) in WAUS and EANT follows Fitzsimmon (2003). Low-latitude position of Tarim next to EANT and WAUS (Huang et al., 2005) is problematic since it has to be removed before collision with India. MBL after Dalziel (1997). However, the location and even existence of MBL at this time is uncertain—oldest MBL rocks are Cambrian but a Proterozoic basement age cannot be excluded (Leat et al., 2005).

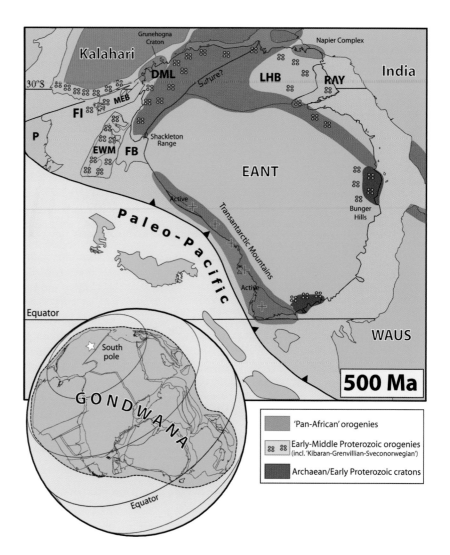

FIGURE 6 Late Cambrian reconstruction of Gondwana (inset globe) and a detailed reconstruction of EANT, SAFR (Kalahari), WAUS, India, and parts of present day SAM (P, FI, MEB). Craton outlines and orogenies mostly after Leat et al. (2005). Position of EWM (Table 3) is based on fitting Late Cambrian paleomagnetic data within error to also allow space for a smaller FB than of today. Terranes and blocks in the Natal Embayment were not affected by Pan-African deformation (see Table 1 for more abbreviations).

Gondwana

Breakup of Rodinia and the subsequent formation of Gondwana at ~550 Ma were marked by protracted Pan-African tectonism, one of the most spectacular mountain-belt building episodes in Earth history. Gondwana incorporated all of Africa, Madagascar, Seychelles, Arabia, India and EANT, most of South America and AUS, and probably some WANT blocks (EWM, FB?). The surface area of Gondwana totaled 95×10^6 km², some 64 percent of today's landmasses or 19 percent of the Earth's surface (Torsvik and Cocks, forthcoming).

In the Late Cambrian, Gondwana (Figure 6, inset globe) stretched from polar (NW Africa) to subtropical northerly latitudes (AUS). EANT covered equatorial to subtropical southerly latitudes. As most reconstructions, we show the Falklands Islands (FI), the Maurice Ewing Bank (MEB), the EWM and FB block located near South Africa, and DML (Figure 6) (Table 3)—the Natal Embayment. The FI block was situated within a ca. 1.1 Ga orogen that also included

the Namaqua-Natal belt (SAFR) and the Maudheim province (DML). We further infer that the EWM terrane belonged to this province (Leat et al., 2005), and we associate all of the above terranes and blocks with Kalahari in Rodinia times (Figure 5). EWM basement is not exposed but ~1.2 Ga Grenvillian Haag Nunataks gneisses (Millar and Pankhurst, 1987) are considered to underlie it (Figures 5 and 6). Paleomagnetic (e.g., Randall and MacNiocaill, 2004), structural and stratigraphic data have been used to argue that EWM was situated near Coats Land (DML) until the Jurassic. In our slightly modified EWM fit (Figure 6) (Table 3) we maintain a similar connection until the early Jurassic (~200 Ma). We allow space for a slightly smaller FB than of today by assuming later Mesozoic extension.

The Pan-African orogenies that stabilized EANT took place in two main zones (Figure 6): (1) in a broad region between the Shackleton Range, the Bunger Hills caused by collision with South Africa (including the Kalahari and Grunehøgna cratons, now part of EANT), and India (including the Napier Complex), and (2) along the Transantarctic

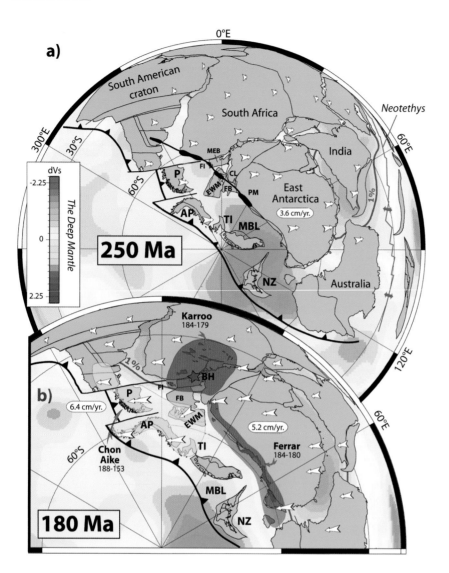

FIGURE 7 (a) 250 Ma reconstruction. Black thick line denote the Permian-Early Mesozoic Gondwanide orogen (Dalziel and Grunow, 1992). PM = Pensacola Mountains (EANT); CL = Coats Land (DML). (b) 180 Ma reconstruction with distribution of the ca. 179-184 Ma Karroo and Ferrar volcanic provinces (LIPs) in SAFR, FI, and EANT and the silicic Chon Aike province (188-153 Ma) located to SAM (P), AP, EWM, and TI (Pankhurst et al., 2000). BH = Bouvet hotspot. White arrows denote absolute plate motion vectors. Mean plate velocities indicated for EANT and Patagonia (P) in (b). The background grid images represent structures in the lower mantle that are long-lived and at regular intervals give rise to plumes and plume-related LIPs (Torsvik et al., 2006). The thick background red line in these images is the potential plume generation zone (e.g., underlying Bouvet in [b]) (see Table 1 for more abbreviations).

Mountains (Ross Orogeny), still active in Late Cambrian times (Leat et al., 2005, and references therein). The hub of Pan-African metamorphism between East Africa and EANT (~580-550 Ma) (Jacobs and Thomas, 2004) is exposed in DML. Conversely, blocks in the adjoining Natal Embayment (e.g., EWM) (Figure 6) escaped both the East Africa-EANT and Ross orogens. Curtis (2001) suggested rifting along the Paleo-Pacific margin, otherwise characterized by active subduction, while Jacobs and Thomas (2004) argued for a lateral-escape scenario.

Pangea

From ~320 Ma onward Gondwana, Laurussia, and intervening terranes merged to form the supercontinent Pangea. Pangea's main amalgamation occurred during the Carboniferous, but Pangea's dimensions were not static. Some continents were subsequently added along the supercontinent margins, and others rifted away (e.g., opening of Neotethys

in Figure 7a) during the Late Paleozoic-Early Mesozoic (Torsvik and Cocks, 2004). Permo-Triassic structures in South America (Argentina), South Africa (Cape Fold Belt), FI, EWM, and ANT (Pensacola Mountains) suggest that an enigmatic Gondwanide orogen (Figure 7a) may have developed in response to subduction-related dextral compression along the convergent SW margin of Gondwana (Johnston, 2000).

Most plate tectonic models assume that the FI block originated off the SE coast of Africa and subsequently rotated ~180° from its current orientation in the Jurassic (e.g., Adie, 1952; Marshall, 1994; Dalziel and Lawver, 2001). This is required both by paleomagnetic data (e.g., Mitchell et al., 1986) from Jurassic dykes (Karroo aged) and by excellent correlations between the basement and the overlying Middle and Late Paleozoic strata of South Africa with the stratigraphy of the FI (Marshall, 1994). Restoration to a position adjacent to SE Africa is also suggested by the structural correlation of the eastern Cape Fold Belt with fold and thrust trends on the Falklands (Figure 7a). The collective data demand that FI have rotated nearly 180°. However, the timing (modeled here between 182 Ma and 160 Ma) and the exact processes responsible for this during separation from southeast Africa remain unclear. The anticlockwise rotation of EWM relative to EANT would be even more difficult to explain if we took into account the presence of another continental crustal block, the Filchner Block (FB) (as defined by Studinger and Miller, 1999) thought to comprise cratonic blocks (Coats Land) and extended continental crust and not oceanic crust.

Pangea ruptured during the Jurassic, preceded by and associated with widespread magmatic activity, including the Karroo flood basalts and related dyke swarms in South Africa and the FI, and the Ferrar province in EANT (Figure 7b). The initial catastrophic outpouring of this deep plume-related LIP event (Torsvik et al., 2006), possibly linked to the Bouvet hotspot, probably triggered the Toarchian (183 ± 1.5 Ma) global warming event (Svensen et al., 2007). Karroo and Ferrar magmatism partly coincided with the more prolonged Chon Aike rhyolite volcanism (Figure 7b), and subduction-related magmatism along the Proto-Pacific margin of Gondwana (Rapela et al., 2005).

Absolute plate motions (Figure 7) show a change from northeast (250 Ma) through southwest (180 Ma) to southward motion from 170 Ma until the end of the Jurassic. A near 90° cusp in the SAPW path (Figure 4a) at around 170 Ma documents an abrupt change in plate driving forces. Unless caused by true polar wander we tentatively link this plate change to a combination of plume activity impinging the south Pangea lithosphere and subduction rollback. Because the subduction angle varied greatly, rollback must have been differential. Thus, we infer Patagonia (P) experienced a strong rollback effect, which we model (Figure 7a) with an offset of about 600 km compared to the present-day location in SAM. In our reconstructions SAM is broken into several discrete blocks whose borders behave as plate boundary scale deformation zones. This is necessary to understand and to reconstruct not only the FI drift story but also the Cretaceous opening history of the South Atlantic.

CIRCUM-ANTARCTIC SEAFLOOR SPREADING SINCE THE LATE JURASSIC

Preserved oceanic crust characterized by distinctive magnetic and gravity signatures allows us to reconstruct the age and extent of oceanic crust through time. However, subduction, complex seafloor spreading or massive volcanism can destroy or overprint this structure. In such cases geological and geophysical data from continental area are the only constraints for plate reconstructions. In the following we present circum-Antarctic reconstructions that take into account both continental and oceanic area evidences of plate motions. For oceanic areas we show the oceanic paleo-age modeled according to magnetic and gravity data of the preserved crust.

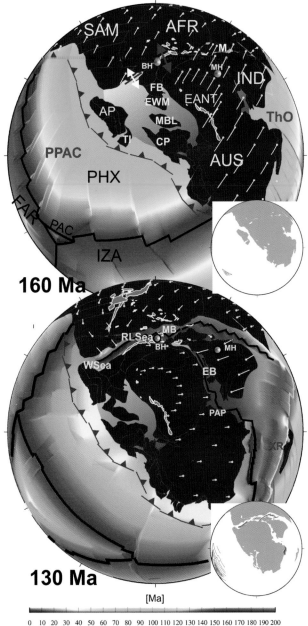

FIGURE 8 Oceanic paleo-age grids and reconstructed continental blocks of Gondwana at 160 Ma and 130 Ma. Continental blocks: CP = Campbell Plateau; M = Madagascar (see Table 1 for more abbreviations). Oceanic plates: FAR = Farallon; IZA = Izanaghi, PAC = Pacific; PHX = Phoenix. Oceanic basins: ThO-Neotethys, PPAC = Paleo-Pacific oceans, Wsea = Weddell Sea, RLSea = Riiser-Larsen Sea; MB = Mozambique basin; EB = Enderby Basin; PAP = Perth Abyssal Plain. Hotspots: BH = Bouvet, MH = Marion. XR indicates location of extinct ridges, toothed grey lines show location of subduction, thick white arrows between southern block of SAM and FB indicate extension. Red arrows indicate first Gondwana breakup in the RLSea. Absolute motion vectors shown as white lines. WANT blocks are connected by uncertain crust type (light brown color); this can be extended continental crust or oceanic crust that has subsequently been subducted or obducted. Inset figures show isochrons based on present-day magnetic and gravity data.

In the case of a single preserved flank we assume symmetric spreading, and in the case of restoring complete oceanic basins, we assume symmetric spreading and rates according to the distances between the two margins whose locations are established by independent data (i.e., not oceanic crust data).

Our reconstructions show vectors of motions for the major continents, which are based on stage poles that indicate the average motion between the continent and underlying mantle for the last 5 million years before the age of the reconstruction. These stage poles are based on our hybrid reference frame and global plate circuit (Torsvik et al., forthcoming) that include finite rotation poles between tectonic plates inferred both quantitatively and qualitatively, based on paleomagnetic data, magnetic and gravity data, and geological data. Due to the complexity of the database, plate circuits, and range of errors involved in our analysis, a method to quantify the resulting errors of our motion vectors is not yet developed, but an estimation of several degrees are expected for a direction deviation.

Isotopic ages of rocks from the southernmost Andes and South Georgia Island, North Scotia Ridge revealed that the formation of oceanic crust in the Weddell Sea region occurred by the Late Jurassic (150 ± 1 Ma) (Mukasa and Dalziel, 1996), but interpretation of new geophysical data indicates that Gondwana breakup probably commenced in the Weddell Sea (Figure 8) at ~160 Ma and propagated clockwise around ANT (Ghidella et al., 2002; Jokat et al., 2003; König and Jokat, 2006). Early AFR-ANT spreading offshore DML has been dated to ~153 Ma (M24) in the Lazarev and Riiser-Larsen Seas (Roeser et al., 1996; Jokat et al., 2003). A new model for the early Indian-ANT spreading system in the Enderby Basin (Figure 8) places the onset of seafloor spreading at ~130 Ma (M9) (Gaina et al., 2003, 2007), consistent with the opening history between India and AUS in the Perth Abyssal Plain.

Early AUS-ANT spreading east of the Vincennes Fracture Zone (~105°E) has been identified by a Late Cretaceous ridge system between Chron 34 (~83.5 Ma) and 31 (~71 Ma) (Tikku and Cande, 1999). In the south Tasman Sea between eastern AUS and the Lord Howe Rise and New Zealand, seafloor spreading began in the late Cretaceous (~83.5 Ma). Spreading propagated northward to the Coral Sea in the Tertiary, terminating at ~52 Ma (Gaina et al., 1998). Seafloor spreading east of AUS is combined with models that include incipient motion between EANT and WANT (Cande et al., 2000; Stock and Cande, 2002; Cande and Stock, 2004).

The evolution of the South Pacific region (Eagles et al., 2004) has been supplemented with reconstructed seafloor formed in the Pacific realm and subducted beneath WANT. The configuration of these "synthetic plates" was established on the basis of preserved magnetic lineations, paleogeography, regional geological data, and the rules of plate tectonics.

Gondwana Breakup

Long-lived subduction in the southern Pacific realm facilitated the amalgamation and accretion of several terranes to westernmost ANT (Vaughan and Storey, 2000). Recent geophysical data and models propose that extension between different continentally affiliated blocks of WANT achieved high degrees of extension but did not develop into seafloor spreading. Rotation, local subduction, and back-arc spreading may first have displaced and later reamalgamated AP blocks (Vaughan et al., 2002).

It has proved difficult to reach a consensus between motion described by paleomagnetic data and other geological and geophysical evidence. We therefore treat the Mesozoic WANT domain as a collection of island arc and continental blocks in a matrix of extended or not well-defined crust until 61 Ma (i.e., when extension between EANT and WANT commenced) (Cande and Stock, 2004). Seafloor spreading in the Pacific region has been quantified using the oldest preserved magnetic anomalies that describe the relative motion between the nascent Pacific plate and neighboring Izanagi (completely subducted under Eurasia), Farallon (partially subducted under North America) and Phoenix (almost completely subducted under WANT, SAM, and AUS) plates. Most of the conjugate plates are now completely subducted and thus we assume symmetric seafloor spreading.

Late Jurassic and Early Cretaceous motion vectors show a general southward trend (Figure 8) that we attribute to subduction rollback. Africa and South America moved more slowly than the block formed by EANT, India, Seychelles, Madagascar, and AUS. Consequently, seafloor spreading started to develop between these two sub-blocks of Gondwana in the Weddell Sea, Riiser-Larsen Sea, Mozambique, and Somali basins. König and Jokat (2006) proposed that a long phase of extension and rifting took place in the southern Weddell Sea before the onset of seafloor spreading dated around 147 Ma (M20). Older magnetic anomalies have been identified in the Riiser-Larsen Sea (M24 at ~154 Ma) by Roeser et al. (1996), who consequently proposed ~165 Ma breakup time between EANT and Africa. At the same time, the southern Lazarev Sea (described as a continental margin) was affected by multiple rifting episodes accompanied by transient volcanism (Hinz et al., 2004).

Our 160 Ma reconstruction (Figure 8) shows the Bouvet plume located at the boundary between EANT and Africa. Although plume-related breakup is controversial, this reconstruction reinforces a possible relationship between breakup, seafloor spreading, and volcanism initiated at the Explora Wedge. This may also explain the multiple rift relocation in the southern Riiser-Larsen Sea and early seafloor spreading that later propagated west into the Weddell Sea.

130 Ma—Eastward Propagation of Seafloor Spreading: Antarctica-India-Australia Breakup

During the mid-Cretaceous, seafloor spreading propagated eastward from the Riiser-Larsen Sea to the Enderby basin between EANT and India (Gaina et al., 2007). At the same time, India broke off from Australia (Figure 8), forming ocean basins west of Australia (Perth, Cuvier, and Gascoyne abyssal plains) (Mihut and Müller, 1998; Heine et al., 2004).

It is unclear what triggered this event. The earliest magmatic activity in the Kerguelen area is dated to ca. 118 Ma (Frey et al., 2000; Nicolaysen et al., 2001) and the EANT margin in the Enderby basin is a nonvolcanic margin. Magmatic activity, however, did occur further to the west (i.e., the ~125 Ma Maud Rise LIP) (offshore DML and EANT) (Torsvik et al. 2006). We link this event with Bouvet hotspot activity (Figure 8, ~130 Ma).

Motion vectors for the Indian-Madagascar-Seychelles triplet at 130 Ma (Figure 8b) show rapid northward movement. We speculate that a possible Tethyan ocean ridge subduction under Eurasia caused the acceleration of India and also a southward ridge jump north of India and northwest of Australia (Heine et al., 2004). A significant cusp (>90°) is recognized in the SAPW path at around 130 Ma signifying a major change in plate driving forces for EANT. Initial exhumation of the Transantarctic Mountains may have begun at this time in the Scott Glacier region (Fitzgerald and Stump, 1992), suggesting onset of extension between EANT and WANT.

99 Ma—Abrupt Change in Relative Velocity

A dramatic acceleration of the Indian (and AFR) plate modified the seafloor spreading geometry north of Enderby Land and west of AUS (Müller et al., 2001). At the same time, the Pacific plate swerved and accelerated (Veevers, 2000) bringing long-lived subduction under the Australian and New Zealand plates to a halt (Figure 9). Transtensional regimes that followed this change in the Pacific plate motion led to the opening of the Tasman Sea east of AUS (Gaina et al., 1998) and rifting of the Chatham Rise, Campbell Plateau (and South New Zealand) from the MBL block (Cunningham et al., 2002). It is noteworthy that increased spreading rates between EANT and India (AFR), from ~3 cm/yr at 100 Ma to ~7 cm/yr at 90 Ma (Figure 10), was not associated with any abrupt changes in the SAPW path (Figure 4a).

50 Ma—Major Change in Plate Driving Forces as a Precursor to Opening of Oceanic Gateway

While related to plume drift as opposed to a change in plate motion, the bend in the Hawaiian-Emperor (recently revised from 43 Ma to 48 Ma) (Sharp and Clague, 2006) apparently does reflect a major tectonic event that affected much of the

Circum-Pacific realm. Besides the inception of rapid northward drift of the Australian plate that caused rapid accretion of oceanic crust on the EANT plate, a major plate tectonic reorganization has been recently reported between Australia and Antarctica (Whittaker et al., 2007). This major event

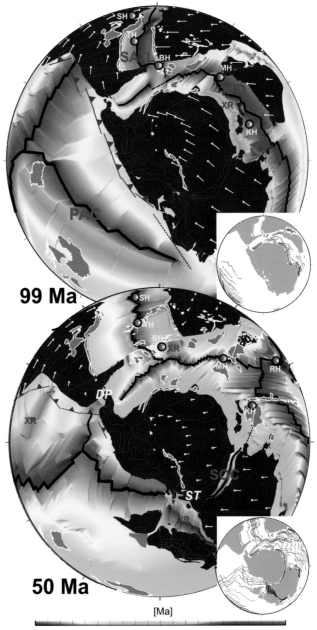

FIGURE 9 Paleo-age for reconstructed oceanic crust at 99 Ma and 50 Ma. Oceanic basins. IND = Indian, SATL = South Atlantic; SOC = Southern Ocean; AT = Adare Trough, ST = South Tasman; DP = Drake Passage. The Southern Hemisphere ocean gateways opened at around 33 Ma and 30 Ma, respectively. Inset figures show isochrons based on present-day magnetic and gravity data (see Figure 8 and Table 1 for explanation of abbreviations).

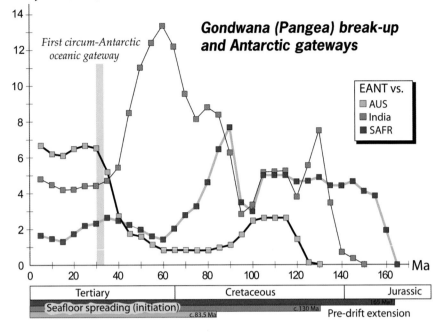

FIGURE 10 Mean plate rates of predrift deformation and seafloor spreading for AUS, India, and SAFR vs. EANT.

(perhaps even global) that coincides with the Hawaiian-Emperor bend time was correlated to the subduction of the Pacific-Izanagi active spreading ridge and subsequent Mariana-Tonga-Kermadec subduction initiation (Whittaker et al., 2007).

Relative extension between EANT and WANT commenced in the Late Cretaceous-early Tertiary, but oceanic crust between these two plates was formed only between 45 Ma and 30 Ma in the Adare Trough of the Ross Sea (Cande et al., 2000) (Figure 9). Rapid extension-related exhumation of the Transantarctic Mountains (TAM) at ~55 Ma is well documented (Fitzgerald and Gleadow, 1988; Fitzgerald, 2002), but the cause of this uplift is still unresolved. Two competing hypotheses seem pertinent: Fitzgerald et al. (1986) suggested a classic asymmetric extension process, while Stern and ten Brink (1989) proposed an elegant model based on the flexural up-warp of a broken, thin lithospheric plate. To date, neither model has been validated nor shown to be wrong. We suggest here that one "shoe" does not necessarily need to fit all: the Stern and ten Brink model (1989) appears to apply well to the Ross Sea sector of the range, outboard of the Wilkes subglacial basin, but may perhaps fit less well in the southern portion of the range. There the sub-ice surface inboard of the TAM achieves greater elevation, and the flexural profile fails. In this region an alternative mechanism—perhaps similar to the one proposed by Fitzgerald et al. (1986)—may become dominant.

In middle to late Eocene times relative motion between microcontinents south and west of Tasmania and the final detachment from ANT led to opening of the first circum-Antarctic oceanic gateway (South Tasman), causing radical changes in oceanic circulation patterns (Brown et al., 2006).

By the dawn of the Oligocene (~33.5 Ma) (Exon 2002) the gateway reached full marine conditions. Seafloor spreading in the Drake Passage and Scotia Sea region is generally considered to have commenced before 26 Ma (Barker, 2001) or 29.7 Ma (Eagles and Livermore, 2002).

EPILOGUE

Far from static, Antarctica has traveled long distances, in both space and time. The most ancient fragments once basked beneath a tropical Precambrian sun, in communion with cratonic West Australia and enveloped in a loosely defined supercontinent, Rodinia. Playing an active role in Rodinia breakup and Gondwana assembly at the dawn of the Paleozoic, Antarctica commenced a long southward drift in Late Ordovician time. During the transit to its present polar position, Antarctica participated in the assembly of yet another supercontinent, Pangea. Jurassic and subsequent divorces left Antarctica surrounded by spreading ridges and marine circum-Antarctic gateways at the beginning of the Oligocene. Once the queen of the continental cotillion, Antarctica has danced away from the heart of it all to a splendid, ice-bound isolation at the bottom of the world—truly the Last Place on Earth.

ACKNOWLEDGMENTS

This paper was funded by the Geological Survey of Norway, the Norwegian Research Council, and StatoilHydro (PETROMAKS Frontier Science and Exploration no. 163395/S30). Reviewers Ian Dalziel, John Gamble, and Steve Cande are thanked for valuable comments.

REFERENCES

Adie, R. J. 1952. The position of the Falkland Islands in a reconstruction of Gondwanaland *Geological Magazine* 89:401-410.

Barker, P. F. 2001. Scotia Sea regional tectonic evolution: Implications for mantle flow and palaeocirculation. *Earth Science Reviews* 55:1-39.

Brown, B., C. Gaina, and R. D. Muller. 2006. Circum-Antarctic palaeobathymetry: Illustrated examples from Cenozoic to recent times. *Palaeogeography, Palaeoclimatology, Palaeoecology* 231:158-168.

Burke, K., and T. H. Torsvik. 2004. Derivation of large igneous provinces of the past 200 million years from long-term heterogeneities in the deep mantle. *Earth and Planetary Science Letters* 227:531-538.

Cande, S. C., and J. M. Stock. 2004. Pacific-Antarctic-Australia motion and the formation of the Macquarie Plate. *Geophysical Journal International* 157:399-414.

Cande, S. C., J. Stock, R. D. Müller, and T. Ishihara. 2000. Cenozoic motion between East and West Antarctica. *Nature* 404:145-150.

Cocks, L. R. M., and T. H. Torsvik. 2002. Earth Geography from 500 to 400 million years ago: A faunal and palaeomagnetic review. *Journal of the Geological Society of London* 159:631-644.

Collins, A. S., and S. A. Pisarevsky. 2005. Amalmagating eastern Gondwana: The evolution of the Circum-Indian Orogens. *Earth Science Reviews* 71:229-270.

Cunningham, A.P., R. D. Larter, P. F. Barker, K. Gohl, and F. O. Nitsche. 2002. Tectonic evolution of the Pacific margin of Antarctica. 2. Structure of Late Cretaceous-early Tertiary plate boundaries in the Bellingshausen Sea from seismic reflection and gravity data. *Journal of Geophysical Research* 107, doi:10.1029/2002JB001897.

Curtis, M. L. 2001. Tectonic history of the Ellesworth Mountains, West Antarctica: Reconciling a Gondwana enigma. *Geological Society of America Bulletin* 113:939-958.

Dalziel, I. W. D. 1992. On the organization of American plates in the Neoproterozoic and the breakout of Lauretia. *GSA Today* 2:237-241.

Dalziel, I. W. D. 1997. Neoproterozoic-Paleozoic geography and tectonics: Review, hypothesis, environmental speculation. *Geological Society of America Bulletin* 109:16-42.

Dalziel, I. W. D., and D. H. Elliot. 1982. West Antarctica: Problem child of Gondwanaland. *Tectonics* 1:3-19.

Dalziel, I. W. D., and A. M. Grunow. 1992. Late Gondwanide tectonic rotations within Gondwanaland. *Tectonics* 11:603-606.

Dalziel, I. W. D., and L. S. Lawver. 2001. The lithospheric setting of the west Antarctic ice sheet. In *The West Antarctic Ice Sheet; Behavior and Environment*, eds. R. B. Alley and R. A. Bindschadler. *Antarctic Research Series* 77:29-44. Washington, D.C.: American Geophysical Union.

Dewey, J. F., and K. Burke. 1973. Tibetan, Variscan and Precambrian basement reactivation: Products of continental collision. *Journal of Geology* 81:683-692.

Eagles, G., and R. A. Livermore. 2002. Opening history of Powell Basin, Antarctic Peninsula. *Marine Geology* 185:195-205.

Eagles, G., K. Gohl, and R. D. Larter. 2004. High resolution animated reconstruction of the South Pacific and West Antarctic margin. *Geochemistry, Geophysics, Geosystems* 5:Q07002, doi:10.1029/2003GC000657.

Exon, N. 2002. Drilling reveals climatic consequence of Tasmanian Gateway opening. *Eos Transactions of the American Geophysical Union* 83:253-259.

Fitzgerald, P. 2002. Tectonics and landscape evolution of the Antarctic plate since the breakup of Gondwana, with an emphasis on the West Antarctic Rift System and the Trans Antarctic Mountains. In *Antarctica at the Close of a Millennium. Royal Society of New Zealand Bulletin* 35:453-469.

Fitzgerald, P. G., and A. J. W. Gleadow. 1988. Fission track geochronology, tectonics and structure of the Transantarctic Mountains in Northern Victoria Land, Antarctica. *Isotope Geoscience* 73:169-198.

Fitzgerald, P. G., and E. Stump 1992. Early Cretaceous uplift in the southern Sentinel Range, Ellsworth Mountains, West Antarctica. In *Recent Progress in Antarctic Earth Science*, eds. Y. Yoshida, K. Kaminuma, and K. Shiraishi, pp. 331-340, TERRAPUB, Tokyo.

Fitzgerald, P. G., M. Sandiford, P. J. Barrett, and A. J. W. Gleadow 1986. Asymmetric extension associated with uplift and subsidence of the Transantarctic Mountains. *Earth and Planetary Science Letters* 81:67-78.

Fitzsimons, I. C. E. 2000. Grenville-age basement provinces in East Antarctica: Evidence for three separate collision orogens. *Geology* 28:879-882.

Fitzsimons, I. C. E. 2003. Proterozoic basement provinces of southern and southwestern Australia, and their correlations with Antarctica. In *Proterozoic East Gondwana: Supercontinent Assembly and Breakup*, eds. M. Yoshida, B. F. Windley, and S. Dasgupta. *Geological Society of London Special Publication* 206:93-130.

Frey, F. A., M. F. Coffin, P. J. Wallace, D. Weis, X. Zhao, S. W. Wise, Jr., V. Waehnert, D. A. H. Teagle, P. J. Saccocia, D. N. Reusch, M. S. Pringle, K. E. Nicolaysen, C. R. Neal, R. D. Müller, C. L. Moore, J. J. Mahoney, L. Keszthelyi, H. Inokuchi, R. A. Duncan, H. Delius, J. E. Damuth, D. Damasceno, H. K. Coxall, M. K. Borre, F. Boehm, J. Barling, N. T. Arndt, and M. Antretter. 2000. Origin and evolution of a submarine large igneous province; the Kerguelen Plateau and Broken Ridge, southern Indian Ocean. *Earth and Planetary Science Letters* 176:73-89.

Gaina, C., D. R. Müller, J.-Y. Royer, J. M. Stock, J. L. Hardebeck, and P. Symonds. 1998. The tectonic history of the Tasman Sea; a puzzle with 13 pieces. *Journal of Geophysical Research* 6:12413-12433.

Gaina, C., R. D. Müller, B. Brown, and T. Ishihara. 2003. Microcontinent formation around Australia. In *Evolution and Dynamics of the Australian Plate*, eds. R. R. Hills and R. D. Müller. *Geological Society of Australia Special Publication* 22 and *Geological Society of America Special Paper* 372:405-416.

Gaina, C., R. D. Müller, B. Brown, T. Ishihara, and K. S. Ivanov. 2007. Breakup and early seafloor spreading between India and Antarctica. *Geophysical Journal International*, doi:10.1111/j.1365-246X.2007.03450.x.

Ghidella, M. E., G. Yáñez, and J. LaBreque. 2002. Raised tectonic implications for the magnetic anomalies of the western Weddell Sea. *Tectonophysics* 347:65-86.

Grunow, A. M. 1993. New paleomagnetic data from the Antarctic Peninsula and their tectonic implications. *Journal of Geophysical Research* 98:13815-13833.

Grunow, A. M., I. W. D. Dalziel, and D. V. Kent. 1987. Ellsworth-Whitmore Mountains crustal block, western Antarctica: New paleomagnetic results and their tectonic significance. In *Gondwana Six: Structure, Tectonics and Geophysics*, ed. G. D. McKenzie, *American Geophysical Union Geophysical Monograph* 40:161-172.

Harley, S. L. 2003. Archaean-Cambrian crustal development of East Antarctica: Metamorphic characteristics and tectonic implication. In *Proterozoic East Gondwana: Supercontinent Assembly and Breakup*, eds. M. Yoshida, B. F. Windley, and S. Dasgupta. *Geological Society of London Special Publication* 206:203-230.

Heine, C., R. D. Müller, and C. Gaina. 2004. Reconstructing the Lost Eastern Tethys Ocean Basin: Convergence history of the SE Asian margin and marine gateways. In *Continent-Ocean Interactions in Southeast Asia*, eds. P. Clift, D. Hayes, W. Kuhnt, and P. Wang. *American Geophysical Union Geophysical Monograph* 149:37-54.

Hinz, K., S. Neben, Y. B. Gouseva, and G. A. Kudryavtsev. 2004. A compilation of geophysical data from the Lazarev Sea and the Riiser-Larsen Sea, Antarctica. *Marine Geophysical Researches* 25:233-245.

Hoffman, P. F. 1991. Did the breakout of Laurentia turn Gondwana inside out? *Science* 252:1409-1411.

Huang, B., B. Xu, C. Zhang, Y. Li, and R. Zhu. 2005. Paleomagnetism of the Baiyisi volcanic rocks (ca. 740 Ma) of Tarim, Northwest China: A continental fragment of Neoproterozoic Western Australia. *Precambrian Research* 142:83-92.

Jacobs, J., and R. J. Thomas. 2004. Himalayan-type indenter-escape tectonics model of the southern part of the late Neoproterozoic-early Paleozoic East African-Antarctic orogen. *Geology* 32:721-724.

Johnston, S. T. 2000. The Cape Fold Belt and Syntaxis and the rotated Falkland Islands: Dextral transpressional tectonics along the southwest margin of Gondwana. *Journal of African Earth Sciences* 31:51-63.

Jokat, W., T. Boebel, M. König, and U. Meyer. 2003. Timing and geometry of early Gondwana breakup. *Journal of Geophysical Research* 108, doi:10.1029/2002JB001802.

König, M., and W. Jokat. 2006. The Mesozoic breakup of the Weddell Sea. *Journal of Geophysical Research* 111, doi:10.1029/2005JB004035.

Leat, P. T., B. C. Storey, and R. J. Pankhurst. 1993. Geochemistry of Palaeozoic-Mesozoic Pacific rim orogenic magmatism, Thurston Island area, West Antarctica. *Antarctic Science* 5:281-296.

Leat, P. T., A. A. Dean, I. L. Millar, S. P. Kelley, A. P. M. Vaughan, and T. R. Riley. 2005. Lithosphere mantle domains beneath Antarctica. In *Terrane Processes at the Margins of Gondwana*, eds. A. P. M. Vaughan, P. T. Leat, and R. J. Pankhurst. *Geological Society of London Special Publication* 246:359-380.

Lythe, M. B., D. G. Vaughan, and the BEDMAP Consortium. 2000. BEDMAP—bed topography of the Antarctic. 1:10,000,000 scale map. BAS (Misc) 9. Cambridge: British Antarctic Survey.

Marshall, J. E. A. 1994. The Falkland Islands: A key element in Gondwana paleogeography. *Tectonics* 13:499-514.

Meert, J. G. 2003. A synopsis of events related to the assembly of eastern Gondwana. *Tectonophysics* 362:1-40.

Mihut, D., and R. D. Müller. 1998. Volcanic margin formation and Mesozoic rift propagators in the Cuvier abyssal plain off Western Australia. *Journal of Geophysical Research* 103:27135-27150.

Millar, I. L., and R. J. Pankhurst. 1987. Rb-Sr geochronology of the region between the Arctic Peninsula and the Transantarctic Mountains: Haag Nunataks and Mesozoic granitoids. In *Gondwana Six: Structure, Tectonics and Geophysics*, ed. G. D. McKenzie, pp. 151-160. *American Geophysical Union Monograph* 40.

Mitchell, C., G. K. Taylor, K. G. Cox, and J. Shaw. 1986. Are the Falklands a rotated microplate? *Nature* 319:131-134.

Mukasa, S. B., and I. W. D. Dalziel. 1996. Southernmost Andes and South Georgia Island, North Scotia Ridge: Zircon U-Pb and muscovite 40Ar/39Ar age constraints on tectonic evolution of Southwestern Gondwanaland. *Journal of South American Earth Sciences* 9(5):349-365.

Müller, R. D., C. Gaina, and W. R. Roest. 2001. A recipe for microcontinent formation. *Geology* 29:203-206.

Nicolaysen, K., S. Bowring, F. Frey, D. Weis, S. Ingle, M. S. Pringle, and M. F. Coffin. 2001. Provenance of Proterozoic garnet-biotite gneiss recovered from Elan Bank, Kerguelen Plateau, southern Indian Ocean. *Geology* 29:235-238.

O'Neill, C., R. D. Müller, and B. Steinberger. 2005. On the uncertainties in hot spot reconstructions and the significance of moving hot spot reference frames. *Geochemistry, Geophysics, Geosystems* 6:Q04003, doi:10.1029/2004GC000784.

Pankhurst, R. J., T. R. Riley, C. M. Fanning, and S. P. Kelley. 2000. Episodic silicic volcanism in Patagonia and the Antarctic Peninsula: Chronology of magmatism associated with the break-up of Gondwana. *Journal of Petrology* 41:605-625.

Randall, D. E., and C. MacNiocaill. 2004. Cambrian palaeomagnetic data confirm a Natal embayment location for the Ellsworth-Whitmore Mountains, Antarctica, in Gondwana reconstructions. *Geophysical Journal International* 157:105-116.

Rapela, C. W., R. J. Pankhurst, C. M. Fanning, and F. Herve. 2005. Pacific subduction coeval with the Karoo mantle plume: The Early Jurassic Subcordilleran belt of northwestern Patagonia. In *Terrane Processes at the Margins of Gondwana*, eds. A. P. M. Leat and R. J. Pankhurst. *Geological Society of London Special Publication* 246:217-239.

Roeser, H. A., J. Fritsch, and K. Hinz. 1996. The development of the crust off Dronning Maud Land, East Antarctica. In *Weddell Sea Tectonics and Gondwana Break-up*, eds. B. C. Storey, E. C. King, and R. Livermore. *Geological Society of London Special Publication* 108:243-264.

Schopf, J. M. 1969. Ellsworth Mountains: Position in West Antarctica due to seafloor spreading. *Science* 164:63-66.

Sharp, W. D., and D. A. Clague. 2006. 50-Ma initiation of Hawaiian-Emperor bend records major change in Pacific plate motion. *Science* 313:1281-1284.

Stern, T. A., and U. S. ten Brink. 1989. Flexural uplift of the Transantarctic Mountains. *Journal of Geophysical Research* 94:10315-10330.

Stock, J. M., and S. C. Cande. 2002. Tectonic history of the Antarctic seafloor in the Australia-New Zealand-South Pacific sector: Implications for Antarctic continental tectonics. In *Antarctica at the Close of a Millennium*, eds. J.A. Gamble, D. N. B. Skinner, and S. Henrys. *Royal Society of New Zealand Bulletin* 35:252-259.

Studinger, M., and H. Miller. 1999. Crustal structure of the Filchner-Ronne shelf and Coats-Land, Antarctica, from gravity and magnetic data: Implications for the break-up of Gondwana. *Journal of Geophysical Research* 104:20379-20394.

Svensen, H., S. Planke, L. Chevallier, A. Malthe-Sørensen, F. Corfu, and B. Jamtveit. 2007. Hydrothermal venting of greenhouse gases triggering Early Jurassic global warming. *Earth and Planetary Science Letters* 256:554-566.

Tikku, A. A., and S. C. Cande. 1999. The oldest magnetic anomalies in the Australian-Antarctic Basin: Are they isochrons? *Journal of Geophysical Research* 104:661-677.

Torsvik, T. H. 2003. The Rodinia jigsaw puzzle. *Science* 300:1379-1381.

Torsvik, T. H., and L. R. M. Cocks. 2004. Earth geography from 400 to 250 million years: A palaeomagnetic, faunal and facies review. *Journal of the Geological Society of London* 161:555-572.

Torsvik, T. H., and L. R. M. Cocks. Forthcoming. The Lower Palaeozoic palaeogeographical evolution of the Northeastern and Eastern peri-Gondwanan margin from Turkey to New Zealand. *Geological Society of London Special Publication*.

Torsvik, T. H., and R. Van der Voo. 2002. Refining Gondwana and Pangea palaeogeography: Estimates of Phanerozoic (octupole) non-dipole fields. *Geophysical Journal International* 151:771-794.

Torsvik, T. H., M. A. Smethurst, J. G. Meert, R. Van der Voo, W. S. McKerrow, M. D. Brasier, B. A. Sturt, and H. J. Walderhaug. 1996. Continental break-up and collision in the Neoproterozoic and Palaeozoic: A tale of Baltica and Laurentia. *Earth Science Reviews* 40:229-258.

Torsvik, T. H., L. M. Carter, L. D. Ashwal, S. K. Bhushan, M. K. Pandit, and B. Jamtveit. 2001. Rodinia refined or obscured: Palaeomagnetism of the Malani Igneous Suite (NW India). *Precambrian Research* 108:319-333.

Torsvik, T. H., M. A. Smethurst, K. Burke, and B. Steinberger. 2006. Large Igneous Provinces generated from the margins of the Large Low Velocity Provinces in the deep mantle. *Geophysical Journal International* 167:1447-1460.

Torsvik, T. H., R. D. Müller, R. Van der Voo, B. Steinberger, and C. Gaina. Forthcoming. Global plate motion frames: Toward a unified model. *Reviews of Geophysics*.

Vaughan, A. P. M., and B. C. Storey. 2000. The Eastern Palmer Land shear zone: A new terrane accretion model for the Mesozoic development of the Antarctic Peninsula magmatic arc. *Journal of the Geological Society of London* 157:1243-1256.

Vaughan, A. P. M., S. P. Kelley, and B. C. Storey. 2002. Mid-Cretaceous ductile deformation on the Eastern Palmer Land Shear Zone, Antarctica, and implications for timing of Mesozoic terrane collision. *Geological Magazine* 139:465-471.

Veevers, J. J. 2000. Change of tectono-stratigraphic regime in the Australian plate during the 99 Ma (mid-Cretaceous) and 43 Ma (mid-Eocene) swerves of the Pacific. *Geology* 28:47-50.

Whittaker, J. M., R. D. Müller, G. Leitchenkov, H. Stagg, M. Sdrolias, C. Gaina, and A. Goncharov. 2007. Major Australia-Antarctic plate reorganization at Hawaiian-Emperor bend time. *Science* 318:83-87.

Wingate, M. T. D., and J. W. Giddings. 2000. Age and palaeomagnetism of the Mundine Well dyke swarm, Western Australia. Implications for an Australia-Laurentia connection at 755 Ma. *Precambrian Research* 100:335-357.

Wingate, M. T. D., S. A. Pisarevsky, and D. A. D. Evans. 2002. Rodinia connections between Australia and Laurentia: NoSWEAT, no AUSWUS? *Terra Nova* 14:121-128.

DVD Contents

The DVD that is included in the back pocket of this book contains a copy of the complete 10th ISAES Online Proceedings website at http://pubs.usgs.gov/of/2007/1047/. The DVD contains Portable Document Format (PDF) version 1.6 (Acrobat 7) files of the 10th ISAES Program book and all publications of the 10th ISAES. A list of these publications is provided below. The disc can be used on any DVD-equipped computer platform that can run Adobe Reader 7 or higher or other software that can translate PDFs. You can get a free copy of the latest version of Adobe Reader from http://www.adobe.com/products/acrobat/readstep2.html. The disc contains a full-text index (index.pdx and associated files in the index directory) that is for use in searching among all of the PDF files at once for words or sets of words using the Search tool in Adobe Reader. To get started: On a Windows-based computer, the file index.html should open automatically. On any platform, open the DVD drive then open the file index.html to view it with your browser.

KEYNOTE PAPERS (KNP – PEER REVIEWED)

KNP 001: *Summary and Highlights of the Tenth International Symposium on Antarctic Earth Sciences* by Wilson et al.

KNP 002: *Antarctic Earth System Science in the International Polar Year 2007-2008* by R. E. Bell

KNP 003: *100 Million Years of Antarctic Climate Evolution: Evidence from Fossil Plants* by J. E. Francis et al.

KNP 004: *Antarctica's Continent-Ocean Transitions: Consequences for Tectonic Reconstructions* by K. Gohl

KNP 005: *Landscape Evolution of Antarctica* by S. S. R. Jamieson and D. E. Sugden

KNP 006: *A View of Antarctic Ice-Sheet Evolution from Sea-Level and Deep-Sea Isotope Changes during the Late Cretaceous-Cenozoic* by K. G. Miller et al.

KNP 007: *Late Cenozoic Climate History of the Ross Embayment from the AND-1B Drill Hole: Culmination of Three Decades of Antarctic Margin Drilling* by T. R. Naish et al.

KNP 008: *A Pan-Precambrian Link Between Deglaciation and Environmental Oxidation* by T. D. Raub and J. L. Kirschvink

KNP 009: *Tectonics of the West Antarctic Rift System: New Light on the History and Dynamics of Distributed Intracontinental Extension* by C. S. Siddoway

KNP 010: *The Significance of Antarctica for Studies of Global Geodynamics* by R. Sutherland

KNP 011: *Antarctica and Global Paleogeography: From Rodinia, Through Gondwanaland and Pangea, to the Birth of the Southern Ocean and the Opening of Gateways* by T. H. Torsvik et al.

SHORT RESEARCH PAPERS (SRP – PEER REVIEWED)

SRP 001: *Advances Through Collaboration: Sharing Seismic Reflection Data via the Antarctic Seismic Data Library System for Cooperative Research (SDLS)* by N. Wardell et al.

SRP 002: *Antarctic Multibeam Bathymetry and Geophysical Data Synthesis: An On-Line Digital Data Resource for Marine Geoscience Research in the Southern Ocean* by S. M. Carbotte et al.

SRP 003: *The Dinosaurs of the Early Jurassic Hanson Formation of the Central Transantarctic Mountains: Phylogenetic Review and Synthesis* by N. D. Smith et al.

EXTENDED ABSTRACTS (EA – NOT PEER REVIEWED)

MEETING AND WORKSHOP REPORTS (MWR – NOT PEER REVIEWED)